THERMAL COMPUTATIONS
FOR
ELECTRONIC EQUIPMENT

THERMAL COMPUTATIONS FOR ELECTRONIC EQUIPMENT

Gordon N. Ellison
Chief Engineer
Tektronix
Computer Science Center
Beaverton, Oregon

KRIEGER PUBLISHING COMPANY
MALABAR, FLORIDA

Original Edition 1984
Reprint Edition 1989

Printed and Published by
KRIEGER PUBLISHING COMPANY
KRIEGER DRIVE
MALABAR, FLORIDA 32950

Library of Congress Cataloging in Publication Data

Ellison, Gordon N.
    Thermal computations for electronic equipment / Gordon N. Ellison.
       p.   cm.
    Bibliography: p.
    Includes index.
    ISBN 0-89464-401-7 (alk. paper)
    1. Electronic apparatus and appliances--Cooling. 2. Heat-
-Transmission. I. Title.
TK7870.25.E4 1989
   621.38--dc20   89-11066
                                          CIP

10   9   8   7   6   5   4   3

To my father and mother,
   for their examples of wisdom and self-discipline.

To my wife and daughters,
   Sharon, Susan, and Cynthia, for their understanding
   and support.

DISCLAIMER

The material herein, digital computer programs included, is supplied as-is and without warranty or representation of any kind. The author and his employer and publisher, past, present, or future make no representations respecting the programs and related material and expressly disclaim any liability for damages from the use of the programs or related material, or any part thereof.

Preface to the Reprint Edition

The author gratefully acknowledges the many gracious and helpful comments concerning this book. I wish to thank Dr. Jack Wilson, California State Polytechnic University at San Luis Obispo and Dr. Kaveh Azar, AT&T Bell Laboratories, and others unknown to me at this date for their selection of this work as a text or required reference for formal course work.

Reader comments or interests concerning the computer code (Appendices V, VI) may be addressed to the publisher:

Krieger Publishing Co.
Attn: Thermal Code
P. O. Box 9542
Melbourne, Florida 32902-9542

Finally, of no small significance, is my appreciation to Robert Krieger and Marie Bowles, Krieger Publishing Co., for their interest in continuing the publication of "Thermal Computations for Electronic Equipment."

Gordon N. Ellison

Preface

It is my intent that this book will be useful to an audience of rather diverse academic preparation. Designers with only a pre-calculus background have been successfully taught to use most of the techniques herein, although those with a more advanced engineering background have typically achieved a greater level of competency in a shorter time.

While most of the book uses algebraic methods, some differential equations have also been necessary, but usually only to derive formulas that are of general use. Although both of the digital computer programs (TNETFA and TAMS) are based on solutions to the partial differential equation for heat conduction, an understanding of the solution methods is not essential for satisfactory application. Successful use of the material in this book does require a quantitative aptitude on the part of the reader.

I have assumed that the reader has a background in some form of physical theory, for example, physics or mechanical or electrical engineering. I do not expect any formal training in heat transfer, but I have found that even engineers who have studied this subject at an advanced level find the analytical methods and basic heat transfer data of considerable value. The book is especially useful in aiding the transition from a purely academic to an industrial, product-oriented environment. I advise the reader that the numerous examples throughout the book are not necessarily recommended as good packaging practice. Rather, the examples have been formulated to illustrate application of some analytical techniques.

In the preparation of this book, I have introduced only those aspects of heat transfer in electronic equipment with which I consider myself to be fully acquainted and that I have applied to practical design problems. No attempt is made to include material on heat pipes, thermoelectric coolers, liquid cooling, and so on; I have confined the material totally to conduction, radiation, and convection by natural or forced airflow. These are the heat transfer mechanisms of greatest interest to designers of electronic test instruments, computer systems, and other ground-based equipment—the industries with which I am most familiar.

Digital computer programs are increasingly being included in scientific and engineering texts and reference books. Those included here (TAMS and TNETFA) are not intended merely as illustrations of numerical methods, but rather are offered as serious analytical tools. The reader who has limited or no experience with such systems is highly encouraged to take the trouble of putting the programs on a modern computing system. Readers who already have a thermal network analyzer may wish to include the recommended radiation and convection formulas in their existing program.

Various features and applications of the TAMS program have been previously reported in the literature, much of which, however, is missed by the practicing engineer. Although geometric constraints of the application problem may be a limitation to the use of the program, its utility has been greatly appreciated by engineers who have taken the small amount of time necessary to become familiar with it.

Both programs are written in FORTRAN IV and should be consistent with ANSI* standard X3.9-1966 for FORTRAN. The programs as included here are identical to those that I am currently using on Control Data Corp. CYBER 175 machines. Conversion to other systems of adequate memory (at least 150000 octal words) should be possible with a minimum of difficulty by an experienced programmer.

The programs have been in use for some years now; thus I believe the possibility of serious "bugs" is somewhat remote, particularly in TAMS. A few features in TNETFA are of somewhat more recent vintage, but these also have been tested by comparing sample problem results from the program with answers obtained from other programs. Also, both programs have been compiled on a minimum of three different compilers, and all indicated diagnostic messages were carefully checked.

I have read the text several times, but I am fully aware that in an effort such as this, a few errors may have been undetected. I welcome any comments from readers concerning text errors or possible program bugs and apologize for any inconvenience they may cause.

The text and figures (exceptions Figs. 7-1, 7-5, 7-12, 9-1, 9-2, and 9-3) of Chapters 7 and 9 were abstracted in whole or in part from the TAMS program manual with permission of Tektronix, Inc., P.O. Box 500, Beaverton, Oregon 97077.

I am gratefully indebted to many people. My unending thanks go to: Don Meier (1), Anthony Kolk (1), Dr. Henry White (1), and Majid Arbab (1) for providing the opportunity for me to enter this field, which has been and continues to be so interesting and challenging; Chris Fischer (1), George Jurkovich (2), and Larry Haroun (3),

---

*American National Standards Institute.
(1) Formerly with NCR, Inc.
(2) Currently with NCR, Inc.
(3) Formerly with Tektronix, Inc.

all of whom provided measurements to verify many of the analytical procedures herein; Imants Golts (4) for his encouragement; Dave Emerick (3) for his help in the preparation of Chapter 10; and certainly to my wife and daughters, who graciously sacrificed many evenings and weekends during the nearly three years in which this book was written.

As a last comment, I gratefully acknowledge the recommendations of Frank Oettinger and Steve Ross of the National Bureau of Standards, Washington, D.C., for their recommendations concerning TAMS code modifications. As a result of these last minute improvements, TAMS output listings produced from the code in Appendix V will have an Aug. 10, 1983 VERSION label rather than the earlier date of the Chapter 7 TAMS output reproductions.

GORDON ELLISON

---

(4) Currently with Tektronix, Inc.

Contents

Chapter 1
Introduction

1.1 PRIMARY MECHANISMS OF HEAT FLOW

Engineering thermal analyses of electronic systems are based on any or all of three methods of thermal-energy transport: conduction, convection, and radiation.

Conduction takes place within a medium, but without obvious transport of the medium itself.

Convective heat transfer also requires a medium for energy flow, but in this instance mass transport of the medium also occurs. One of the most visible examples of this material transport is water heated in an open kettle. The warm water rises in the center of the kettle and falls to the bottom as it becomes cooled upon transferring heat to the kettle walls. The liquid flow not only mixes the fluid, but actually aids in the rate of heat transfer in the vicinity of the walls. A very careful examination of the fluid immediately adjacent to the wall would show negligible fluid flow. In this very thin layer, heat is transferred by conduction.

In a manner similar to the heated water in a kettle, circulating, convective air currents expected in the interior of sealed electronic enclosures aid the transfer of heat energy to the cabinet wall. Upon conduction through the metal or plastic enclosure, external convective heat transfer would then aid in removing thermal energy from the system.

Radiation heat transfer is totally unique when compared with conduction and convection in that no medium of energy transfer is required because thermal radiation transport occurs via the propagation of an electromagnetic radiation field. Although this field typically covers the entire electromagnetic spectrum, most of the thermal radiation encountered from conventional microelectronic components and systems is located in the infrared region.

All three heat transfer mechanisms obey the second law of thermodynamics in the sense that a net energy transfer flows only from a higher temperature to a lower temperature region.

1.2 CONDUCTION

The pertinent aspects of heat conduction are demonstrated by a one-dimensional solid element, as in Fig. 1-1, for which it is assumed that convection and radiation are not present. The convention of heat

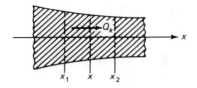

Fig. 1-1. Heat conduction in a one-dimensional solid element.

flow in the positive x-axis direction is used. The heat flows through a path length $x_2 - x_1$, and the cross-sectional area A_k is perpendicular to the axis. In this case, therefore, the temperature $T(x_1)$ at x_1 is greater than the temperature $T(x_2)$ at x_2.

Fourier's law is used to quantify conductive heat flow. In one dimension:

$$Q_k = -kA_k(dT/dx)|_x$$

E1.1
Fourier's Law

where

Q_k = heat transferred (watts)
A_k = cross-sectional area of heat flow path, cm^2 or $in.^2$
dT/dx = temperature gradient, $°C/cm$ or $°C/in.$
k = thermal conductivity, watts/$°C \cdot cm$ or watts/$°C \cdot in.$

The gradient and the cross-sectional area are defined at the same point x. The convention of a positive heat flow through a decreasing temperature, i.e., a negative temperature gradient, necessitates the minus sign immediately preceding the thermal conductivity factor.

The thermal conductivity may be considered a constant of proportionality that is dependent only on the specific material involved. While it is not uncommon for k to be temperature-dependent, this usually causes only a minor source of error in engineering design problems. When the temperature dependence of k becomes important, the digital computation techniques discussed in Chapter 8 may be applied.

Thermal conductivities for various materials are given in Table 1-1.

Generalization of Fourier's law to several dimensions is accomplished by applying the one-dimensional law to an element of volume $\Delta V = \Delta x \Delta y \Delta z$ as shown in Fig. 1-2, for which:

$$q_x = -k_x \Delta y \Delta z (\partial T/\partial x) \quad \text{at } c$$

$$q_{x+\Delta x} = -k_x \Delta y \Delta z (\partial T/\partial x) \quad \text{at } d$$

$$q_y = -k_y \Delta x \Delta z (\partial T/\partial y) \quad \text{at } a$$

$$q_{y+\Delta y} = -k_y\Delta x\Delta z(\partial T/\partial y) \quad \text{at } b$$

$$q_z = -k_z\Delta x\Delta y(\partial T/\partial z) \quad \text{at } e$$

$$q_{z+\Delta z} = -k_z\Delta x\Delta y(\partial T/\partial z) \quad \text{at } f$$

The element is also provided with an internal heat-generating source of Q_V watts/unit volume.

Application of conservation of energy requires that in a steady state condition, i.e., no energy storage, heat in = heat out, so that for the elemental volume ΔV, adopting the convention that heat in is positive:

$$[(-k_x\Delta y\Delta z\partial T/\partial x)|_c - (-k_x\Delta y\Delta z\partial T/\partial x)|_d]$$

$$+ [(-k_y\Delta x\Delta z\partial T/\partial y)|_a - (-k_y\Delta x\Delta z\partial T/\partial y)|_b]$$

$$+ [(-k_z\Delta x\Delta y\partial T/\partial z)|_e - (-k_z\Delta x\Delta y\partial T/\partial z)|_f]$$

$$+ Q_V\Delta x\Delta y\Delta z = 0$$

Dividing both sides by $\Delta x\Delta y\Delta z$,

Table 1-1. Thermal conductivities for various materials.

Material	k (watts/in. $\cdot$ °C)
Metals	
Aluminum	5.0
Copper	10.0
Gold	8.0
Nickel	1.5
Alloys	
Aluminum	3.5–4.5
Brass	1.5–3.0
Bronze, 75% Cu, 25% Sn	0.7
Cast iron	1.5
Steel	1.1
Steel, stainless 18-8	
Type 304	0.4
Type 347	0.4
Insulators	
Asbestos	0.004
Bakelite	0.006
Plate glass	0.02
Borosilicate	0.03
Rubber, hard	0.004
Mica	0.02

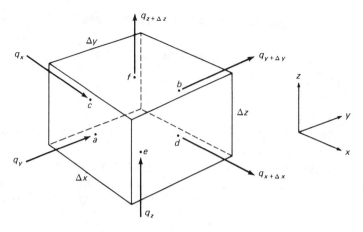

Fig. 1-2. Volume element used in heat balance.

$$(1/\Delta x)(k_x \partial T/\partial x)|_c^d$$

$$+ (1/\Delta y)(k_y \partial T/\partial y)|_a^b$$

$$+ (1/\Delta z)(k_z \partial T/\partial z)|_e^f$$

$$= -Q_V$$

Taking the limit of $\Delta V \to 0$ and recalling from basic calculus that:

$$\lim_{\Delta x \to 0} (1/\Delta x)(k_x \partial T/\partial x)|_c^d = \lim_{\Delta x \to 0} (1/\Delta x)(k_x \partial T/\partial x)_{x,y,z}^{x+\Delta x,y,z}$$

$$= \frac{\partial}{\partial x}\left(k_x \frac{\partial T}{\partial x}\right)$$

then:

$$\boxed{\frac{\partial}{\partial x}\left(k_x \frac{\partial T}{\partial x}\right) + \frac{\partial}{\partial y}\left(k_y \frac{\partial T}{\partial y}\right) + \frac{\partial}{\partial z}\left(k_z \frac{\partial T}{\partial z}\right) = -Q_V} \qquad \text{E1.2}$$

Three-dimensional heat conduction, steady state

which is the three-dimensional steady state heat conduction equation in cartesian coordinates.

Not all practical problems are steady state. The temperature/time relationship of a component or instrument may be of interest during a warmup phase, or the heat source(s) Q_V may operate in some time-dependent fashion. The theoretical basis of quantifying this is accomplished by recognizing that the steady state energy balance must be modified to account for energy storage or discharge, i.e.:

heat in – heat out = heat stored

The differential form of heat stored is $\rho \Delta V C_p (\partial T / \partial t)$ for the volume element ΔV of density ρ and specific heat C_p. Thus the partial differential equation for time-dependent heat conduction is:

$$\frac{\partial}{\partial x}\left(k_x \frac{\partial T}{\partial x}\right) + \frac{\partial}{\partial y}\left(k_y \frac{\partial T}{\partial y}\right) + \frac{\partial}{\partial z}\left(k_z \frac{\partial T}{\partial z}\right)$$

$$= -Q_V + \rho C_p (\partial T / \partial t)$$

E1.3

Three-dimensional heat conduction, time-dependent

Any consistent set of units for ρ and C_p may be used, e.g., $\rho = \rho$ (gm/in.^{3}), $C_p = C_p$ (cal/gm $\cdot$ °C). Use of the cal unit when heat sources are quantified in watts requires C_p to be multiplied by the factor 4.18 joules/cal, since 1 watt = 1 joule/sec.

Note that particular care has been taken to maintain the generality of an anisotropic thermal conductivity. Solutions to the three-dimensional partial differential equations are rather complex even in the simplest of geometries and require a working knowledge and experience-level of mathematical physics that are outside the reach of many readers of this book. Furthermore, although there is a vast amount of literature containing multi-dimensional mathematical solutions, very few of them are useful in practical engineering problems. Therefore, further discussion of problems requiring more than one spatial coordinate is deferred to Chapters 7 and 8, where digital computer techniques are offered that bring rather complex problem solutions within the capability of more individuals. Both computer methods used in Chapters 7 and 8 are useful in steady state problems. The thermal network method applied in Chapter 8 is applicable to time-dependent problems from very simple to very complex.

Fourier's law may also be directly integrated to result in a common application formula. Rearranging the variables:

$$dT = -(Q_k / kA_k)\, dx$$

$$\int_{T_1}^{T_2} dT = -Q_k \int_{x_1}^{x_2} dx / (kA_k)$$

If A_k and k do not vary over the path length $L = x_2 - x_1$, then:

$$\Delta T = (L / kA_k) Q_k$$

where $\Delta T = T_1 - T_2$, the temperature difference over L. The quantity preceding Q_k is a thermal resistance for conduction.

$$R_k = L/kA_k, \, °C/\text{watt}$$

where R_k is the temperature drop over the length L due to a heat transfer of one watt.

The following example illustrates the application of Fourier's law in the simplest form. The heat transfer through an aluminum bracket is calculated for a 20°C temperature difference between the ends. The bracket is 5 in. long, 1 in. wide, and 0.05 in. thick. Use $k = 5$ watts/in. · °C.

$$Q_k = kA_k\Delta T/L$$

$$= (5)(0.05)(1.0)(20)/(5.0)$$

$$= 1 \text{ watt}$$

The conduction resistance is clearly 20°C/watt.

More generally then:

$$R_k = \int_0^L dx/(kA_k)$$

which may be used in problems consisting of several different materials stacked in series. Consider the problem of a transistor base attached to a planar heat sink with a mica wafer used as an electrical insulator. "Thermal paste" is used to fill air voids between the mica and the metal parts. Figure 1-3 illustrates the geometry and resulting temperature gradient.

This interface resistance may be adequately estimated by assuming that the transistor base and heat sink are at a uniform temperature and the cross-sectional area is invariant over the total length of the flow path.

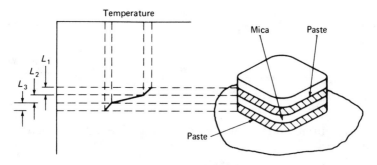

Fig. 1-3. Interface resistance geometry for a transistor package attached to a heat sink.

$$R_k = \int_0^L dx/(kA_k)$$

E1.4
General thermal
resistance of a solid

$$R_k = (L_1/k_1A_1) + (L_2/k_2A_2) + (L_3/k_3A_3)$$

which may be recognized as a formula for adding thermal resistances in series.

$$R = \sum_{i=1}^N R_i$$

E1.5
Series addition

In this application, $k_1 = k_3 = 0.02$ watt/in. $\cdot$ $^\circ$C, the thermal conductivity of a typical filled thermal paste; $k_2 = 0.02$ watt/in. $\cdot$ $^\circ$C for mica. The component path lengths are given as $L_1 = 0.0005$ in., $L_2 = 0.003$ in., and $L_3 = 0.0005$ in. A value of $A_k = 1.0$ in.^{2} is representative of a TO-3 transistor base.

$$R_1 = R_3 = 0.0005/(0.02)(1.0)$$

$$= 0.025^\circ C/watt$$

$$R_2 = 0.003/(0.02)(1.0)$$

$$= 0.150^\circ C/watt$$

$$R_k = 0.025 + 0.150 + 0.025$$

$$= 0.2^\circ C/watt$$

which may not seem very large except that applications requiring 50 to 100 watts through an interface of this type are possible.

Interface resistances may be encountered for which a precise thickness and thermal conductivity are not well defined. For example, the interface formed by two similar metal brackets bolted together is not a continuum of the same metal, but rather is a complex mixture of metal–metal contacts and air voids (assuming the absence of thermal paste, silicone grease, or vacuum). The actual thermal resistance of a contact is a function of any or all physical surface characteristics such as oxide thickness, surface roughness, material hardness, surface temperature, and finally the contact pressure holding the two pieces together. Figures 1-4(a), (b) provide contact resistances for two materials, aluminum and stainless steel, commonly

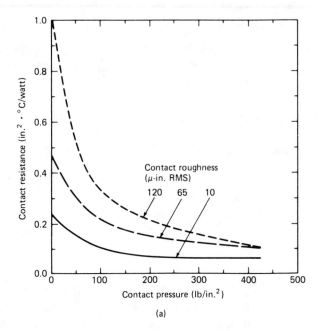

(a)

Fig. 1-4(a). Contact resistance of 75S-T6 aluminum to aluminum joint with air interface. Mean temperature of joint = 93°C. Data plotted from [1].

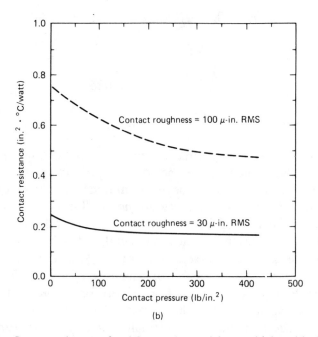

(b)

Fig. 1-4(b). Contact resistance of stainless steel to stainless steel joint with air interface. Mean temperature of joint = 93°C. Data plotted from [1].

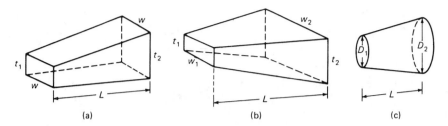

Fig. 1-5. One-dimensional conduction geometries.

encountered in electronic packaging. Note that the accepted dimensions are in.$^2 \cdot °C/$watt or the equivalent. Thus if r_c is the ordinate of Fig. 1-4 and A_k is the apparent cross-sectional flow path area, the resistance in $°C/$watt is:

$$R = r_c/A_k$$

so that the larger the area, the smaller the resistance R.

Thermal design problems are sometimes encountered where A_k is not uniform over the path length, but may be approximated by one of a few simple shapes. For example, suppose that the one-dimensional geometry is represented by a uniform width but linearly diverging height, as shown in Fig. 1-5(a). The area as a function of x is:

$$A_k = w[t_1 + (t_2 - t_1)x/L]$$

Applying E1.4:

$$R_k = \int_0^L dx/(kA_k)$$

$$= (1/kw) \int_0^L dx/[t_1 + (t_2 - t_1)x/L]$$

Substituting:

$$u = t_1 + (t_2 - t_1)x/L$$

$$du = (t_2 - t_1)\, dx/L$$

$$R_k = [L/kw(t_2 - t_1)] \int_{t_1}^{t_2} \frac{du}{u}$$

$$R_k = L \ln(t_2/t_1)/kw(t_2 - t_1)$$

E1.6
Uniform
width wedge

For the geometry in Fig. 1-5(b):

$$R_k = (1/k) \int_0^L dx/A_k$$

$$A_k = [t_1 + (t_2 - t_1)x/L][w_1 + (w_2 - w_1)x/L]$$

$$= t_1 w_1 + t_1(w_2 - w_1)x/L + w_1(t_2 - t_1)x/L$$

$$+ (t_2 - t_1)(w_2 - w_1)x^2/L$$

$$= t_1 w_1 + (t_1 \Delta w + w_1 \Delta t)x/L + \Delta t \Delta w x^2/L^2$$

$$= a + bx + cx^2$$

where:

$$a = t_1 w, \quad b = (t_1 \Delta w + w_1 \Delta t)/L, \quad c = \Delta t \Delta w/L^2$$

and:

$$\Delta w = w_2 - w_1, \quad \Delta t = t_2 - t_1$$

$$kR_k = \int_0^L dx/(a + bx + cx^2)$$

$$= (1/\sqrt{-q}) \ln\left(\frac{2cx + b - \sqrt{-q}}{2cx + b + \sqrt{-q}}\right)\Bigg|_{x=0}^{x=L}, \quad q = 4ac - b^2$$

Finally:

$$R_k = [L/k(t_1 \Delta w - w_1 \Delta t)] \ln\left[\frac{(\Delta w/w_1) + 1}{(\Delta t/t_1) + 1}\right]$$

E1.7

Variable width wedge

The integration is more complex, but straightforward for the geometry illustrated in Fig. 1-5(c):

$$A_k = \pi r^2$$

$$r = r_1 + (r_2 - r_1)x/L$$

$$R_k = (1/\pi k) \int_0^L dx/[r_1 + (r_2 - r_1)x/L]^2$$

$$u = r_1 + (r_2 - r_1)x/L$$

$$du = (r_2 - r_1)\, dx/L$$

$$R_k = [L/\pi k(r_2 - r_1)] \int_{r_1}^{r_2} \frac{du}{u^2}$$

$$= [L/\pi k(r_2 - r_1)][-1/u]_{r_1}^{r_2}$$

$$\boxed{R_k = 4L/(\pi k D_1 D_2)}$$

E1.8
Cone

1.3 CONVECTION

The basic relation that describes heat transfer by convection from a surface (see Fig. 1-6) presumes a linear dependence on surface temperature rise and is referred to as Newtonian cooling.

$$\boxed{Q_c = \bar{h}_c A_s (T_s - T_A)}$$

E1.9
Newtonian cooling

where

Q_c = heat transferred from a surface to ambient by convection, watts
A_s = area of surface, cm^2 or $in.^2$
T_s = temperature of surface, $°C$
T_A = temperature to which heat is being transferred
$\bar{h}_c$ = average convective heat transfer coefficient, watts/$cm^2 \cdot °C$ or watts/$in.^2 \cdot °C$

If the formula for Newtonian cooling is rewritten as:

$$\Delta T = (1/\bar{h}_c A_s) Q_c$$

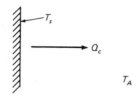

Fig. 1-6. Surface heat transfer.

it becomes clear that a surface resistance may be identified as

$$R_s = 1/\bar{h}_c A_s$$

E1.10
Convective surface
resistance

with the dimensions of °C/watt. Expected values of $\bar{h}_c$ in air are 0.0015 to 0.015 watt/in.$^2 \cdot$°C for natural convection and 0.015 to 0.15 watt/in.$^2 \cdot$°C for forced convection. Methods of estimating $\bar{h}$ for application purposes will be elaborated on in Chapter 2.

Several comments are in order before we proceed to a sample calculation. First, the reader should recognize that E1.9 is not really a law of heat transfer in the sense of Fourier's law. Rather, it is a definition of the heat transfer coefficient. This definition appears rather simple and straightforward: it specifies the quantity of heat transferred through a temperature difference. The oversimplification is that $\bar{h}_c$ may depend quite significantly on the surface and surrounding temperatures, fluid velocity in the case of forced convection, fluid properties such as viscosity and density, and finally the surface geometry. A detailed mathematical analysis of the "boundary layer" phenomena is given in many standard college-level texts on heat transfer. The results of these studies will be summarized in Chapter 2, but with emphasis placed on relatively simply stated formulas and graphs useful for application to real problems.

The actual convective heat transfer from a surface varies with location. Thus it would be more accurate to specify:

$$dQ_c = h_c \, dA_s (T_s - T_A)$$

where both T_s and h_c are a function of position on the surface. Heat transfer coefficients for most surfaces encountered in electronic equipment are not, however, known with sufficient accuracy to warrant use of anything other than an h averaged over the geometry of the surface if not the temperature.

A physical understanding of h is aided by considering the case shown in Fig. 1-7, of an air-moving device such as a small fan blowing air in the x-direction parallel to a flat plate. As the air approaches the plate, it has a uniform frontal velocity v_0. As the air flows past

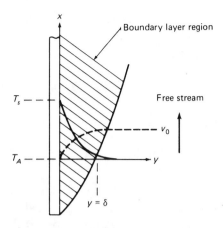

Fig. 1-7. Boundary layer geometry.

the leading edge, the molecules immediately adjacent to the plate
($y = 0$) adhere to the surface, and as the flow progresses in the
x-direction, succeeding layers of low-velocity air molecules build
up along the y-axis. The result is a hydrodynamic boundary layer
with a velocity profile such that $v = 0$ at $y = 0$, the plate surface, and
$v = v_0$, the free stream velocity just outside the boundary layer.

A temperature gradient also exists in the air between the plate
surface and the free stream flow. The characteristics of air are such
that this thermal boundary layer is developed over approximately the
same y distance as the hydrodynamic boundary layer. The convective heat transfer coefficient may then be considered as the conductance per unit area obtained by dividing the average thermal conductivity of the boundary layer by the thickness, i.e.:

$$h \sim k/\delta$$

A more complete physical description of boundary layer physics is
included in Chapter 2.

Even the temperature T_A to which heat is convected is not always well specified. Consider, for example, the situation of closely
spaced circuit cards within an electronic enclosure. The air temperature measured in a direction perpendicular to the cards may vary so
much that a T_A representative of a fixed ambient is really only an
idealization of a rather cumbersome concept. However, analysis of
the heat flow from the cabinet exterior is another matter. The ambient temperature T_A is taken for the surrounding air at a sufficiently
remote distance as to be uniform.

Solutions of differential equations for conductive heat flow with
convective surface losses may require boundary conditions specified
by Newtonian cooling, in which case Fourier's law is also used. Just
inside a surface specified as x_s:

$$\Delta Q_k = -k\Delta A_k (dT/dx)\big|_{x_s - \xi}$$

and just outside the surface:

$$\Delta Q_c = \bar{h}_c \Delta A_s (T - T_A)\big|_{x_s + \xi}$$

In the limit $\xi \to 0$, $\Delta Q_k \to \Delta Q_c$; thus:

$$k(dT/dx)\big|_{x_s} = -\bar{h}_c (T - T_A)\big|_{x_s}$$

A sample problem illustrates the use of the convective heat transfer coefficient. A rectangular 10 in. × 12 in. cabinet top (Fig. 1-8) is at an average temperature of 30°C in a room at 20°C. Compute the heat transferred to the room. An $\bar{h}_c = 0.0022$ watt/in.$^2 \cdot$ °C is consistent for a horizontal plate of these dimensions.

$$Q_c = \bar{h}_c A_s (T_s - T_A)$$

$$= 0.0022(10)(12)(30 - 20)$$

$$= 2.6 \text{ watts}$$

1.4 SIMPLE RADIATION

Newtonian cooling may also be used to define a radiation heat transfer coefficient, h_r:

$$Q_r = \mathcal{F} h_r A_s (T_s - T_A)$$

$\mathcal{F}$ is a factor that includes both surface finish and geometry effects. A common, but elementary, case is that of the exterior surface of a cabinet radiating to the surrounding room walls which are very nearly the same temperature as the ambient air. In this case $\mathcal{F} = \epsilon$ where ϵ is defined as the cabinet surface emissivity, a quantity that is less than 1.0 for all real surfaces ($\epsilon \simeq 0.8$ for many painted surfaces). For T_s within 20°C of T_A and T_A between 0 and 100°C:

$$h_r \simeq 1.463 \times 10^{-10} (T_A + 273)^3$$

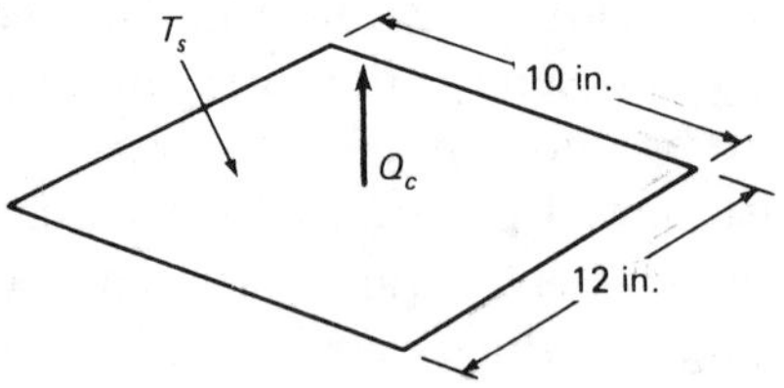

Fig. 1-8. Natural convection heat transfer from a cabinet top.

with dimensions of watts/in.$^2 \cdot °$C. This quantity is derived in Section 3.4.3, where it is given by E3.25.

A comparison with E1.9 and E1.10 indicates radiation may be quantified by a resistance using the radiation heat transfer coefficient:

$$R_r = 1/\mathfrak{F}h_r A_s$$

Further discussion of radiation is postponed until Chapter 3, which is totally devoted to the subject.

1.5 COMBINED CONDUCTION, CONVECTION, AND RADIATION

In the preceding paragraphs and examples conduction and convection have been treated separately. Unfortunately, these processes usually take place simultaneously. Suppose, for example that four transistor cans are attached in the manner illustrated in Fig. 1-9(a). The heat sinking bracket is constructed of aluminum with a cross-sectional area large enough to conduct a significant amount of heat to a bulkhead that convects and radiates to the internal cabinet air via a resistance R_{Bk}. Both the bracket and transistor can surface areas are

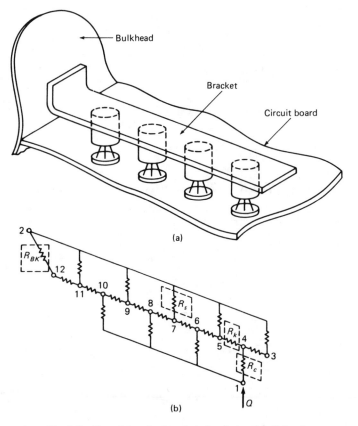

Fig. 1-9. Transistors heat sunk to bracket and bulkhead.

lumped into the set of distributed convection resistances R_s. Conduction from the transistor header to the bracket is included in each R_c, and conduction along the bracket is accounted for by the R_k. The simple circuit in Fig. 1-8 consists of only 12 "nodes," and is probably too complex to solve except with digital computation techniques.

Many multi-element problems may, however, be analyzed by simple circuit techniques using the previously given formula E1.4 for series resistance addition and the easily proven parallel resistance formula. Using the illustration in Fig. 1-10 for reference, a total heat flow of Q watts is split into two parallel flow paths represented by R_1 and R_2. It is desired to find a resistance R equivalent to R_1 and R_2. The temperature drop across R_1, R_2, or R is $T_1 - T_2$, and the heat flow through each leg is:

$$Q_1 = (T_1 - T_2)/R_1$$

$$Q_2 = (T_1 - T_2)/R_2$$

Also:

$$Q = (T_1 - T_2)/R$$

From conservation of energy:

$$Q = Q_1 + Q_2$$

or:

$$(T_1 - T_2)/R = (T_1 - T_2)/R_1 + (T_1 - T_2)/R_2$$

$$\boxed{\frac{1}{R} = \frac{1}{R_1} + \frac{1}{R_2}}$$

E1.11
Parallel resistances

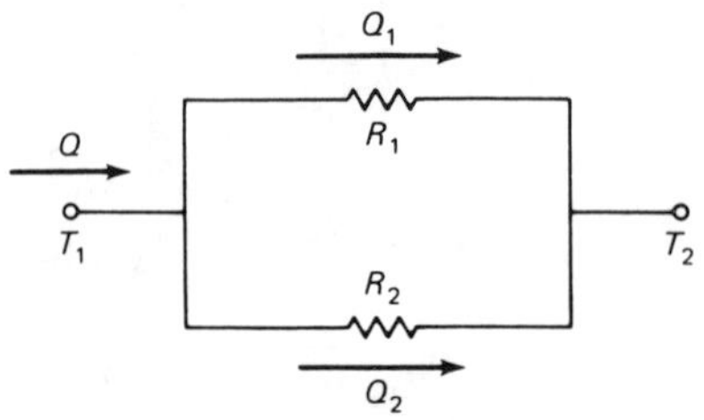

Fig. 1-10. Simple parallel resistance circuit.

It is often convenient to work with conductance, the reciprocal of resistance:

$$C = \frac{1}{R}$$

E1.12
Conductance

$$\frac{1}{C} = \frac{1}{C_1} + \frac{1}{C_2}$$

E1.13
Series conductances

$$C = C_1 + C_2$$

E1.14
Parallel conductances

The series and parallel element formulae are illustrated in the following example of heat dissipation through a plastic cabinet with convection and radiation effects. Figure 1-11 illustrates the cabinet and equivalent thermal circuit. The cabinet dimensions are $L = W = H = 10$ in. Electronic components distributed throughout the interior dissipate 12 watts. The cabinet wall is 0.125 in. thick plastic with a thermal conductivity of 0.01 watt/in. $\cdot$ °C. Reasonable interior and exterior convection coefficients are given as a surface average of approximately 0.002 watt/in.^{2} $\cdot$ °C, whereas the radiation $h_r = 0.0039$ watt/in.^{2} $\cdot$ °C, and $\epsilon \simeq 1.0$. The thermal circuit illustrates the

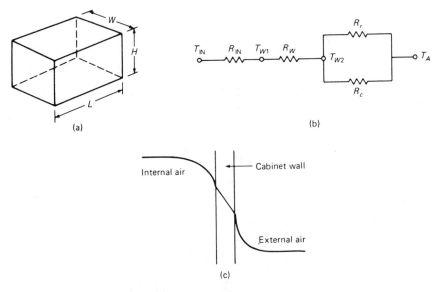

Fig. 1-11. Heat dissipation from a plastic cabinet. (a) Cabinet. (b) Thermal circuit. (c) Temperature gradient.

assumption that internal radiation is negligible because of the multiple shielding effects caused by several circuit cards.

The thermal circuit elements are:

Interior convection:

$$R_{IN} = 1/h_c A_s$$

$$= 1/(0.002)(6)(10)(10)$$

$$= 0.833°C/watt$$

Wall conduction:

$$R_w = L/kA_k$$

$$= 0.125/(0.01)(6)(10)(10)$$

$$= 0.021°C/watt$$

External convection/radiation:

$$C_{EX} = C_r + C_c$$

$$= (1.0)(0.0039)(6)(10)(10) + (0.002)(6)(10)(10)$$

$$= 2.340 + 1.200$$

$$= 3.540 \text{ watts}/°C$$

$$R_{EX} = 1/C_{EX}$$

$$= 0.283°C/watt$$

Total:

$$R = R_{IN} + R_w + R_{EXT}$$

$$= 1.137°C/watt$$

The wall temperature rise above the external ambient is:

$$T_{w2} - T_A = R_{EX} Q$$

$$= (0.283)(12)$$

$$= 4.4°C$$

The rise across the wall thickness is:

$$T_{w1} - T_{w2} = R_w Q$$

$$= (0.021)(12)$$

$$= 0.3°C$$

The internal air temperature rise above external ambient is:

$$T_{IN} - T_A = RQ$$

$$= (1.137)(12)$$

$$= 13.6°C$$

Note how small the temperature rise is across the plastic wall—less than 1°C!

Quite frequently even a modest internal temperature rise of 14°C is excessive if a component must be operated with a large power dissipation. Suppose, for example, that the instrumentation housed in the cabinet in Fig. 1-11(a) is to be guaranteed by the manufacturer to operate in a maximum room temperature of 35°C. The designer is concerned about an integrated circuit that dissipates 1.2 watts and must not exceed a chip junction temperature of 125°C–conditions to ensure an adequate lifetime for the device. A heat sink of some type may be required for the IC package, but even then there is the possibility that a heat sink may not provide adequate cooling. The designer must estimate the magnitude of the problem before proceeding. A quick reference to a heat sink manufacturer's catalog indicates a sink of the style illustrated in Fig. 1-12 with tested thermal characteristics shown in Fig. 1-13. The required minimum case to ambient thermal resistance may be calculated starting with the basic equation for a junction temperature, T_J:

$$T_J = (R_{JC} + R_{CA}) Q + T_A$$

where

R_{JC} = junction to case resistance
R_{CA} = case to ambient resistance
T_A = ambient air temperature in vicinity of component

The local worst-case ambient is the sum of the maximum room temperature and the predicted rise from the preceding example, i.e.:

$$T_A = 35°C + 14°C$$

$$= 49°C$$

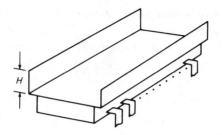

Fig. 1-12. Component with a simple heat sink.

R_{JC} is determined from the IC manufacturer's data sheet. For a ceramic dual-in-line package, $R_{JC} = 25°C/watt$ is a good estimate. Then the required minimum case to ambient resistance is:

$$R_{CA} \leqslant [(T_J - T_A)/Q] - R_{JC}$$

$$= [(125 - 49)/1.2] - 25$$

$$R_{CA} \leqslant 38°C/watt$$

The heat sink manufacturer's data sheet is referred to (illustrated in Fig. 1-13). The specifications also include a resistance plot for the unsinked device. Note that at 1.2 watts the unsinked resistance is

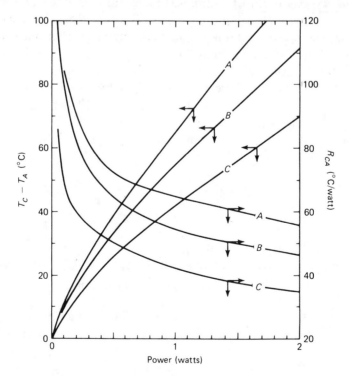

Fig. 1-13. Component thermal characteristics. *A*—Package only. *B*—Heat sink only, $H = 0.25$ in. *C*—Heat sink only, $H = 0.50$ in.

63°C/watt, and only a sink with a height of 0.5 in. may be adequate. The plot shows a sink resistance of 40°C/watt, slightly in excess of the required 38°C/watt. At this point the designer would be well advised not to take either his calculations or the sink manufacturer's curves too literally, but rather to use the analysis as an estimate of the magnitude of the problem. If an improved heat sink is not found, at least a quick bench test of the selected cooling scheme is recommended.

1.6 NUMERICAL SOLUTION OF STEADY STATE PROBLEMS

Most complex problems are solvable by large thermal circuits built of many resistance or conductance elements as defined in Section 1.5. Temperatures and heat flow for such circuits are easily determined by digital computation methods. This section provides the basis for the program TNETFA, which is discussed in Chapters 8 and 10.

The basic network equation may be derived by examining only a small portion of a circuit. Consider the example in Fig. 1-11. The node (2) indicated by temperature T_{w2} is connected to nodes (1) and (A). Conservation of energy applied to (2) requires that:

$$\text{heat in} = \text{heat out}$$

for steady state conditions. Suppose that a heat source (e.g., a surface-mounted transistor) dissipates Q_2 watts. Conservation of energy may be written in terms of temperature, conductance, and heat source dissipation:

$$Q_2 = C_w(T_{w2} - T_{w1}) + C_r(T_{w2} - T_A) + C_c(T_{w2} - T_A)$$

where:

$$C_w = 1/R_w, \ C_r = 1/R_r, \text{ and } C_c = 1/R_c$$

Solving for T_{w2}:

$$T_{w2} = \frac{(C_w T_{w1} + C_r T_A + C_c T_A) + Q_2}{C_w + C_r + C_c}$$

This is readily generalized to any node i connected to N other nodes:

$$T_i = \frac{\sum_{\substack{j \neq i}}^{N} C_{ij} T_j + Q_i}{\sum_{\substack{j \neq i}}^{N} C_{ij}}$$

E1.15
Steady state
numerical solution

The indication $j \neq i$ is used as a reminder that a node is not connected to itself.

E1.15 is used to solve for a set of temperatures in an iterative fashion. A set of equations defined by E1.15 must contain one or more fixed-temperature nodes. The ambient T_A of Fig. 1.11 is a fixed temperature. Clearly T_A is one of the set T_j, but not T_i. Fixed temperatures are not iterated!

The solution method starts with a guess of all T_j that are not fixed. Each T_i is in turn computed using the guessed T_j on the right side of E1.15. As each T_i is computed, it may be used as a T_j in succeeding computations; i.e., the most recent best estimate should be used for each T_j. Nodes that do not have associated heat-dissipating devices naturally have $Q_i = 0$. One iteration is said to have been made when E1.15 has been applied to each nonfixed node i once and only once. After each successive iteration, the temperature change for each node should become smaller. Sometimes this occurs with an initial overshoot followed by convergence to a solution after many iterations.

One particularly important feature of the iteration method is that temperature-dependent conductances may be successively recomputed for each iteration. Elements representing natural convection, radiation, and temperature-dependent thermal conductivities are typical examples.

There is no generally reliable method of predicting the number of iterations that will be required of any particular problem. A rough rule of thumb, however, is to start with as many iterations as there are nodes. Some problems will require far fewer iterations, and some surprisingly small problems will require many, many more iterations than the number of nodes. A good practice is to observe the size of temperature change between each iteration. It should be at least two orders of magnitude less than the difference between the highest-temperature node and the fixed ambient temperature at problem termination.

A second quantity that is useful in recognizing a good solution (in the mathematical sense) is the total residual energy of the system. This is just the heat energy unaccounted for by an imperfect mathematical solution. The residual r_i for node i is:

$$r_i = \sum_{\substack{j \neq i}}^{N} C_{ij}(T_i - T_j) - Q_i$$

If the iterations have converged to a perfect mathematical solution, each r_i for the nonfixed temperature nodes is zero. In most real situations $r_i \neq 0$. The sum of all r_i for a complete thermal network is the system residual energy. A good system residual is less than about 2% of the total heat input (sum of all r_i). Both the temperature changes and system residual energy may be monitored by

TNETFA (Chapters 8 and 10). In TNETFA the system residual is identified as energy balance.

A method of overrelaxation that extrapolates two successive iterations may sometimes be applied to reduce the number of required iterations. Defining T_0 for node i as the previously iterated temperature, and T_N as the newly iterated temperature, the overrelaxed, or extrapolated, temperature T_R is computed with an overrelaxation factor β:

$$T_R = T_0 + \beta(T_N - T_0)$$

E1.16
Overrelaxed
temperature

$$1.0 \leqslant \beta < 2.0$$

E1.17
Overrelaxation
factor

T_0 in E1.16 may actually be an overrelaxed result from the previous iteration. Note that $\beta = 1.0$ gives $T_R = T_N$, the nonrelaxed result. It is very important to have β never equal to, but always less than 2.0. A divergent solution will result for $\beta = 2.0$.

As in the case of estimating the required number of iterations, the optimum β is determined only by experience, and any two problems may require different optimal β values. Some problems require $\beta = 1.0$, and any other values will result in divergent iterative solutions. The author recommends particular care for problems with many nonlinear elements (i.e., temperature-dependent conductances, etc.).

1.7 NUMERICAL SOLUTION OF TIME-DEPENDENT PROBLEMS

Analytical solutions to time-dependent thermal problems are well covered by the vast number of good texts now available. Therefore, only theory sufficient to support the time-dependent features of TNETFA (Chapters 9 and 10) is included here.

The formulation of a time-dependent equation for a nodal temperature is based on an extension of the steady state energy balance equation:

$$Q_i - \sum_{j \neq i}^{N} C_{ij}(T_i - T_j) = \frac{\rho C_p \Delta V_i (T_i^\Delta - T_i)}{\Delta t}$$

The last term on the right quantifies energy storage for an energy in greater than out of node i. This is just the finite difference form of E1.3 where T_i is the temperature of node i at time t, and T_i^Δ the temperature at time $t + \Delta t$. The temperatures T_j are taken at time t. This is a "forward" finite difference in time. The quantities ρ (mass/volume), C_p (calories/degree · mass), and ΔV_i (volume) are density, specific heat at constant pressure, and volume of the node i, respectively. Any set of consistent units may be used. Care must be taken to use a mechanical equivalent of heat conversion for C_p when watts are used. TNETFA allows selection of three different sets of units. The sample time-dependent problem in Section 8.5 should be carefully studied.

T_i^Δ is obtained following a little algebraic manipulation:

$$T_i^\Delta = T_i(1 - STAB_i) + \frac{\Delta t}{CAP_i}\left(Q_i + \sum_{\substack{j \neq i}}^{N} C_{ij}T_j\right)$$

E1.18

Time-dependent temperature

The quantity CAP_i is the thermal capacitance of node i:

$$CAP_i = \rho C_p \Delta V_i$$

E1.19
Node capacitance

$STAB_i$ is the stability constant of node i:

$$STAB_i = \frac{\Delta t}{CAP_i} \sum_{\substack{j \neq i}}^{N} C_{ij}$$

E1.20
Stability constant

The stability constant sets an absolute upper limit to the time step. Note that by E1.18 $STAB_i \leqslant 1.0$. A $STAB_i > 1.0$ would mean that the higher the temperature T_i at time t, the lower T_i^Δ will be at time $t + \Delta t$, which is nonsense. It would also result in instabilities in the numerical solution. The upper limit to the time step is given by E1.21:

$$\Delta t \leqslant \frac{CAP_i}{\displaystyle\sum_{\substack{j \neq i}}^{N} C_{ij}}$$

E1.21
Time-step limit

A network with several different values of CAP_i and $\Sigma_{j \neq i}^{N} C_{ij}$ for the nodes must have a Δt that is smaller than the smallest $CAP_i/\Sigma_{j \neq i}^{N} C_{ij}$ for the entire system.

Readers interested in further details on consideration of numerical stability are advised to start with a good text such as [2].

Chapter 2
Convection

2.1 INTRODUCTION

The Newtonian cooling law is used to define:

E2.1
Newtonian
cooling, convection

$$Q_c = \bar{h}_c A_s (T_s - T_A)$$

as previously written in Section 1.3. The bar in $\bar{h}_c$ is used to indicate an average over a surface at temperature T_s. The quantity T_A is the temperature of the ambient air to which the surface is convecting. In most applications to electronic equipment, the surface A_s is not at a uniform temperature, and h_c is both temperature- and position-dependent. Consideration of these factors in calculating h_c tends to make problems more complex than necessary for adequate accuracy. The procedure used in most of this book is to assume an average surface temperature for T_s and an $\bar{h}_c$ averaged over the surface.

If the thermal conductance concept is used, simultaneous radiation and convection transport from a surface to ambient is very conveniently quantified by a total surface conductance, C_s:

$$C_s = C_c + C_r$$

$$= (\bar{h}_c + \mathcal{F} \bar{h}_r) A_s$$

$$= \bar{h} A_s$$

or:

$$\bar{h} = \bar{h}_c + \mathcal{F} \bar{h}_r$$

E2.2
Combined heat
transfer coefficient

As an example, the 10 in. × 12 in. surface in Fig. 1-8 was characterized by an $\bar{h}_c = 0.0022$ watt/in.$^2 \cdot$ °C. If the same surface has an emissivity of 1.0 and radiates to a perfectly absorbing 20°C ambient such that $\mathcal{F} = 1.0$, then:

$$\bar{h}_r = 1.463 \times 10^{-10}(T_A + 273)^3$$

$$= 1.463 \times 10^{-10}(293)^3$$

$$= 0.0037 \text{ watt/in.}^2 \cdot {}^\circ C$$

and:

$$\bar{h} = \bar{h}_c + 1.0(\bar{h}_r)$$

$$= 0.0022 + 0.0037$$

$$= 0.006 \text{ watt/in.}^2 \cdot {}^\circ C$$

Note that for a highly emissive surface, radiative heat transfer is greater than convective heat transfer by the ratio $0.0037/0.0022 = 1.7$. This refutes the often erroneous belief that radiation is only important at very large temperatures.

The fundamental problem associated with convective heat transfer is to obtain the appropriate convective heat transfer coefficient. This will be accomplished for a few significant cases by providing formulae and/or graphs, relating $\bar{h}_c$ to air velocity (forced air convection), temperature rise (natural convection), and a geometric parameter. However, much of the heat-transfer literature uses a set of dimensionless numbers to correlate the various physical quantities. The reader who wishes to have a capability beyond the bounds of this volume should be aware of which quantities are relevant.

2.2 NUSSELT AND PRANDTL NUMBERS

This chapter will be associated with one of the most common convection systems, i.e., heat transfer across a solid to air interface as illustrated in Fig. 2-1. A thin thermal boundary layer exists such

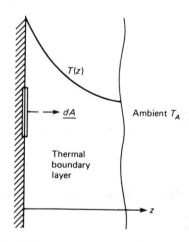

Fig. 2-1. Heat convection at a solid–air interface.

that very close to the wall surface the dominant heat transfer mechanism in the air is molecular conduction. Then very close to the wall Fourier's law may be applied to a surface area dA:

$$dQ_c = -kdA(dT/dz)\big|_{z=0}$$

Also, by definition of h_c:

$$dQ_c = h_c dA(T_s - T_A)$$

where T_s is the surface temperature, and T_A is the air temperature at a sufficiently large z such that no significant temperature gradient exists. The quantity k is the thermal conductivity of the boundary layer.

Setting the right sides of the preceding two equations equal:

$$-kdAdT/dz = h_c dA(T_s - T_A)$$

$$-kdT/dz = h_c(T_s - T_A)$$

Since T_s is a constant, $dT = d(T - T_s)$ and:

$$kd(T_s - T)/dz = h_c(T_s - T_A)$$

Multiplying through by some appropriate parameter P with the dimensions of a length, and rearranging:

$$\frac{h_c P}{k} = \frac{\dfrac{d(T_s - T)}{dz}}{\left(\dfrac{T_s - T_A}{P}\right)}$$

The quantity on the left is the dimensionless manner of specifying h_c and is referred to as the Nusselt number:

$$\boxed{Nu_p = \frac{h_c P}{k}}$$

E2.3
Nusselt number

Two physical interpretations of the Nusselt number stem from:

$$Nu_p = \frac{d(T_s - T)/dz}{(T_s - T_A)/P} \quad \text{or} \quad \frac{d[(T_s - T)/(T_s - T_A)]}{d(z/P)}$$

both taken at $z = 0$, of course. The first form of Nu_p is the ratio of the temperature gradient at the wall to a reference gradient. The second form of Nu_p may be interpreted as a dimensionless temperature gradient at the wall.

The nature of the parameter P does not become clear until more detail is provided concerning the geometry of the convecting surface. As an example, a vertical, free convecting flat plate uses $P = H$, the plate height. The recommended P will be indicated at the appropriate sections.

Another dimensionless combination required to quantify all types of convective heat flow is the Prandtl number, Pr.

$$Pr = \mu C_p / k$$

E2.4
Prandtl number

where

μ = absolute viscosity, gm/in. · sec
C_p = specific heat at constant pressure, joules/gm · °C
k = thermal conductivity of air, watts/in. · °C

Division of the numerator and denominator by ρ (air density) lb_m/ft^3 gives:

$$Pr = \frac{\mu/\rho}{k/C_p\rho} = \frac{\nu}{a}$$

where

ν = kinematic viscosity, a form of momentum diffusivity
a = thermal diffusivity

Thus:

$$Pr \sim \frac{\text{momentum diffusivity}}{\text{thermal diffusivity}}$$

which permits a physical interpretation of Pr as the relative rates with which the velocity and temperature profiles, i.e., the aerodynamic and thermal boundary layers, develop, as for instance at the surface on a flat plate. The following therefore applies to all fluids (including air):

$Pr < 1$:   Temperature profile develops more rapidly than velocity profile.

$Pr = 1$: Temperature and velocity profiles develop at same rate.

$Pr > 1$: Velocity profile develops more rapidly than temperature profile.

Prandtl number vs. temperature is tabulated for air in Table 2-1. Note that $Pr = 0.72$ for temperatures of 0–100°C. This is sufficiently close to unity that velocity and thermal boundary layers develop such that they are of about equal thickness. Thus when both the velocity and temperature are uniform at the approach to the plate or duct, boundary layer thickness estimates refer to both thermal and velocity profiles.

2.3 FREE CONVECTION

Thermal resistances characteristic of heat transfer by free convection in air are often the dominant elements in the overall resistance of a thermal system. However, the ultrasimplicity of a free convection cooling system without failure-prone electric fans or blowers has resulted in its widespread use.

Free convective air currents occur when a solid and the surrounding air are at different temperatures. Heat flows to or from the fluid and produces a density decrease or increase, respectively, which in turn causes an air displacement (convection current). Gravity is the acting force. Figure 2-2 illustrates convective heat transfer Q from a vertical plate at a temperature T_s in air at a temperature T_A. Figure 2-3 indicates the boundary layer development on a single sided plate.

A dimensionless quantity unique to free convection is the Grashof number, a ratio of buoyant to viscous forces:

$$Gr_p = \frac{g\rho^2}{\mu^2}\,\beta(T_s - T_A)\,P^3$$

E2.5
Grashof number

where ρ, μ, and P are as defined in Section 2.2. In addition:

$g = 386.0$ in./sec²

β = coefficient of thermal expansion of air, 1/°C

T_s = surface temperature, °C

T_A = ambient fluid temperature, °C

If air is approximated as an ideal gas:

$$\beta = \frac{1}{T_A + 273.15}$$

E2.6
Coeff. thermal
expansion, ideal gas

Table 2-1. Physical properties of air at atmospheric pressure with units converted from [2]. Table A-3 (p. 636) from *Principles of Heat Transfer*, 3rd Edition by Frank Kreith. Copyright © 1958, 1965, 1973 by Harper & Row, Publishers, Inc. Reprinted by permission of the publisher.

T (°C)	T (°F)	ρ (gm/in.³)	c_p (joule/gm·°C)	μ (10^{-4} gm/in.·sec)	ν (in.²/sec)	k (10^{-4} watt/in.·°C)	Pr	β (10^{-3}/°C)	$g\rho^2/\mu^2$ (10^6/in.³)	$g\beta\rho^2/\mu^2$ (10^3/in.³·°C)
-18	0	0.023	1.00	4.195	0.0187	5.846	0.73	3.916	1.12	4.38
0	32	0.021	1.01	4.403	0.0209	6.153	0.72	3.661	0.899	3.29
38	100	0.019	1.01	4.856	0.0259	6.769	0.72	3.216	0.569	1.83
93	200	0.016	1.01	5.442	0.0344	7.648	0.72	2.729	0.324	0.885
149	300	0.014	1.02	6.085	0.0441	8.483	0.71	2.370	0.195	0.462
204	400	0.012	1.03	6.614	0.0544	9.318	0.689	2.094	0.128	0.269
260	500	0.0108	1.03	7.143	0.0655	10.15	0.683	1.876	0.243	0.166

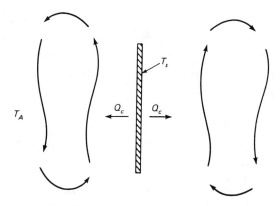

Fig. 2-2. Illustration of natural convection heat transfer and air currents for a vertical plate with $T_s > T_A$.

The quantities β, $g\rho^2/\mu^2$, and $g\rho^2\beta/\mu^2$ are tabulated in Table 2-1 vs. temperature to facilitate computation.*

Most of the literature concerning free convection heat transfer coefficients is written in terms of a correlation of Nu_p vs. Gr_p and Pr, or more specifically:

$$Nu_p = C(Gr_p Pr)^n$$

E2.7
Nusselt number,
free convection

The configurations most seriously addressed in this chapter are ver-

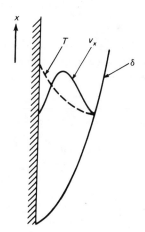

Fig. 2-3. Velocity and temperature profiles for a heated, vertical flat plate in free convection.

---

*Formulae useful for scientific calculator use are:

$$g\rho^2/\mu^2, \, 1/in.^3 = 8.674 \times 10^5 \exp[-1.023 \times 10^{-2} T(^\circ C)]$$

$$g\rho^2\beta/\mu^2, \, 1/in.^3 \cdot {}^\circ C = 3.139 \times 10^3 \exp[-1.311 \times 10^{-2} T(^\circ C)]$$

both to within 10% of Table 2-1, $T = -18^\circ C$ to $150^\circ C$.

tical and horizontal flat plates, and vertical and horizontal cylinders. Table 2-2 defines the dimensional parameter P and the constants for these configurations. $\overline{Nu}_p$ specifically denotes a Nusselt number, and therefore an $\overline{h}_c$ also averaged over the plate dimensions. The average $\overline{h}_c$ is determined from:

$$\overline{h}_c = \frac{k}{P}\,\overline{Nu}_p$$

The properties k, Gr_p, and Pr for air are evaluated at the mean between the surface and ambient temperatures, i.e., approximately the mean boundary layer temperature.

Additional comments are in order concerning horizontal plates. The dimensional parameter in Table 2-2 for a horizontal plate is actually the surface area of one side divided by the surface perimeter. The limit for a square plate, $P = W/4$, appears contradictory if carefully compared with other references, e.g., [2] and [4]. However, the more recent data of [6] indicate advances in technique and accuracy. The practice of using $C = 0.27$, consistent with [2] and [4], is used in this book because of a lack of any alternative at this time. For horizontal, rectangular plates:

E2.8
Dimensional parameter, horizontal, rectangular plates

$$P = WL/[\,2(W + L)\,]$$

Table 2-2. Free convection parameters for average Nusselt numbers for cylinders and flat plates. H = height; D = diameter; W = width; L = length; T_s = surface temperature; T_A = air temperature.

Configuration	P	C	n	Comment
Vertical plate	H	0.59	0.25	$10^4 < GrPr < 10^9$
		0.13	0.33	$10^9 < GrPr < 10^{12}$
				Results from [4]
Vertical cylinder	H		Same as vertical plate	
Horizontal plate lower surface $T_s < T_A$ or upper surface $T_s > T_A$	$\dfrac{WL}{2(W+L)}$	0.54	0.25	$2.2 \times 10^4 \leqslant GrPr \leqslant 8 \times 10^6$
		0.15	0.33	$8 \times 10^6 \leqslant GrPr \leqslant 1.6 \times 10^9$
	from [5]	from [6]	from [6]	from [6]
Horizontal plate upper surface $T_s < T_A$ or lower surface $T_s > T_A$	$\dfrac{WL}{2(W+L)}$	0.27	0.25	$3 \times 10^5 < GrPr < 3 \times 10^{10}$
	from [5]	from [4]	from [4]	from [4]
Horizontal cylinder	D	0.53	0.25	$10^3 < Gr < 10^9$
	from [4]	from [4]	from [4]	from [4]

For a vertical plate:

$$P = H$$

E2.9
Dimensional
parameter,
vertical plate

The appearance of a *GrPr* criterion for *C* and *n* is due to the laminar or turbulent nature of the resulting boundary layer on the plate. The smaller and larger of the *GrPr* products are for laminar and turbulent flow, respectively. This appears to make matters more complicated, but in most applications such is not the case. A simple calculation of two extreme, but practically possible, situations adequately allows resolution of this issue.

Consider a vertical plate. For a 2.0 in. height, a 70°C ambient, and a 1°C surface rise, a somewhat minimal value of *GrPr* is obtained:

$$GrPr = (g\beta\rho^2/\mu^2)(H)^3(\Delta T)Pr$$

$$= (1.25 \times 10^3)(2.0)^3(1)(0.72)$$

$$= 7.2 \times 10^3$$

A rather maximal *GrPr* is obtained for a 10 in. high plate in a 0°C ambient, but with a 100°C rise:

$$GrPr = (1.6 \times 10^3)(10)^3(100)(0.72)$$

$$= 1.2 \times 10^8$$

Reference to Table 2-2 indicates that the larger value, *GrPr* = 1.2×10^8, is still within the range of a laminar boundary layer. The value *GrPr* = 7.2×10^3 is just slightly below the 10^4 lower limit indicated. Referral to Fig. 2-4, from [4], is required. For log (*GrPr*) = log (7.2×10^3) = 3.86, *Nu* = 6.0. If the correlation from Table 2-2 were used, one would obtain:

$$Nu = 0.59(GrPr)^{0.25}$$

$$= 0.59(7.2 \times 10^3)^{0.25}$$

$$= 5.4$$

which is only 10% in error. This is within the accuracy of any of the correlations. Most standard applications in electronic equipment appear to fall within the laminar flow region.

It is worthwhile to use the preceding example to illustrate the

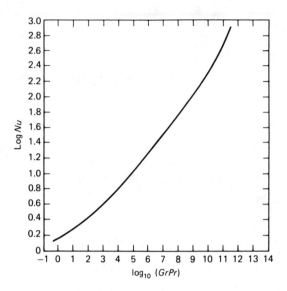

Fig. 2-4. Natural convection from short vertical plates to air. [4]. From *Heat Transmission*, 3rd Edition, by W. H. McAdams, copyright © 1954 by McGraw-Hill, Inc. Used with the permission of McGraw-Hill Book Company.

effect of the significant dimension on the heat transfer coefficient. Suppose both 2.0 in. and 10 in. high plates are at a temperature of 75°C in 25°C air. Then $\bar{T}_s = 50°C$, and:

$$\Delta T_s = 75 - 25 = 50°C \quad \text{and} \quad g\beta\rho^2/\mu^2 = 1.6 \times 10^3$$

Using E2.3:

$H = 2.0$ in.:

$$h_c = \frac{k}{H} Nu_H$$

$$= \frac{k}{H} (0.59)(GrPr)^{0.25}$$

$$= \frac{k}{H} (0.59)[(g\beta\rho^2/\mu^2)(\Delta T)(H)^3 Pr]^{0.25}$$

$$= \frac{6.96 \times 10^{-4}}{2.0} (0.59)[(1.6 \times 10^3)(50)(2.0)^3(0.72)]^{0.25}$$

$$= 0.0054 \text{ watt/in.}^2 \cdot °C^*$$

---

*An adequate formula for k for air is:

$$k \text{ (watts/in.} \cdot °C) = 6.145 \times 10^{-4} + 1.61 \times 10^{-6} T \text{ (°C)}$$

$H = 10.0$ in.:

$$h_c = \frac{6.96 \times 10^{-4}}{10} (0.59)[(1.6 \times 10^3)(50)(10)^3(0.72)]^{0.25}$$

$$= 0.0036 \text{ watt/in.}^2 \cdot °C$$

The convective heat transfer coefficients differ by nearly a factor of two. If the plates had the same width, the thermal resistances would differ by:

$$\frac{R(H = 2.0)}{R(H = 10.0)} = (10.0/2.0)(0.0036/0.0054) = 3.3$$

as opposed to a ratio of 5 if only area effects were considered.

Variations of h_c along the height are of interest in heat sinking effects of vertical plates and perhaps circuit cards. Schmidt and Beckman, [11], analytically determined that:

$$h_{cx} \propto 1/x^{0.25}$$

where x is the distance from the plate leading edge. This is consistent with:

$$\boxed{h_{cx} = 0.75 \frac{k}{x} Nu_x}$$

E2.10
x variations of free convection h_c

which may be used to show that:

$$\frac{h_L}{\bar{h}_L} = 0.75$$

The 0.25 exponent dependence of $GrPr$ actually results in only a slight dependence of h_c on average surface temperature. Using the recommended formulae for k and $g\beta\rho^2/\mu^2$, Table 2-3 was constructed using:

$$0.54k \left(\frac{g\beta\rho^2}{\mu^2} Pr \right)^{0.25}$$

for a horizontal plate. The result definitely indicates that for the temperature range 0–100°C, the free convection plate and cylinder formulae may be greatly simplified for air at sea level pressure.

Table 2-3. Temperature dependence of horizontal plate.

$\bar{T}\,(°C)$	$0.54k\left(\dfrac{g\beta\rho^2}{\mu^2}Pr\right)^{0.25}$
0	0.0023
25	0.0023
50	0.0022
75	0.0021
100	0.0021

$$h_c = 0.0022(\Delta T/P)^{0.25}$$

E2.11
Horizontal plate

$$h_c = 0.0011(\Delta T/P)^{0.25}$$

E2.12
Horizontal plate

$$h_c = 0.0024(\Delta T/P)^{0.25}$$

E2.13
Vertical plate

The arrows for each of the three plate orientations indicate the direction of heat transfer based on the plate temperature with respect to the surrounding air. In the first case (E2.11), the upper surface of a plate with a temperature greater than the ambient above it would convect heat in the indicated direction of the arrow. The lower surface of the same plate with a temperature less than the ambient beneath it would again transfer heat in the direction indicated by the arrow.

In the second case (E2.12), the upper surface of a plate with a temperature less than the ambient above it would convect heat in the indicated direction of the arrow. The lower surface of the same plate with a temperature greater than the ambient beneath it would similarly transfer heat in the direction of the arrow.

Heat is transferred to or from ambient by the vertical plate when the ambient temperature is less than or greater, respectively, than that of the plate.

An application example is deferred until after some experiments performed in the author's laboratory have been described. The impetus for the experimental work was the rather poor correlation of classical convection formulas when used with small devices such as a bracket with a 0.5 in. height or perhaps even smaller components.

The test apparatus was simple. A flat plate was used as the convecting surface with one TO-3 power transistor centrally attached. The plate was thick enough to cause only a modest temperature gradient from the transistor case to the plate edge. Thermocouples were used to determine a temperature profile, from which an average surface temperature was determined. One vertical and two horizontal plates were used. The dimensions and orientations are illustrated in Fig. 2-5, nine configurations in all. Each was powered at three different values to provide additional data.

The convective conductance is the difference between the total and radiative conductances:

$$C_c = C - C_r$$

where

$$C = \frac{\text{power dissipated}}{\bar{T}_s - T_A}$$

$$C_r = \epsilon h_r A_s$$

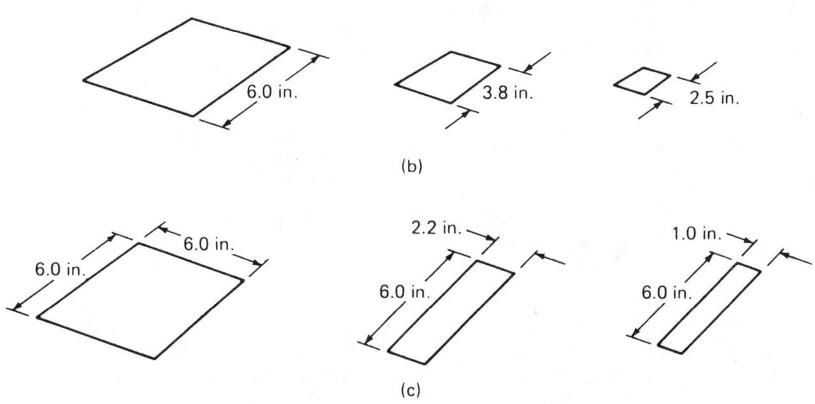

Fig. 2-5. Orientation and geometry of flat plate used in natural convection correlation. (a) Vertical plate–fixed width, variable height. (b) Horizontal plate–square. (c) Horizontal plate–fixed length, variable width.

The emissivity ϵ was pre-established by spraying the plates with a flat black paint.

It was assumed that the h_c for the horizontal heated upper and lower surfaces differed by a factor of two, consistent with Table 2-2 and [4]. The horizontal surface test results plotted in Fig. 2-6 are for the heated surface facing upward. The parameter P is H and $WL/[2(W + L)]$ for the vertical and horizontal cases, respectively.

Curve-fitting the author's empirical data results in a small-device formula:

$$h_c = 0.0018f\left(\frac{\Delta T}{P}\right)^n$$

E2.14
Small device

where

$f = 1.22, n = 0.35$:   Vertical plate

$f = 1.00, n = 0.33$:   Horizontal plate, upper surface $T_s > T_A$, lower surface $T_s < T_A$

$f = 0.5, n = 0.33$:   Horizontal plate, lower surface $T_s > T_A$, upper surface $T_s < T_A$

It is tempting, upon examining the exponents 0.33 and 0.35, to speculate that turbulent conditions existed. However, none of the $Gr_p Pr$ products for the test data indicated turbulent conditions according to the criteria of Table 2-2.

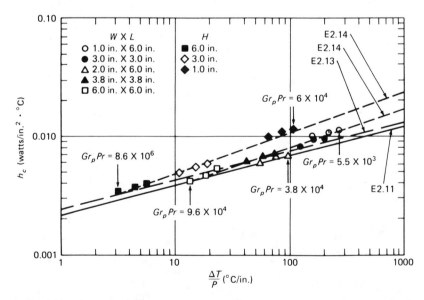

Fig. 2-6. Flat plate convective heat transfer coefficients. $Gr_p Pr$ products correspond to appropriate experimental values.

A rather large surface such as a cabinet side panel of 15 in. height and a rather typical surface temperature rise of 10–20°C result in $\Delta T/P \sim 1$–2. In this range the empirical formula:

$$h_c = 0.0022(\Delta T/H)^{0.35}$$

$$= 0.0022(20/15)^{0.35}$$

$$= 0.0024$$

compares favorably with the classical formula:

$$h_c = 0.0024(\Delta T/H)^{0.25}$$

$$= 0.0024(20/15)^{0.25}$$

$$= 0.0026$$

A different result is obtained with small devices. Consider a 0.5 in. high vertically oriented bracket at 113°C in an air ambient of 73°C:

$$\bar{T} = (\bar{T}_s + T_A)/2 = 93°C$$

From Table 2-1:

$$g\beta\rho^2/\mu^2 = 0.885 \times 10^3$$

$$k = 7.648 \times 10^{-4}$$

$$Pr = 0.72$$

Referring to Fig. 2-4 and using:

$$\log(GrPr) = \log\left[\left(\frac{g\beta\rho^2}{\mu^2}\right)(\Delta T)(H)^3 Pr\right]$$

$$= \log[(0.885 \times 10^3)(43)(0.5)^3(0.72)]$$

$$= \log(3.42 \times 10^3) = 3.53$$

then:

$$\log Nu = 0.64$$

$$Nu = 4.37$$

and:

$$h_c = \frac{k}{H} Nu = \frac{7.648 \times 10^{-4}}{0.5} \quad (4.37)$$

$$= 0.0067 \text{ watt/in.}^2 \cdot {}^\circ\text{C}$$

Using the classical formula:

$$h_c = 0.0024(40/0.5)^{0.25} = 0.0072$$

By the empirical formula:

$$h_c = 0.0022(40/0.5)^{0.35} = 0.010$$

The correct classical result should be 0.0067 watt/in.$^2 \cdot {}^\circ$C, whereas the experimental systems indicated in Fig. 2-5 require $h_c = 0.010$. This much error, or discrepancy, can have serious consequences. Suppose the bracket is 4.0 in. wide. Then:

$$A_s = 2(0.5)(4.0) = 4.0 \text{ in.}^2$$

$$h_c = 0.0067, \ Q = h_c A_s \Delta T = 1.07 \text{ watts}$$

$$h_c = 0.0072, \ Q = 1.15 \text{ watts}$$

$$h_c = 0.010, \ Q = 1.60 \text{ watts}$$

Both Fig. 2-4 and E2.13 significantly underpredict the heat transfer from the small surface. In summary, E2.14 should be used for small devices. For cabinet surfaces with heights greater than about 6 in., E2.13 should be used.

An example of a natural convection and radiation cooled heat sink is illustrated in Fig. 2-7. The 2 in. × 1 in. device is oriented with the 1 in. dimension as the height in order to obtain the maximum heat transfer coefficient. The plate is sufficiently thick to make the

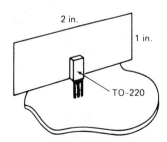

Fig. 2-7. Natural convection/radiation cooled TO-220 transistor on simple heat sink.

Table 2-4.   Watts transferred from heat sink surface
in Fig. 2-7.

$\bar{T}_s$ (°C)	Q_r (watts)	Q_c (watts)	Q (watts)	R_s (°C/watt)
50	0.39	0.68	1.07	23
75	0.89	1.73	2.62	19
100	1.51	2.99	4.50	17

temperature gradient rather minimal.   A flat black spray paint on
both surfaces is used to obtain an emissivity $\epsilon = 0.9$.   The transistor
leads are soldered into an epoxy-glass card.   The ambient is 25°C.
The heat sink temperature is required.

The total heat transferred is the sum of the radiation and convection effects:

$$Q = Q_r + Q_c$$

if lead losses are neglected.   E3.24 is used to compute Q_r:

$$Q_r = 3.657 \times 10^{-11} \epsilon A_s [(\bar{T}_s + 273)^4 - (T_A + 273)^4]$$

$$Q_c = h_c A_s (\bar{T}_s - T_A), h_c = 0.0022 (\Delta T/H)^{0.35}$$

An explicit solution of $\bar{T}_s$ is not possible, but this difficulty may be
circumvented by using several values of $\bar{T}_s$ to calculate Q, as given
in Table 2-4.   A $\bar{T}_s$ of 100°C results in a $Q = 4.50$ watts or a total
resistance:

$$R_s = (100 - 25)/4.50 = 17°C/\text{watt}$$

The results in Table 2-4 as well as a few additional values are plotted
in Fig. 2-8.   This permits graphical extrapolation to $\bar{T}_s = 83 + 25 =$
108°C at 5 watts.

The difference between the experimental and theoretical may be
explained by the lead to board losses through a resistance R_L in
parallel with R_s:

$$\frac{1}{R} = \frac{1}{R_s} + \frac{1}{R_L}$$

where R_s is taken as the computed radiation/convection contribution.   At 5 watts $R_s = 16.0°C/\text{watt}$, and from experiment $R = 72/5 =$
14.4°C/watt.   Thus:

$$R_L = 1/(1/R - 1/R_s)$$

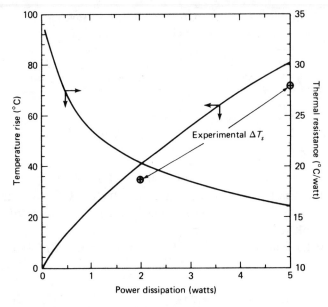

Fig. 2-8. Theoretical/experimental characteristics of 2 in. (width) × 1 in. (height) vertical plate. $\epsilon = 0.9$.

$$= 1/(1/14.4 - 1/16.0)$$

$$= 144°C/watt$$

The reader may show that this is also consistent with the measured data at 2 watts.

An alternate method that illustrates a true iterative technique is also applicable to the nonlinear radiation and convection problem. The surface temperature rise is:

$$\bar{T}_s - T_A = QR$$

$$\bar{T}_s^N - \bar{T}_A = Q \bigg/ \left\{ \sigma \epsilon A_s \frac{(\bar{T}_s'^4 - \bar{T}_A'^4)}{\bar{T}_s - T_A} \right.$$

$$\left. + 0.0022 A_s \left(\frac{\bar{T}_s - T_A}{H} \right)^{0.35} + \frac{1}{144} \right\},$$

$$T' = T(°C) + 273$$

The procedure in using such a formula is to first select an estimated (guessed) value of $\bar{T}_s$ for use in the right side of the equation. Any value of $\bar{T}_s$ may be used for this initial estimate, although the closer it is to the true value, the fewer the number of iterations that will be required. A new value, designated here as $\bar{T}_s^N$, is computed. This

Table 2-5.   Temperature rise
from 5-watt heat sink in
Fig. 2-7.   Computed using
iterative method.

$\bar{T}_s$ (°C)	$\bar{T}_s^N$ (°C)	$\bar{T}_s^N - \bar{T}_s$ (°C)
50	139	89
139	99	−40
99	111	12
111	107	−4
107	108	1

then becomes the new trial value for the right side. Several trials or iterations are run, finally arriving at a satisfactory solution. A set of results is shown in Table 2-5, where $\bar{T}_s = 50°C$ was used initially. Iterative methods are ideal for digital computer applications.

The cube-shaped, enclosed cabinet problem in Chapter 1, Fig. 1-11, was used as an illustration of the thermal circuit method. Estimates of the radiative and convective heat transfer coefficients were supplied as known values. An iterative method is now applied.

Suppose that the exterior cabinet surfaces have an emittance $\epsilon = 0.9$, the ambient air and surrounding room walls are at 25°C, and, as given in Section 1.5, $Q = 12$ watts power dissipation within the cabinet. An approximation that will not drastically affect the final computed interior air temperature but greatly simplifies the calculations is that all the walls are at the same temperature.

The external conductance is the sum of the side, top, and bottom panel convection and radiation conductances. The external convection conductance for $W = L$ is:

$$C_c = \left\{ 0.0024 \left(\frac{\Delta T}{H} \right)^{0.25} (4WH) + 0.0022 \left[\frac{\Delta T}{\frac{WL}{2(W+L)}} \right]^{0.25} (W^2) \right.$$

$$\left. + 0.0011 \left[\frac{\Delta T}{\frac{WL}{2(W+L)}} \right]^{0.25} (W^2) \right\}$$

$$= \left\{ 0.0024 \left(\frac{\Delta T}{10} \right)^{0.25} (4)(10)(10) + 0.0022 \left[\frac{\Delta T}{\frac{(10)(10)}{2(10+10)}} \right]^{0.25} (10)^2 \right.$$

$$+ 0.0011 \left. \left[\frac{\Delta T}{\frac{(10)(10)}{2(10 + 10)}} \right]^{0.25} (10)^2 \right\}$$

$$= 0.802(\Delta T)^{0.25} \text{ watt/°C}$$

The external radiation conductance is:

$$C_r = \epsilon(4WH + 2W^2)(1.463 \times 10^{-10})(T_A + 273)^3$$

$$= (0.9)[4(10)(10) + 2(10)^2] \times (1.463 \times 10^{-10})(T_A + 273)^3$$

$$= 2.09 \text{ watts/°C}$$

The total exterior conductance is:

$$C_{EX} = C_c + C_r$$

$$= 0.802(\Delta T)^{0.25} + 2.09$$

The exterior wall temperature is computed from:

$$\bar{T}_w^N = T_A + Q/C_{EX}$$

where C_{EX} is a function of ΔT. The interior air temperature is computed assuming interior radiation exchange is negligible, owing to low emissivity surfaces as well as surface shielding effects; i.e., the various $\mathcal{F}_{ij} \simeq 0$. Also, using the result from Section 1.5 that showed a negligible temperature gradient through the wall:

$$T_{IN}^N = \bar{T}_w + Q/C_{IN}$$

$$\bar{T}_w = 29°C, Q = 12 \text{ watts}$$

$$C_{IN} = 0.802(T_{IN} - \bar{T}_w)^{0.25}$$

The wall and interior air temperature results, as shown in Table 2-6(a), (b), illustrate the iterative scheme in considerable detail. Note that the significant variation of T_{IN}^N occurs first in the ones, then the tenths, and finally the hundredths.

The net thermal resistance is:

$$R = R_{IN} + R_{EX}$$

$$= 0.7 + 0.33 = 1.0°C/\text{watt}$$

Table 2-6(a). Iterative results for wall temperature
of 10 in. × 10 in. enclosure in 25°C ambient.

Iteration No.	$\bar{T}_w$ (°C)	$\bar{T}_w^N$ (°C)	$R_{EX} = (\bar{T}_w^N - T_A)/Q$ (°C/watt)
1	100	28	0.25
2	28	29	0.33
3	29	29	0.33

Table 2-6(b). Iterative results for interior air
temperature of 10 in. × 10 in. enclosure.

Iteration No.	T_{IN} (°C)	T_{IN}^N (°C)	$R_{IN} = (T_{IN}^N - \bar{T}_w)/Q$ (°C/watt)
1	100.000	34.155	0.43
2	34.155	38.930	0.83
4	38.930	37.429	0.79
6	37.429	37.781	0.73
8	37.781	37.692	0.72
10	37.693	37.714	0.73
12	37.714	37.709	0.73

2.4 FORCED CONVECTION OVER PLATES

In Section 2.3 a dimensionless quantity, the Grashof number, Gr_p, unique to free convection heat transfer, was discussed. Gr_p was shown to be physically useful when computing convective heat transfer coefficients. Forced convection heat transfer has an analogous quantity, the Reynold's number, $Re_p = VP\rho/\mu$, that is required to calculate Nusselt numbers. Quantities ρ and μ, the density and absolute viscosity, are as defined in Section 2.2. P is an appropriate length parameter, and V is the fluid (air) speed.

The dimensions of the various quantities must be consistent. If μ and ρ have dimensions as used in Table 2-1 (gm, in., sec), several variations in units are possible for V and P. The following are convenient for most engineering work in the United States today:

$$Re_p = VP(\rho/\mu)(1/5)$$

E2.15
V ft/min., P in.

$$Re_p = VP(\rho/\mu)$$

E2.16
V in./sec, P in.

$$Re_p = VP(\rho/\mu)(144)$$

E2.17
V ft/sec, P ft

$$Re_p = VP(\rho/\mu)(12/5)$$

E2.18
V ft/min., P ft

The quantity $\nu = (\mu/\rho)$ is defined as the kinematic viscosity and is tabulated in Table 2-1.* The Reynold's number represents a ratio of inertia forces to viscous forces in the fluid.

The literature is abundant with theoretically and experimentally based formulae and graphs for Nusselt numbers characterizing convective heat transfer for forced fluid flow over a variety of shapes and sizes. Unfortunately, very little of this information is readily applicable to the complexity of geometry encountered in electronic equipment. The procedure adopted here is to state a few of the more common Nu vs. Re relationships that best meet the requirements of the electronic designer. Because complete credit to the original sources would require nearly as much space as is devoted to explaining the material, the reader is urged to consult one of the excellent general references [2] or [12] for more detail.

Perhaps the most basic of all surface shapes encountered is the flat plate (see Fig. 2-9). A boundary layer may be laminar, turbulent, or in an intermediate stage (called the transition region), depending on the distance from the leading edge of the plate, the free stream velocity, and the amount of upstream turbulence. Flow in the vicinity of the leading edge tends to have a velocity profile with the boundary layer of a laminar nature; i.e., the fluid particles follow streamlines. As the flow proceeds farther from the leading edge, small disturbances start to become more significant in the "transition" region until finally inertia forces dominate viscous effects, and the boundary layer becomes turbulent with only a very thin laminar sublayer at the plate surface. The location at which the boundary becomes turbulent is called the critical length, x_c, and is calculated from the critical Reynold's number Re_{x_c}. Empirical studies reported in the literature indicate Re_{x_c} may vary from 8×10^4 to 5×10^6, depending on the nature of the surface and the degree of turbulence in the flow upstream from $x = 0$. Typically, the standard treatises indicate:

$$Re_{x_c} = 5 \times 10^5$$

E2.19
Flat plate

where the free stream velocity V_∞ is used to estimate Re_{x_c}.

---

*A formula satisfactory for scientific calculator use is:

$$\nu, \text{in.}^2/\text{sec} = 0.02134 \exp [0.00454T(°C)]$$

to within 5% of Table 2-1, $T = -18°C$ to $150°C$.

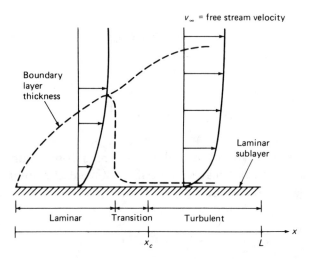

v_∞ = free stream velocity

Boundary layer thickness

Laminar sublayer

Laminar | Transition | Turbulent

x_c | L

Fig. 2-9. Flat plate airflow, [2]. Fig. 6-1 (p. 312) from *Principles of Heat Transfer*, 3rd Edition, by Frank Kreith. Copyright © 1958, 1965, 1973 by Harper & Row, Publishers, Inc. Reprinted by permission of the publisher.

The critical Reynold's number is used to judge whether a laminar or turbulent Nusselt number should be used. For laminar flat plate flow, the classical correlation for a local heat transfer coefficient h_x is:

$$Nu_x = 0.332 Re_x^{1/2} Pr^{1/3}$$

E2.20
Flat plate, local

This assumes a surface temperature T_s that is uniform the length of the plate. When this condition is not met, an average temperature, $\bar{T}_s$, must be used. Temperature-dependent physical properties (k, v, Pr) are evaluated at the mean boundary layer temperature, taken as $(\bar{T}_s + T_A)/2$.

A surface average is more convenient. For a plate of length L:

$$\overline{Nu}_L = \frac{1}{L} \int_0^L Nu_x \, dx$$

$$= k \, 0.332 Pr^{1/3} \left(\frac{V}{v}\right)^{1/2} \frac{1}{L} \int_0^L x^{-1/2} \, dx$$

$$\overline{Nu}_L = 0.332 Pr^{1/3} \left(\frac{V}{v}\right)^{1/2} \frac{2}{L} x^{1/2} \Big|_{x=0}^{x=L}$$

$$= 2(0.332 Re_L^{1/2} Pr^{1/3})$$

$$\boxed{\overline{Nu}_L = 2Nu_L}$$

E2.21
Flat plate average

$$\boxed{\overline{h}_L = 2h_L}$$

E2.22

E2.22 can be very useful in evaluating forced convection effects for shapes resembling flat plates. One merely needs to obtain h_x at the trailing edge $x = L$ and multiply by two. h_x is readily computed from

$$h_x = 0.332 \frac{k}{x} Re_x^{1/2} Pr^{1/3}$$

$$= \left(\frac{V}{x}\right)^{1/2} \left(0.332 \frac{k}{\nu^{1/2}} Pr^{1/3}\right)$$

If h_x is in units of watts/in.$^2 \cdot °C$, x in inches, and V in the very common ft/min., the above must be multiplied by a units conversion factor of $(12/60)^{1/2}$. The coefficient to $(V/x)^{1/2}$ is slightly temperature-dependent and is tabulated in Table 2-7. This tabulation indicates that for h_x at $\overline{T} = 50°C$, the maximum error at 0–100°C is less than 5%. Then:

$$\boxed{h_x = 0.000546\sqrt{V/x}}$$

E2.23
Laminar flat
plate flow

where h_x, V, and x have units of watts/in.$^2 \cdot °C$, ft/min., and in., respectively. For more convenient reference, h_x is plotted in Fig. 2-10.

Table 2-7. Temperature
variation of h_x for
laminar flat plate flow.
h_x watts/in.$^2 \cdot °C$;
V ft/min.; x in.

$\overline{T}$	$h_x/\sqrt{V/x}$
0	0.000572
25	0.000559
50	0.000546
75	0.000541
100	0.000537

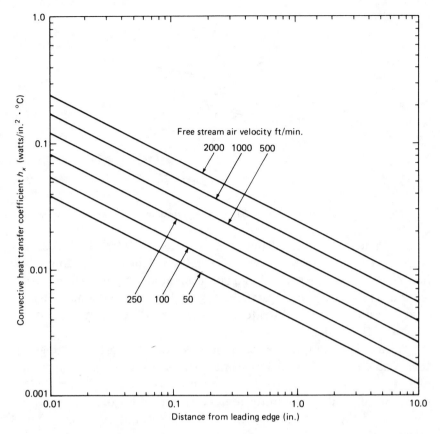

Fig. 2-10. Theoretical heat transfer coefficient for flat plate laminar flow. Should be used with Fig. 2-12.

The maximum Re_x associated with Fig. 2-10 is for $x = 10$ in., $V = 2000$ ft/min.:

$$Re_x = VP(\rho/\mu)(1/5)$$

$$= (2000)(10)(1/0.0268)(1/5)$$

$$= 1.5 \times 10^5$$

which is within the realm of laminar flow.

The average convective heat transfer coefficient for laminar flow over a flat plate from E2.22 and E2.23 is:

$$\bar{h}_L = 0.001092\sqrt{V/L}$$

E2.24
Laminar flat
plate flow

Once again the author wishes to summarize the essential aspects

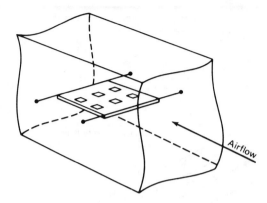

Fig. 2-11. Illustration of 2 in. × 2 in. ceramic substrate in air duct test system.

of experiments conducted in his laboratory. A 0.025 in. thick, 2.0 in. long, ceramic substrate with several small heat-dissipating thick film resistors was suspended in a component airflow test facility, one section of which is illustrated in Fig. 2-11. Small-diameter nylon cord was used as the substrate supporter. Airflow was measured upstream of the device.

A detailed computer model of this conducting/convecting system was developed with one variable used as an unknown parameter: $\bar{h}_L$. Radiation effects were taken into account. Several values of $\bar{h}_L$ were used to compute temperature profiles on the substrate. These were then compared with profiles measured using an infrared microscope.

The results of the experiments were that the theoretical, laminar flat plate flow heat transfer coefficients led to a significant overprediction of temperature unless the $\bar{h}_L$ values were increased significantly. This correction is indicated in Fig. 2-12.

An independent but similar set of experiments was used with an extruded aluminum heat sink as illustrated in Fig. 2-13, with resultant $\bar{h}_L$ indicated in Fig. 2-14. The lines of various densities are for different fin spacings. Since the data for $S = 0.182$ and 0.102 tend to be somewhat indistinguishable, it may be presumed that the $S = 0.182$ is equivalent to a flat plate.

Figure 2-14 and Figs. 2-10 and 2-12 should be checked for consistency. This is done in Table 2-8 for $L = 2$ in. The maximum discrepancy is 12%. Figure 2-14 is recommended for $\bar{h}_L$ for flat plate flow at altitudes in the region of sea level, although the re-plot, with some curve smoothing, in Fig. 2-15 is usually more convenient. As a sample application, suppose the device in Fig. 2-7 is in a forced convection airstream of 100 ft/min. This airspeed is not within the realm (solid lines) of the recommended $\bar{h}_L$ in Fig. 2-15 and requires extrapolation. This is a typical example of how the thermal designer must often work with an incomplete set of facts. The transistor dis-

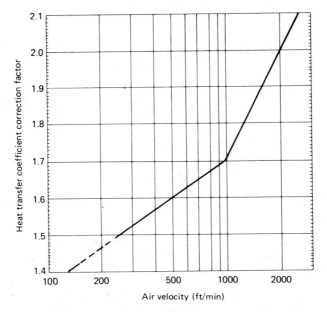

Fig. 2-12. Correction factor applicable to heat transfer coefficient for laminar flow over a 2.0 in. × 2.0 in. (0.025 in. thick) ceramic substrate. Used with Fig. 2-10.

sipates 5 watts, the heat sink has an $\epsilon = 0.9$, and $T_A = 25°C$. Use $R_L = 144°C/watt$.

   Case (1): Flow is parallel to the 2 in. dimension:

From Fig. 2-15:

$$\bar{h}_L = 0.011 \text{ watt/in.}^2 \cdot °C$$

From Figs. 2-10 and 2-12:

$$\bar{h}_L = 0.0038(2)(1.37) = 0.010 \text{ watt/in.}^2 \cdot °C$$

Hence both methods agree.

$$R_c = 1/\bar{h}_L A = 1/(0.01)(2)(2)(1) = 25°C/watt$$

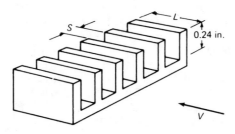

Fig. 2-13. Extruded heat sink used for experimental determination of forced convection heat transfer coefficient.

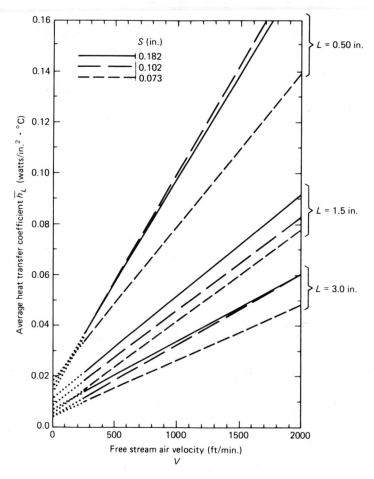

Fig. 2-14. Average convective heat transfer coefficient for extruded heat sink illustrated in Fig. 2-13. From [10], copyright © 1976 by the Institute of Electrical and Electronics Engineers, Inc. Reprinted by permission of the publisher.

Table 2-8. Comparison of $\bar{h}_L$.

V (ft/min.)	$\bar{h}_L$–Fig. 2-10 (watts/in.$^2 \cdot °$C)	f–Fig. 2-12	$f\bar{h}_L$ (watts/in.$^2 \cdot °$C)	$\bar{h}_L$–Fig. 2-14 (watts/in.$^2 \cdot °$C)
250	0.0122	1.5	0.018	0.017
500	0.0173	1.6	0.028	0.025
1000	0.0244	1.7	0.042	0.043
2000	0.0345	2.0	0.069	0.078

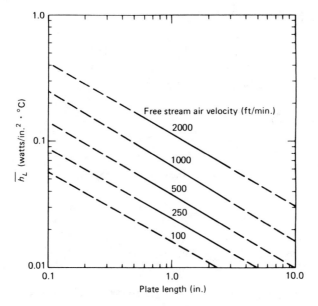

Fig. 2-15. Average convective heat transfer coefficient for a flat plate.

Neglecting radiation for a first approximation:

$$R = R_c R_L/(R_c + R_L)$$

$$= (25)(144)/(25 + 144) = 21°C/\text{watt}$$

$$T_s = RQ + T_A = (21)(5) + 25$$

$$= 130°C$$

The large difference between T_s and T_A precludes use of the simple formula in Section 1.4 for h_r. Figure 3-16 must be used.

$$R_r = 1/\epsilon h_r A$$

$$= 1/(0.9)(0.0065)(2)(2)(1)$$

$$= 43°C/\text{watt}$$

Now adding R_r into the total:

$$R = (21)(43)/(20 + 43) = 14°C/\text{watt}$$

Making a second iteration of the surface temperature and h_r:

$$T_s = RQ + 25 = (14)(5) + 25$$

$$= 95°C$$

From Fig. 3-16, $h_r \simeq 0.0054$.

$$R_r = 1/(0.9)(0.0054)(4)$$

$$= 51°C/watt$$

Settling for a final result:

$$\frac{1}{R} = \frac{1}{R_c} + \frac{1}{R_L} + \frac{1}{R_r}$$

$$= \frac{1}{25} + \frac{1}{144} + \frac{1}{51}$$

$$R = 15°C/watt$$

Case (2): This same device is oriented such that airflow is parallel to the 1 in. dimension:

From Fig. 2-15:

$$\bar{h}_L = 0.016 \text{ watt/in.}^2 \cdot °C$$

$$R_c = 1/(0.016)(4) = 16°C/watt$$

First iteration:

$$R = R_c R_L/(R_c + R_L) = 14°C/watt$$

$$T_s = RQ + 25 = 95$$

$$h_r = 0.0054 \text{ watt/in.}^2 \cdot °C$$

$$R_r = 1/(0.9)(0.0054)(4) = 51°C/watt$$

$$\frac{1}{R} = \frac{1}{16} + \frac{1}{144} + \frac{1}{51}$$

$$R = 11°C/watt$$

Second iteration:

$$T_s = RQ + 25$$

$$= (11)(5) + 25$$

$$= 80°C$$

$$h_r = 0.0051$$

$$R_r = 1/(0.9)(0.0051)(4)$$

$$= 55°C/\text{watt}$$

$$\frac{1}{R} = \frac{1}{R_c} + \frac{1}{R_L} + \frac{1}{R_r}$$

$$= \frac{1}{16} + \frac{1}{144} + \frac{1}{55}$$

$$R = 11°C/\text{watt}$$

Note that a simple change in orientation indicates a potential temperature decrease of $\Delta T = (15 - 11)(5) = 20°C$! If only half of this were realized, the increase in the transistor lifetime would be significant.

2.5 FORCED CONVECTION IN DUCTS

The remainder of this chapter will be devoted to forced airflow in confined spaces, i.e., ducts. In practice, nearly all airflow is contained within a space that may be defined by boundaries, thus forming a duct of some sort. Two parallel flat plates with the spaces on either side closed form a duct. Aerodynamically and thermally, however, boundary layer effects must be considered. For example, air with a uniform velocity profile at the entrance to a duct will, upon entering the confined system, develop a hydrodynamic boundary layer on each interior wall. Initially, boundary layers develop as on flat plates until the boundary layers on opposing surfaces become sufficiently thick to come into contact with one another. Once this happens, the flow soon becomes fully developed; i.e., the velocity profile remains unchanged the length of the duct. The boundary layer thickness for the entrance region may be estimated from flat plate theory.

$$\delta = 5x/Re_x^{1/2}$$

E2.25
Laminar boundary
layer thickness

$$\delta = 0.376x/Re_x^{1/5}$$

E2.26
Turbulent boundary
layer thickness

The two preceding formulae for δ use the flat plate Reynold's number criteria for predicting laminar or turbulent effects.

Before we proceed to convection characteristics of duct systems, the concepts of hydraulic diameter and bulk mean fluid temperature must be addressed. The significant dimension in duct flow (particularly fully developed) is the hydraulic diameter, defined by:

$$D = \frac{4 \times \text{cross-sectional area of flow}}{\text{wetted perimeter}}$$

$$\boxed{D = 4A_c/P_w}$$

E2.27
Hydraulic diameter
for airflow

A few illustrations clarify this point:

Tube with circular cross section with diameter d:

$$D = \frac{4\pi d^2/4}{\pi d} = d$$

Two concentric tubes with flow in annular region between:

$$D = \frac{4\pi \left(\dfrac{d_O^2}{4} - \dfrac{d_I^2}{4} \right)}{\pi(d_O + d_I)} = d_O - d_I$$

Rectangular duct with width W, height H:

$$D = \frac{4WH}{2(W + H)} = \frac{2WH}{W + H}$$

$$= 2H, \quad H \ll W$$

A note of caution is given to readers seeking more detail in the convective heat transfer literature. A different equivalent diameter D may be required in Nusselt vs. Reynold's number correlations. For example, an empirical investigation may correlate Nu_D vs. Re_D for a four-sided channel with heat flux from only two surfaces, in which case the thermal calculation should use a perimeter defined by only the two surfaces transferring heat.

In fully developed duct flow the confined air temperature increases as it travels the length of the duct.

$$\Delta T = Q/\dot{m}C_p$$

where

    Q = heat transferred into air, watts
    $\dot{m}$ = mass flow rate, gm/sec
    C_p = specific heat of air, joules/gm · °C
    ΔT = fluid temperature rise, °C, due to heat input Q

A more convenient form is obtained from the volumetric flow rate $G = \dot{m}/\rho$:

$$\Delta T = Q/\rho C_p G$$

C_p for air is nearly independent of temperature, but ρ varies with temperature. Thus:

$$\Delta T = \frac{5.997 \times 10^{-3}}{G} Q(\bar{T}_B + 273) \qquad \text{E2.28}$$

Air temperature rise at atmos. press.

where

    ΔT = temperature rise of well-mixed air, °C
    Q = heat transferred into air, watts
    G = airflow, ft³/min.
    $\bar{T}_B$ = well-mixed bulk temperature averaged from inlet to exit, °C

Clearly T_B is not uniform throughout the duct length over which Q is transferred to the air. An average of the air temperature over the duct length may also be used. If $\bar{T}_B \simeq 30°C$ is used as typical:

$$\Delta T = 1.82 Q/G \qquad \begin{array}{l} \text{E2.29} \\ \bar{T}_B = 30°C \end{array}$$

Another form of the equation for air temperature rise may be developed that is slightly more complex, but removes any uncertainty concerning the average bulk temperature. Denoting $C = 5.997 \times 10^{-3}$, T_I = inlet air temperature (°C), and T_E = exit air temperature (°C):

$$\Delta T = \frac{CQ}{G} (\bar{T}_B + 273)$$

$$= \frac{CQ}{G}\left(\frac{T_E + T_I}{2} + 273\right)$$

$$= \frac{CQ}{G}\left(\frac{\Delta T + 2T_I}{2} + 273\right)$$

$$\left(\frac{G}{CQ}\right)\Delta T = \frac{1}{2}\Delta T + T_I + 273$$

$$\Delta T\left(\frac{G}{CQ} - \frac{1}{2}\right) = T_I + 273$$

$$\Delta T = \frac{2(T_I + 273)}{\left(\frac{2G}{CQ} - 1\right)}$$

$$\boxed{\Delta T = \frac{2(T_I + 273)}{\left(333.5\,\dfrac{G}{Q} - 1\right)}}$$

E2.30
Air temperature
rise based on inlet
temperature

Applications are common in which heat is convected from the interior of a duct wall at a nearly uniform temperature over the duct length to an airstream that has a considerable temperature increase as it proceeds down the length of the duct. For example, a 12 in. long extruded aluminum heat sink with a fan on one end represents such a system. The most approximate estimate of the heat sink temperature rise would use the air temperature averaged over duct length as an ambient. This approximation may be refined.

Referring to Fig. 2-16(a), the heat ΔQ_{wf} transferred from a short element Δx of wall is:

$$\Delta Q_{wf} \simeq hP\Delta x[T_w - T(x)]$$

for a duct perimeter P. The heat ΔQ_a absorbed by the air element with a mass flow rate $\dot{m}$ is:

$$\Delta Q_a \simeq \dot{m}C_p[T(x + \Delta x) - T(x)]$$

In the limit $\Delta x \to 0$:

$$dQ_{wf} = hP(T_w - T)\,dx$$

$$dQ_a = \dot{m}C_p\,dT$$

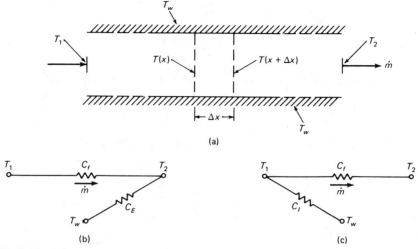

Fig. 2-16. Representation of duct flow heat transfer models. (a) Continuous model of duct section. (b) Two-element model referenced to exit temperature. (c) Two-element model referenced to inlet temperature.

Conservation of energy requires that:

$$dQ_{wf} = dQ_a$$

$$hP(T_w - T)\, dx = \dot{m} C_p\, dT$$

$$\int_{T_1}^{T_2} \frac{dT}{T - T_w} = -\int_0^L \frac{hP}{\dot{m} C_p}\, dx$$

$$\ln\left(\frac{T_2 - T_w}{T_1 - T_w}\right) = -\frac{hA_s}{\dot{m} C_p}$$

$$T_w - T_2 = (T_w - T_1)\, e^{-\beta}, \quad \beta = \frac{hA_s}{\dot{m} C_p}$$

$$T_2 - T_1 = (T_w - T_1)(1 - e^{-\beta})$$

The total heat transferred to the air over a length L is:

$$Q_a = \dot{m} C_p (T_2 - T_1)$$

$$= \dot{m} C_p (T_w - T_1)(1 - e^{-\beta})$$

For a conductance C_E defined by reference to T_2 at the exit Fig. 2-16(b):

$$C_E = Q_{wf}/(T_w - T_2)$$

$$Q_{wf} = C_E(T_w - T_2)$$

$$= C_E(T_w - T_1)e^{-\beta}$$

The heat absorbed by the air equals the total heat Q_{wf} convected:

$$Q_a = Q_{wf}$$

$$\dot{m}C_p(T_w - T_1)(1 - e^{-\beta}) = C_E(T_w - T_1)e^{-\beta}$$

$$\boxed{\frac{C_E}{hA_s} = \frac{e^\beta - 1}{\beta}}$$

E2.31
Conductance referred
to exit air,
$\beta = hA_s/\dot{m}C_p$

If an overall conductance C_I is defined with reference to an inlet temperature T_1 [Fig. 2-16(c)]:

$$C_I = Q_{wf}/(T_w - T_1)$$

$$Q_{wf} = C_I(T_w - T_1)$$

and if we use:

$$Q_a = \dot{m}C_p(T_w - T_1)(1 - e^{-\beta})$$

$$\dot{m}C_p(T_w - T_1)(1 - e^{-\beta}) = C_I(T_w - T_1)$$

then:

$$\boxed{\frac{C_I}{hA_s} = \frac{1 - e^{-\beta}}{\beta}}$$

E2.32
Conductance
referred to inlet air

Both C_E/hA_s and C_I/hA_s are plotted in Fig. 2-17. These formulae have application in both desk calculator and computer simulation studies. In the latter case either C_E or C_I may be used, with the former probably preferred. Note that if the element size is sufficiently small, $A_s \ll \dot{m}C_p/h$, then $C_E = C_I = hA_s$.

The duct Nusselt numbers for fully developed laminar flow are readily addressed because of an independence of the Reynold's number. Theoretical solutions have indicated a significant dependence on the aspect ratio of rectangular ducts. This dependence is listed in Table 2-9, from [12].

For duct systems where entry length effects are significant, the

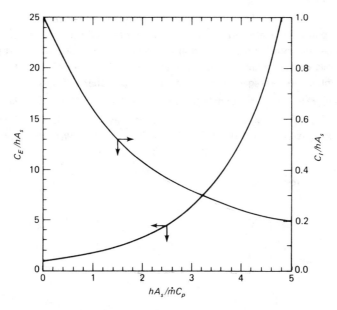

Fig. 2-17. Wall to fluid conductances for flow in a duct.

results in Fig. 2-18 for round tubes may be applicable. Note that as L is decreased, the curve seems to asymptotically approach the fully developed $Nu_{\textcircled{T}} = 3.66$. These results suggest a duct length L_{FD} criterion for assuming fully developed laminar flow for the complete duct length. From Fig. 2-18:

$$\frac{Re_D \, Pr D}{L_{FD}} \times 10^{-2} \lesssim 0.1$$

$$\boxed{L_{FD} \gtrsim VD^2/14\nu}$$

E2.33
Fully developed
laminar duct flow

The boundary layer thickness for flat plate laminar flow may be used to develop an estimate L_{FD} for which flow is fully developed. Since boundary layers develop slowly at distances remote from the surface, the criterion of two opposing boundary layers interfering significantly when $\delta \gtrsim D$ is reasonable. E2.25 is used:

$$\delta = 5L_{FD}\sqrt{Re_{L_{FD}}}$$

$$D \lesssim 5L_{FD}/\sqrt{Re_{L_{FD}}}$$

$$L_{FD} \gtrsim VD^2/25\nu$$

Which is about half of E2.33.

Suppose an extruded heat sink viewed endwise is treated as a

Table 2-9. Nusselt numbers for fully developed velocity and temperature profiles in tubes of various cross sections with laminar flow. The constant-heat-rate solutions are based on constant axial heat rate, but with constant temperature around the tube periphery. Nusselt numbers are averages with respect to tube periphery. $Nu_{(H)}$ for constant heat rate, $Nu_{(T)}$ for constant temperature. From *Convective Heat and Mass Transfer*, 2nd Edition, by W. M. Kays and M. E. Crawford. Copyright © 1980, 1966 by McGraw-Hill, Inc. Used with the permission of McGraw-Hill Book Company.

Cross Section Shape	b/a	$Nu_{(H)}$	$Nu_{(T)}$
○		4.364	3.66
□	1.0	3.63	2.98
▭	1.4	3.78	
▭	2.0	4.11	3.39
▭	3.0	4.77	
▭	4.0	5.35	4.44
▭	8.0	6.60	5.95
═	∞	8.235	7.54
⫛		5.385	4.86
△		3.00	2.35

series of rectangular U-channels. A typical fin spacing and height (perpendicular to sink base) would be $S = 0.25$ in. and $H = 1.0$ in. The hydraulic diameter for a thermal calculation is:

$$D = \frac{4A_c}{P_w} = \frac{4SH}{(S + 2H)}$$

$$= 4(0.25)(1.0)/[0.25 + 2(1.0)]$$

$$= 0.44 \text{ in.}$$

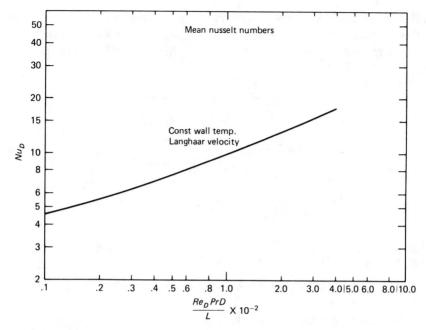

Fig. 2-18. Mean Nusselt numbers with respect to tube length. From [13], reprinted by permission of The American Society of Mechanical Engineers.

At an airflow of 100 ft/min., the Reynold's number is:

$$Re_D = VD/\nu$$

$$= \frac{(100 \text{ ft/min.})(12 \text{ in./ft})\left(\dfrac{1}{60 \text{ sec/min.}}\right)(0.44 \text{ in.})}{0.026 \text{ in./sec}}$$

$$= 339$$

The minimum heat sink length for which a fully developed laminar flow approximation may be used is found from E2.33:

$$L_{FD} \geqslant \frac{VD^2}{14\nu} = \frac{D}{14} Re_D = \left(\frac{0.44}{14}\right)(339) = 11 \text{ in.}$$

If the sink length is 12 in.:

$$\frac{Re_D PrD}{L} \times 10^{-2} = \frac{(339)(0.72)(0.44)}{12} \times 10^{-2}$$

$$= 0.09$$

Reference to Fig. 2-18 or a curve fit recommended by Kays in [13],

E2.34, indicates $\overline{Nu}_D = 4.6$ for cylinders. Consideration of entrance effects ($\overline{Nu}_D = 4.6$) results in better accuracy than a fully developed flow approximation ($\overline{Nu}_D = 3.66$), even for this relatively long duct ($L/D = 12.0$).

$$\overline{Nu}_D = 3.66 + \frac{0.104\left(\dfrac{Re_D Pr}{L/D}\right)}{1 + 0.016\left(\dfrac{Re_D Pr}{L/D}\right)^{0.8}} \qquad \text{E2.34}$$

Laminar duct flow in circular tubes

Table 2-9 shows that a rectangular duct with an aspect ratio of $1/0.25 = 4$ has a fully developed $Nu_{\widehat{T}}$ that is greater than that of a circular duct by a factor of $4.44/3.66 = 1.21$. Thus for $L = 12$ in., $\overline{Nu} = (1.21)(4.6) = 5.6$.

If either the airflow or duct diameter is increased, the corresponding Reynold's number increases. Re_D is used to surmise the nature of the flow:

$$Re_D < 2100 \qquad \begin{array}{l}\text{E2.35}\\ \text{Laminar duct flow}\end{array}$$

$$Re_D > 10{,}000 \qquad \begin{array}{l}\text{E2.36}\\ \text{Turbulent duct flow}\end{array}$$

Although these criteria are not absolute, they are recommended throughout the literature. Intermediate Re_D's are taken as indicative of transition flow, i.e., a mix of laminar and turbulent effects. Heat transfer in this region is difficult to predict, and no recommendations will be made here except to indicate that the reader may calculate bounds based on the preceding laminar and the following turbulent flow formulae.

For fully developed turbulent air flow in a smooth circular duct, [2]:

$$Nu_D = 0.023 Re_D^{0.8} \qquad \begin{array}{l}\text{E2.37}\\ \text{Turbulent flow}\end{array}$$

Rectangular ducts require the use of a hydraulic diameter in the Nusselt correlation with the duct perimeter restricted to the region dissipating heat into the airstream.

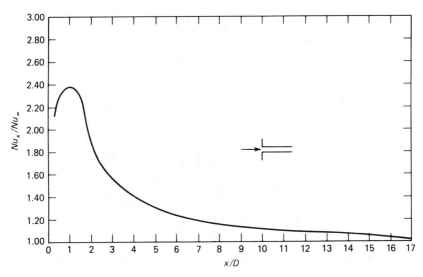

Fig. 2-19. Local Nusselt number for turbulent flow in the entry length of a circular tube with a right angle edge entrance, [12]. From *Convective Heat and Mass Transfer*, 2nd Edition, by W. M. Kays and M. E. Crawford, Fig. 13-10, copyright © 1980 by McGraw-Hill, Inc. Used with the permission of McGraw-Hill Book Company.

Turbulent flow in a duct that is sufficiently short to have significant heat transfer in the entrance where the flow is not yet fully developed has been evaluated by Boelter et al., [14], for several entrance configurations. The ratio of local Nusselt numbers for an abrupt contraction as computed by Kays, [12], from [14] is shown in Fig. 2-19. The thermal and hydrodynamic boundary layers both start at the entrance. The average $\overline{Nu}_L$ is obtained by integrating Fig. 2-19.

$$\frac{\overline{h}_L}{h_D} = 1 + 1.68 \left(\frac{D}{L}\right)^{0.58}$$

E2.38
Turbulent flow,

$$2 \leqslant \frac{L}{D} \leqslant 20$$

$$\frac{\overline{h}_L}{h_D} = 1 + 6\left(\frac{D}{L}\right)$$

E2.39
Turbulent flow,

$$20 \leqslant \frac{L}{D}$$

The case $2 \leqslant L/D \leqslant 20$ was obtained directly from Fig. 2-19 by this author.

The heat sink illustrated in Fig. 2-20 is a realistic application of forced air in a duct. Calculation of the convective heat transfer coefficients for duct flow requires the use of Re_D where D is the hydraulic diameter. D is computed as $D = 4(0.23)(1.0)/[(0.23)(2) + (1.0)(2)] = 0.37$ in. for determining if the flow is laminar or turbu-

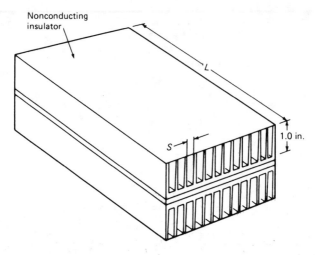

Fig. 2-20. Ducted, forced air cooled heat sink. $S = 0.23$ in., 24 passages.

lent. Nusselt number calculations use the heat transfer perimeter for $D = 4(0.23)(1.0)/[0.23 + (1.0)2] = 0.41$ in. Both methods yield very similar results for this problem; thus a nominal $D = 0.4$ in. is used throughout.

$$Re_D = \left(\frac{VD}{\nu}\right)\left(\frac{1}{5}\right) = \frac{(G/A_c)}{\nu} D\left(\frac{1}{5}\right)$$

for G ft³/min., A_c ft², D in., ν in.²/sec. Assuming an average $(T_w + T_A)/2$ of 50°C, $\nu = 0.027$ in.²/sec. The total cross-sectional area for the 24 channels is $A_c = 24(0.23$ in.$)(1$ in.$)(0.0069$ ft²/in.²$) = 0.038$ ft². Then:

$$Re_D = \frac{(0.4 \text{ in.})}{5} \frac{G}{(0.027 \text{ in.}^2/\text{sec})(0.038 \text{ ft}^2)}$$

$$= 78\,G$$

where G is the total flow through the system in ft³/min. The system is evaluated for flow up to 30 ft³/min., which means $Re_D = 0$–2340, i.e., just into the laminar–turbulent transition region at the greater airflows. D/L for the 12 in. long sink is 0.033.

The laminar flow Nusselt numbers are determined from Fig. 2-18 for a circular tube. Table 2-9 is used to correct the circular tube Nu to a rectangular tube with an aspect ratio of $1/0.23 \simeq 4$. The circular tube Nu should be increased by $4.44/3.66 = 1.21$.

Turbulent flow Nusselt numbers are first computed for fully developed duct flow using $Nu_D = 0.023\,Re_D^{0.8}$ and multiplied by the entry length correction, $1 + (6D/L)$. No correction for a noncircular cross section is required. The laminar and turbulent Nu's are tabulated in Table 2-10.

The heat sink surface temperature is best computed using C_I referred to the air inlet temperature. This requires evaluation of $\dot{m}C_p$ at 50°C:

$$\dot{m}C_p = \rho C_p G$$

$$= 0.018\,(\text{joules/in.}^3 \cdot \text{°C})(12 \text{ in./ft})^3 (1/60 \text{ sec/min.})\,G\,(\text{ft}^3/\text{min.})$$

$$= 0.518\,G$$

which is also tabulated in Table 2-10. C_I is computed from:

$$\frac{C_I}{hA_s} = \frac{1 - e^{-\beta}}{\beta}, \quad \beta = hA_s/\dot{m}C_p$$

or Fig. 2-17, and is plotted in Fig. 2-21(a). A_s for the 24 channels is:

$$A_s = 24[0.23 + 2(1.0)]L$$

$$= 642 \text{ in.}^2 \text{ for } L = 12 \text{ in.}$$

The experimental heat sink surface temperature rise is also indicated in Fig. 2-21(a). The theoretical calculations are based on a uniform surface temperature, but, as expected, the measurements reflect a temperature that increases along the length of the sink. The variation in sink temperature is indicated by the vertical bars in Fig. 2-21(a). The plotted conductance C_I shows a break at 25 ft³/min. or $Re_D \simeq 2000$, the upper limit of the laminar flow range.

The agreement between the experimental and theoretical temperature rise is quite good and could be improved even more if natural convection losses on the external surfaces of the duct system were accounted for.

The same forced air cooled heat sink was also tested for $L = 6$ in. and $P = 30$ watts. The maximum Re_D ($G = 12$ ft³/min.) was 936, indicative of laminar flow conditions. The experimental and theoretical results are plotted in Fig. 2-21(b). Once again, the temperature rise predictions are somewhat in excess of, but sufficiently close to the test data.

A commonly occurring mechanical system that at first glance appears amenable to duct flow analysis is the card cage, i.e., a set of closely spaced parallel epoxy glass circuit boards with heat-dissipating components on at least one side of each card. The author has not had sufficient success with duct flow theory in such systems to be

Table 2-10. Nusselt numbers for forced-air-cooled heat sink (Fig. 2-20). L = 12 in.

Flow			Laminar					Turbulent				
G ft³/min.	Re_D	$\dot{m}C_p$ joules/°C·sec	$\overline{Nu_D}$ Fig. 2-18	$\overline{Nu_D}$ ×1.21	$\overline{h_D}$ watts/°C·in.²	hA_s watts/°C	C_I Fig. 2-17 watts/°C	Nu_D	$\overline{Nu_D}$	$\overline{h_D}$ watts/°C·in.²	hA_s watts/°C	C_I Fig. 2-17 watts/°C
3	234	1.55	3.7	4.5	.0078	5.01	1.49					
5	390	2.6	4.5	5.5	.010	6.24	2.36					
10	780	5.2	5.4	6.5	.011	7.06	3.86					
15	1170	7.8	6.0	7.3	.013	8.35	5.13					
20	1560	10.4	6.8	8.2	.014	8.99	6.02					
25	1950	13.0	7.4	9.0	.016	10.27	7.10	9.9	11.9	.021	13.48	8.39
30	2340	15.5	7.8					11.4	13.7	.024	15.41	9.76

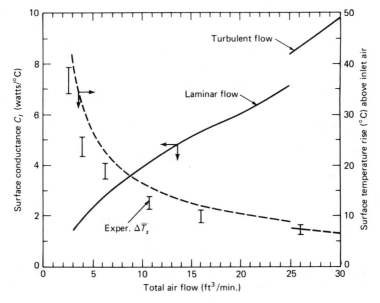

Fig. 2-21(a). Theoretical and experimental characteristics of forced air cooled heat sink. $P = 62.8$ watts, $L = 12$ in.

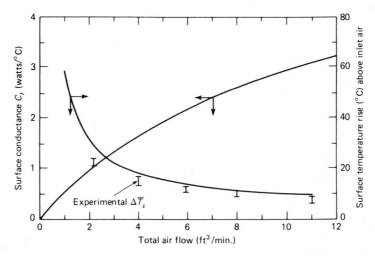

Fig. 2-21(b). Theoretical and experimental characteristics of forced air cooled heat sink. $P = 30$ watts, $L = 6$ in.

able to recommend any particular formula as helpful in design analysis. A component level approach is preferred.

Figures 2-22, 2-23, and 2-24 illustrate the thermal characteristics of some 16 lead dual-in-line packages. The curves may be considered somewhat typical of these package styles, but structural characteristics such as lead-frame geometry, particularly in epoxy packages, can significantly effect the thermal resistance. The following example illustrates the use of component thermal resistance data.

An array of 10 circuit cards, dissipating 100 watts each, is con-

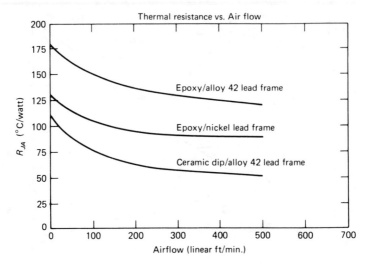

Fig. 2-22. Thermal resistance, junction to ambient, of some 16 lead dual-in-line packages, [15]. Reprinted from *Electronics*, October 31, 1974. Copyright © McGraw-Hill Inc. 1982. All rights reserved.

tained in an assembly designed for 225 ft³/min. of airflow. The temperature rise of the cooling air is:

$$\Delta T \simeq \frac{1.8P}{G}$$

$$= \frac{1.8(10)(100)}{225}$$

$$= 8°C$$

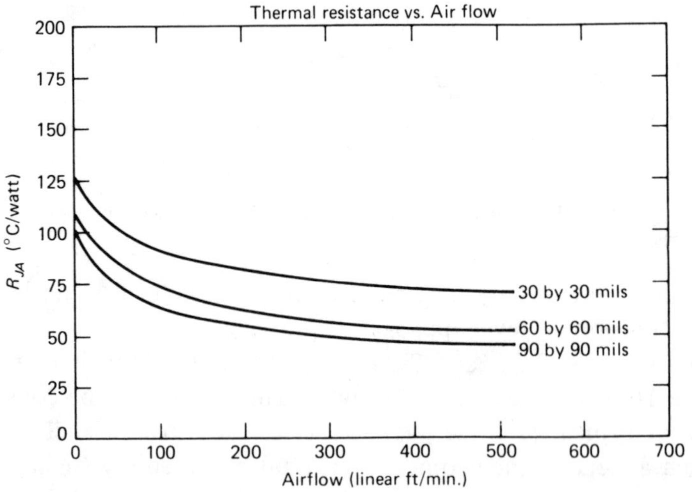

Fig. 2-23. Thermal resistance, junction to ambient, of a 16 lead ceramic dual-in-line package, [15]. Reprinted from *Electronics*, October 31, 1974. Copyright © McGraw Hill Inc. 1982. All rights reserved.

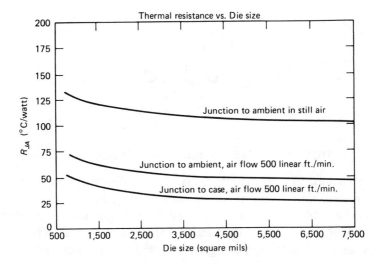

Fig. 2-24. Thermal resistance, junction to ambient, of a 16 lead ceramic dual-in-line package, [15]. Reprinted from *Electronics*, October 31, 1974. Copyright © McGraw-Hill Inc. 1982. All rights reserved.

A common occurrence in practice is to measure several air temperature rises at the last couple of rows of components. It is not uncommon to find that they are approximately twice the value computed from the standard air temperature rise formula. Thus:

$$\max \Delta T_{\text{AIR}} \simeq 16°\text{C}$$

The frontal cross-sectional area of 10 in. × 10 in. for the card cage is sufficiently obstructed by components to cause a 50% reduction in area. The average linear velocity of the flow is then:

$$V = \frac{G}{A} = \frac{225}{\left[\dfrac{0.5(10)(10)}{144}\right]}$$

$$= 648 \text{ ft/min.}$$

Each card has about 200 epoxy, 16 lead DIPS. Referring to Fig. 2-22 and assuming a nickel–iron (alloy 42) lead frame, R_{JA} (junction to local ambient) = 90°C/watt. The component junction temperature rise above the room ambient is then:

$$T_J - T_A = R_{JA}P + \Delta T_{\text{AIR}}$$

$$= (90°\text{C/watt})\left[\frac{1000}{10(200)}\right] + 16$$

$$= 61°\text{C}$$

Chapter 3
Thermal Radiation

3.1 RADIATION CHARACTERISTICS OF BLACKBODY SURFACES

Thermal energy transport by radiation is unique compared to conduction and convection in that a transport medium is not required. In fact, heat transfer between two surfaces is greatest when there is no intervening material. Fortunately the absorption of radiation between surfaces within most ground-based electronic equipment is sufficiently small that it will be neglected throughout this text. The theoretical approximations used to quantify the radiation heat exchange between a cabinet exterior and ambient are actually improved by atmospheric absorption.

Radiation from solid objects such as components, cabinet walls, and heat sinks may be considered as a totally surface-related phenomenon. The radiation is electromagnetic in character with a theoretical basis found in the realm of quantum physics of solids. Briefly, however, it may be thought of as originating at microscopic electronic oscillators emitting radiation over an extremely broad bandwidth. The maximum monochromatic radiation flux per unit wavelength interval into a half space, or monochromatic emissive power, is described by Planck's radiation law:

$$E_{\lambda b} = (c_1/\lambda^5)/(e^{c_2/\lambda T'} - 1)$$

E3.1
Planck's radiation law

where $c_1 = 3.7403 \times 10^4$, $c_2 = 1.4387 \times 10^4$, and T' is the temperature of the radiating surface on an absolute scale. If T is °C, then $T' = T + 273.15$ (degrees Kelvin). The radiation wavelength λ is measured in microns ($1\ \mu = 10^{-4}$ cm). The radiation flux $E_{\lambda b}$ has dimensions of watts/cm² · μ.

Plots of $E_{\lambda b}$ for high and low temperatures, Figs. 3-1 and 3-2, respectively, illustrate that a radiation maximum occurs. This particular wavelength, λ_M, is found by setting the derivative of $E_{\lambda b}$ equal to zero and solving for λ, i.e., λ_M:

$$\frac{\partial E_{\lambda b}}{\partial \lambda} = - \frac{5c_1}{\lambda^6 (e^{c_2/\lambda T'} - 1)}$$

$$+ \frac{c_1 c_2 e^{c_2/\lambda T'}}{\lambda^7 T'(e^{c_2/\lambda T'} - 1)^2} = 0$$

$$5(e^{c_2/\lambda_M T'} - 1) = \frac{c_2 e^{c_2/\lambda_M T'}}{\lambda_M T'}$$

Setting $z = c_2/\lambda_M T'$:

$$(1 - z/5) = e^{-z}$$

which is a transcendental equation and is not explicitly soluble. A simple procedure is to plot $(1 - z/5)$ and e^{-z} vs. z. The two functions intersect at $z = 4.965$ or $c_2/\lambda_M T' = 4.965$.

$$\boxed{\lambda_M T' = 2897.7}$$

E3.2
Wien's
displacement law

which is Wien's displacement law. This formula is also indicated in Figs. 3-1 and 3-2 and quantifies the shift of the radiation peak toward shorter wavelengths as the temperature is increased.

A few words are in order concerning the spectral radiation plots. Figure 3-1, for temperatures of 500–1000°C, is typical of radiation discussions in most heat transfer texts. Very few applications in electronic package design are concerned with such large temperatures, but, Fig. 3-1 is useful in the sense that it is an aid in explain-

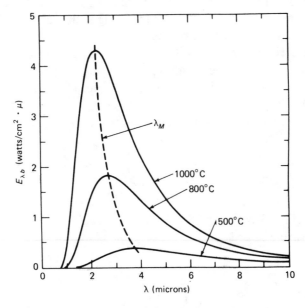

Fig. 3-1. High temperature spectral distribution of monochromatic emissive power.

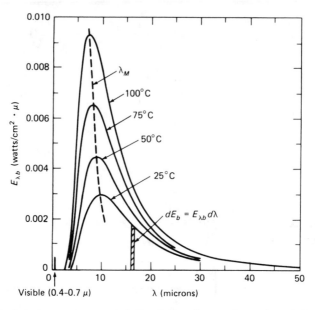

Fig. 3-2. Low temperature spectral distribution of monochromatic emissive power.

ing the color shift of metals from red to blue as they are heated with a torch or in a furnace. Only at rather high temperatures does the leading edge of the curve overlap into the very narrow 0.4–0.7 μ region to such an extent as to be visible to the eye.

Surface temperatures encountered in practical applications with which this book is concerned are typically in the realm of 25–100°C, as indicated in Fig. 3-2. A rather hot component may reach 100°C, and an external surface of a cabinet may be between 25°C and 50°C. Note that in all cases most of the radiation, about 85%, is between about 5 μ and 25 μ. Particularly interesting is the rather well defined region of 7 μ to 10 μ, where the peak radiation occurs. This relationship, as described by Wien's displacement law, will be used later to account for less than ideal radiators (gray bodies) that may deviate from the characteristics in Fig. 3-2 by an amount that is wavelength-dependent.

The monochromatic emissive power at any wavelength for a small element $d\lambda$ is:

$$dE_b = E_{\lambda b} d\lambda$$

as seen in Fig. 3-2 at $T = 25°C$. The total emissive power or black-body emissive power is:

$$E_b = \int_{\lambda=0}^{\lambda=\infty} dE_b$$

$$= \int_0^\infty E_{\lambda b} d\lambda$$

$$= (\pi^4/15)(c_1/c_2^4) T'^4$$

$$\boxed{E_b = \sigma T'^4}$$

where σ is the Stefan-Boltzmann constant, $\sigma = 3.657 \times 10^{-11}$ watt/in.$^2 \cdot {}^\circ$K^4. E3.3 states that the radiant emittance or radiant energy emitted by a surface of unit area per unit time is proportional to the fourth power of the Kelvin temperature.

3.2 GEOMETRIC SHAPE EFFECTS

This development of the spatial aspects of thermal radiation begins with the definition of the radiation field intensity, I, which is the radiation energy emanating per unit time from an infinitesimal area element normal to the direction of I and contained within a cone that is subtended by an area dA_2 at the originating surface dA_1. This definition is typically described by a hemispherical surface with dA_{2s} as an element at radius r (Fig. 3-3). Thus if dQ_{1-2} is the radiation heat transfer (watts), passing through the area dA_p normal to the beam direction, and $d\omega$ is the cone of radiation subtended by dA_2:

$$I = dQ_{1-2}/dA_p d\omega$$

$$dQ_{1-2} = I dA_p d\omega$$

Figure 3-4 shows the geometry required to obtain dA_p in terms of the actual emitting surface area dA_1. Note that the significance of dA_p is that this is the element of area actually "seen" by dA_{2s}, the area element of the hemisphere. Clearly, then:

$$dA_p = \cos \phi \, dA_1$$

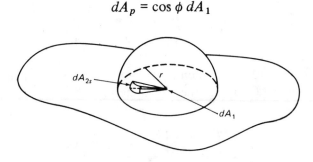

Fig. 3-3. Hemispherical surface used to define radiation field intensity.

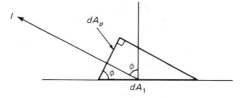

Fig. 3-4. Magnified view of dA_1 projected as dA_p into direction $\hat{I}$ of I.

The definition of an element of a solid angle as the area element sub-tended by a sphere (in this case hemisphere) of unit radius:

$$d\omega \equiv dA_{2s}, \quad r = 1$$

provides then:

$$d\omega = dA_{2s}/r^2$$

for any dA_{2s} at any r, since the area of a sphere is proportional to r^2. Finally, then:

$$dQ_{1\text{-}2} = \frac{I \cos \phi \, dA_1 dA_{2s}}{r^2}$$

The element $dQ_{1\text{-}2}$ integrated over the hemisphere of which dA_{2s} is an element is related to the total emissive power E_b of dA_1:

$$E_b = (1/dA_1) \int_{A_{2s}} dQ_{1\text{-}2}$$

$$= \int_{A_{2s}} \frac{I \cos \phi \, dA_{2s}}{r^2}$$

At this point it is advantageous to restrict consideration to diffuse surfaces, i.e., those that emit radiation independent of direction.

$$E_b = I \int_{A_{2s}} \frac{\cos \phi \, dA_{2s}}{r^2}$$

In a spherical coordinate system, $dA_{2s} = r^2 \sin \phi \, d\theta d\phi$. If the preceding integration is taken over the hemisphere:

$$E_b = I \int_{\theta=0}^{2\pi} d\theta \int_{\phi=0}^{\pi/2} \cos \phi \sin \phi \, d\phi$$

then:

$$I = E_b/\pi$$

An element of receiving area dA_2 *not* on a hemispherical surface is seen from dA_1 as a projected element $\cos \phi_2 \, dA_2$ in a manner similar to Fig. 3-4. This situation is illustrated in Fig. 3-5 where $\underline{dA_1}$ and $\underline{dA_2}$ are vector areas in the direction of the normals to the area elements. The radiation transfer between the two diffuse elements is then:

$$dQ_{1\text{-}2} = E_{b1} \, dA_1 \left(\frac{\cos \phi_1 \, \cos \phi_2 \, dA_2}{\pi r^2} \right)$$

$$dQ_{2\text{-}1} = E_{b2} \, dA_2 \left(\frac{\cos \phi_1 \, \cos \phi_2 \, dA_1}{\pi r^2} \right)$$

The net rate of thermal radiation exchange between surface elements is:

$$dQ_{\text{NET}} = dQ_{1\text{-}2} - dQ_{2\text{-}1}$$

$$= (E_{b1} - E_{b2}) \, \frac{\cos \phi_1 \, \cos \phi_2 \, dA_1 dA_2}{\pi r^2}$$

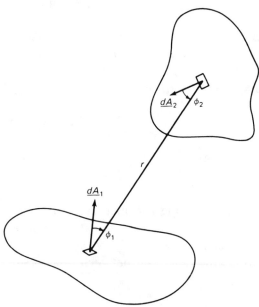

Fig. 3-5. Radiation exchange geometry.

or:

$$Q_{NET} = (E_{b1} - E_{b2}) \int_1 \int_2 \frac{\cos \phi_1 \cos \phi_2 \, dA_1 dA_2}{\pi r^2}$$

for surfaces 1 and 2 at uniform temperature. The quantity within the double integral is written as:

$$A_1 F_{1-2} = \iint \frac{\cos \phi_1 \cos \phi_2 \, dA_1 dA_2}{\pi r^2} \qquad \text{E3.5}$$

Shape factor, area product

and clearly:

$$A_1 F_{1-2} = A_2 F_{2-1} \qquad \begin{array}{c} \text{E3.6} \\ \text{Reciprocity} \end{array}$$

so that:

$$Q_{NET} = (E_{b1} - E_{b2}) F_{1-2} A_1$$

or:

$$Q_{NET} = (E_{b1} - E_{b2}) F_{2-1} A_2$$

F_{1-2} is called a shape factor, view factor, or configuration factor and represents the fraction of radiation emitted by surface 1 that is intercepted by surface 2.

By conservation of energy:

$$A_1 F_{1-1} + A_1 F_{1-2} + \cdots + A_1 F_{1-N} = A_1$$

$$\sum_{j=1}^{N} F_{ij} = 1$$

where all space surrounding and including A_i is composed of N surfaces. The term F_{1-1} accounts for radiation from a surface to itself. When the surface is not concave, $F_{1-1} = 0$.

Computation of radiation between black surfaces is accomplished via:

$$Q_{ij} = \sigma F_{ij} A_i (T_i'^4 - T_j'^4) \qquad \begin{array}{c} \text{E3.7} \\ \text{Blackbody} \\ \text{radiation exchange} \end{array}$$

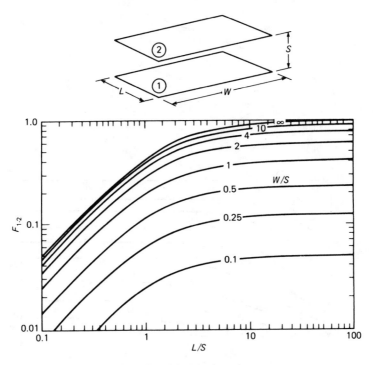

Fig. 3-6. Parallel plate shape factors.

The major part of any such calculation is the determination of the shape factors. Fortunately, tables and graphs are found throughout the literature documentating some very common shapes. Two of the most commonly occurring problems in electronic chassis are parallel and perpendicular plates.

For parallel plates with nomenclature following Fig. 3-6:

$$x = L/S, \qquad y = W/S$$

$$\left(\frac{\pi xy}{2}\right) F_{1\text{-}2} = \ln \left[\frac{(1+x^2)(1+y^2)}{1+x^2+y^2}\right]^{1/2}$$

$$+ y\sqrt{1+x^2} \ \tan^{-1}\left(\frac{y}{\sqrt{1+x^2}}\right)$$

$$+ x\sqrt{1+y^2} \ \tan^{-1}\left(\frac{x}{\sqrt{1+y^2}}\right)$$

$$- y \ \tan^{-1} y - x \ \tan^{-1} x$$

For perpendicular plates according to Fig. 3.7:

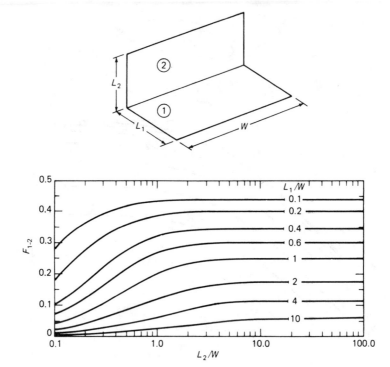

Fig. 3-7. Perpendicular plate shape factors.

$$x = L_2/W, \qquad y = L_1/W, \qquad z = x^2 + y^2$$

$$(\pi y) F_{1\text{-}2} = \frac{1}{4} \left\{ \ln \left[\frac{(1 + x^2)(1 + y^2)}{1 + z} \right] \right.$$

$$+ y^2 \ln \left[\frac{y^2(1 + z)}{(1 + y^2)z} \right] + x^2 \ln \left[\frac{x^2(1 + z)}{z(1 + x^2)} \right] \right\}$$

$$+ y \tan^{-1}(1/y) + x \tan^{-1}(1/x) - \sqrt{z} \tan^{-1}(1/\sqrt{z})$$

Shape factors for more complex systems can be obtained from known values. Consider the problem illustrated in Fig. 3-8 where $A_1 = A_2$, and $A_3 = A_4$. Using conservation of radiation flux, the cross term $F_{2\text{-}3}$ may be found:

$$(A_1 + A_2) F_{(1+2)\text{-}(3+4)} = (A_1 + A_2)(F_{(1+2)\text{-}3} + F_{(1+2)\text{-}4})$$

$$F_{(1+2)\text{-}(3+4)} = 2 F_{(1+2)\text{-}3}$$

$$= 2(F_{1\text{-}3} + F_{2\text{-}3})$$

$$F_{2\text{-}3} = \tfrac{1}{2} F_{(1+2)\text{-}(3+4)} - F_{1\text{-}3}$$

Both terms on the right are obtainable from Fig. 3-7.

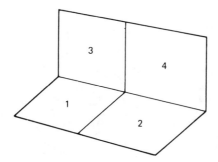

Fig. 3-8. Shape factor geometry for four surfaces.

A simple method for computing the shape factor between two flat, very long surfaces is known as Hottel's crossed string method, [4], because of the relationship of the variables as shown in Fig. 3-9, for which:

$$F_{1-2} = [(L_3 + L_4) - (L_5 + L_6)]/2L_1$$

E3.8

Crossed strings

The accuracy of this method in real applications depends on the actual end losses, which are not accounted for here.

Typical of significant applications of radiation heat transfer and shielding effects is the U-channel heat sink illustrated in Fig. 3-10. Mounting two devices to the same sink ensures minimum temperature difference between the transistors, sometimes a significant feature. The heat sink is painted black to ensure near-blackbody conditions. The radiation transferred to ambient for a sink temperature of 75°C in an ambient of 25°C is calculated.

The two fins are identified as surfaces 1 and 3, whereas the base is surface 2. The exterior surface radiation is assumed to be totally absorbed by the surrounding structure, e.g., an instrument cabinet. The heat transferred to ambient from the exterior surfaces is:

$$Q_{EXT} = \sigma(A_1 + A_3)(T_1'^4 - T_A'^4)$$

$$= 3.657 \times 10^{-11}(4)[(75 + 273)^4 - (25 + 273)^4]$$

$$= 0.99 \text{ watt}$$

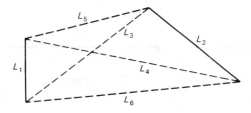

Fig. 3-9. Geometry for crossed string method.

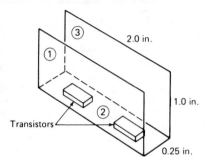

Fig. 3-10. Heat sinking two transistors.

Calculation of the interior radiation to ambient requires incorporation of the angle factors for each radiating surface. Surfaces 1, 2, and 3 radiate to ambient through both end openings as well as the top gap. Notice that shielding conditions for 1 and 3 are identical. This fact is used to simplify the analysis. Using the notation:

$$F_{i-E} = \text{shape factor for surface } i \text{ to the end opening}$$

$$F_{i-T} = \text{shape factor for surface } i \text{ to the top opening}$$

then:

$$Q_{INT} = \sigma(2A_1F_{1-T} + 4A_1F_{1-E} + 2A_2F_{2-E} + A_2F_{2-T})(T_1'^4 - T_A'^4)$$

and:

$$F_{1-T}: \quad L_1/W = 1.0/2.0 = 0.5$$
$$L_2/W = 0.25/2.0 = 0.125$$

$$F_{1-T} = 0.10 \text{ from Fig. 3-7}$$

$$F_{1-E}: \quad L_1/W = 2.0/1.0 = 2.0$$
$$L_2/W = 0.25/1.0 = 0.25$$

$$F_{1-E} = 0.05 \text{ from Fig. 3-7}$$

$$F_{2-E}: \quad L_1/W = 2.0/0.25 = 8.0$$
$$L_2/W = 1.0/0.25 = 4.0$$

$$F_{2-E} = 0.06 \text{ from Fig. 3-7}$$

$$F_{2-T}: \quad L/S = 2.0/1.0 = 2.0$$
$$W/S = 0.25/1.0 = 0.25$$

$$F_{2-T} = 0.087 \text{ from Fig. 3-6}$$

so that:

$$Q_{INT} = 3.657 \times 10^{-11}[2(1)(2)(0.10) + 4(1)(2)(0.05)$$

$$+ 2(2.0)(0.25)(0.06) + (2.0)(0.25)(0.087)]$$

$$\times [(75 + 273)^4 - (25 + 273)^4]$$

$$= 0.22 \text{ watt}$$

Thus the total radiation transfer is 1.21 watts, and the radiative resistance of the heat sink at this temperature is $50/1.21 = 41°C/\text{watt}$.

3.3 RADIATION CHARACTERISTICS OF NONBLACK SURFACES

As indicated earlier, surfaces encountered in practice do not radiate precisely in the manner described by the blackbody equations. The actual monochromatic emissive power of a real surface is always less, though sometimes not by much, than $E_{\lambda b}$. The ratio of actual to blackbody monochromatic emissive powers defines the monochromatic emissivity:

$$\epsilon_\lambda = E_\lambda / E_{\lambda b}$$

The total emissivity is:

$$\epsilon = E/E_b = E/\sigma T'^4$$

The term "emissivity" should be confined to the description of a material property. Radiation properties are affected by surface conditions such as roughness and surface oxidation, for which circumstances the term "emittance" should be used. The two terms are, however, used interchangeably by engineers. Table 3-1 lists emittances for some common materials as selected from a variety of sources. Where values for several wavelengths were available, the region of about 10 μ was selected. A particularly interesting example of the importance of recognizing the interrelationship between monochromatic emissive power and monochromatic emittance is illustrated by the plot of ϵ_λ for anodized aluminum in Fig. 3-11. Although this material has an ϵ_λ that varies considerably over the region, Fig. 3-2 indicates that at temperatures commonly encountered in electronic equipment (25–100°C), the dominant spectral region is 5–15 μ. This would suggest according to Fig. 3-11, that anodized aluminum heat sinks may be characterized by total emittance in the vicinity of 0.4 to 0.9.

The author of the original source, [16], of data used in Fig. 3-11 computed a total emittance using an equivalent of:

$$\epsilon = \int_0^\infty \epsilon_\lambda E_{\lambda b}\, d\lambda \bigg/ \int_0^\infty E_{\lambda b}\, d\lambda$$

to obtain $\epsilon = 0.697$. Additional spectral data are available in [17].

All surfaces satisfy the following conditions based on conservation of energy. In Fig. 3-12:

$$Q_I = \text{incident radiation}$$
$$Q_R = \text{reflected radiation}$$
$$Q_T = \text{transmitted radiation}$$
$$Q_A = \text{absorbed radiation}$$

and:

$$Q_R + Q_T + Q_A = Q_I$$

or if $\alpha_\lambda = Q_A/Q_I$, $\tau_\lambda = Q_T/Q_I$, and $\rho_\lambda = Q_R/Q_I$, then:

$$\rho_\lambda + \tau_\lambda + \alpha_\lambda = 1$$

For opaque surfaces, $\tau_\lambda = 0$ and:

$$\alpha_\lambda = 1 - \rho_\lambda$$

Table 3-1. Emittance of some common materials used in electronic equipment.

Surface	Emittance
Metals	
Aluminum, polished	0.05
Aluminum, lightly oxidized	0.1
Aluminum, anodized	0.7–0.9
Copper, polished	0.06
Copper, oxidized	0.25–0.7
Gold	0.02
Iron, polished	0.06
Iron, bright cast	0.2
Iron, oxidized cast	0.6
Paints	
Varnish	0.89
Lacquer, clear on bright Cu	0.07
Lacquer, white	0.9
Lacquer, flat black	0.95
Enamel, most colors	0.9
Oil, most colors	0.9

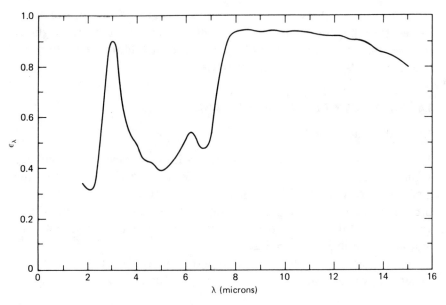

Fig. 3-11. Monochromatic emittance for anodized aluminum at 282°C. Plotted from [16], Table 2.

The monochromatic absorptivity α_λ is related to ϵ_λ by Kirchhoff's law (see [2]):

$$\alpha_\lambda = \epsilon_\lambda$$

for a directional polarized beam, but is of more use written as:

$$\rho_\lambda = 1 - \epsilon_\lambda$$

E3.9
Monochromatic
reflectivity

The following version:

$$\rho = 1 - \epsilon$$

E3.10
Total reflectivity

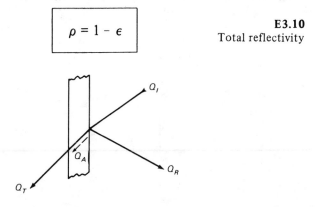

Fig. 3-12. Application of conservation of energy.

is valid only if ϵ does not vary with wavelength. Kirchhoff's law requires thermal equilibrium, and identical temperatures throughout the system of surfaces. The latter condition seldom exists in practice; thus E3.9 and E3.10 are approximations.

3.4 GRAY BODY RADIATION EXCHANGE

3.4.1 Algebraic Method

The algebraic method reviewed in the following pages is particularly applicable to problems involving many surfaces and is the technique incorporated into TNETFA (Chapters 8 and 10). The simpler two or three surface problems for which one might wish to obtain nonmatrix analytical solutions are best evaluated by the radiation network method given in Section 3.4.2.

It was shown in Section 3.2 that the net radiation exchange between any two surfaces at absolute temperatures T'_i, T'_j is:

$$Q_{ij} = \sigma F_{ij} A_i (T'^4_i - T'^4_j)$$

This equation requires that no surface reflection take place, i.e., $\epsilon_i = \epsilon_j = 1 - \rho_i = 1.0$. In actual practice, radiation may not be a very significant effect unless the gray body emittance ϵ is rather close to unity. There are also situations, however, in which the geometry is such that although ϵ may be significantly less than one, there are enough multiple reflections to make radiation heat transfer non-negligible. In cases where a natural-convection-only analysis indicates a very marginal component operating temperature, the addition of radiation effects to the calculations may indicate that the design is indeed worth pursuing.

The situation of a particular surface emitting radiation and being irradiated by one or more other surfaces is illustrated in Fig. 3-13. E is the emissive power, H is the sum total of the irradiance from all other interacting surfaces, and ρH is the reflected irradiance. The

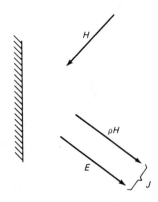

Fig. 3-13. Emission and reflection from an opaque surface.

total radiation energy per unit surface area leaving the surface in a unit of time interval is called the *radiosity*. The surface illustrated in Fig. 3-13 is a gray, diffusely reflecting surface of uniform temperature and area A_1.* If A_1 is irradiated by only one other surface, A_2, the irradiance is:

$$H_1 = F_{2-1} A_2 J_2$$

where J_2, the radiosity of surface 2, is the total radiation density leaving A_2.

$$J_1 A_1 = E_1 A_1 + \rho_1 H_1$$

$$J_1 A_1 = \epsilon_1 E_{b1} A_1 + \rho_1 F_{2-1} J_2 A_2$$

For a system of N surfaces this is easily generalized to:

$$J_i A_i = \epsilon_i E_{bi} A_i + \sum_{j=1}^{N} \rho_i F_{ji} J_j A_j$$

Using the reciprocity relation:

$$F_{ji} A_j = F_{ij} A_i$$

$$J_i A_i = \epsilon_i E_{bi} A_i + \rho_i A_i \sum_{j=1}^{N} F_{ij} J_j$$

$$\boxed{J_i = \epsilon_i E_{bi} + \rho_i \sum_{j=1}^{N} F_{ij} J_j} \qquad \begin{array}{l} \textbf{E3.11} \\ \text{Radiosity equation} \end{array}$$

Naturally $E_{bi} = \sigma T_i'^4$, and if surface i is not concave, $F_{ii} = 0$. The radiosity equation is actually a system of N linear equations where J_i are the unknown.

The *net radiation* for any surface i is the difference between the total outgoing and incoming radiation. Therefore for an arbitrary surface i:

$$Q_{\text{NET } i} = \text{total outgoing} - \text{total incoming}$$

$$= J_i A_i - \sum_{j=1}^{N} F_{ji} J_j A_j$$

---

*These assumptions will be used from this point on, unless it is otherwise noted.

$$= J_i A_i - \sum_{j=1}^{N} A_i F_{ij} J_j$$

$$Q_{\text{NET } i} = A_i \left(J_i - \sum_{j=1}^{N} F_{ij} J_j \right)$$

E3.12
Net surface
radiation

This may be further simplified. The radiosity equation may be
solved for:

$$\sum_{j=1}^{N} F_{ij} J_j = \frac{J_i - \epsilon_i E_{bi}}{\rho_i}$$

Then:

$$\frac{Q_{\text{NET } i}}{A_i} = J_i - \left(\frac{J_i - \epsilon_i E_{bi}}{\rho_i} \right)$$

$$= \frac{\rho_i J_i - J_i + \epsilon_i E_{bi}}{\rho_i}$$

$$= \frac{\epsilon_i E_{bi} - (1 - \rho_i) J_i}{\rho_i}$$

$$\frac{Q_{\text{NET } i}}{A_i} = \left(\frac{\epsilon_i}{1 - \epsilon_i} \right) (E_{bi} - J_i)$$

E3.13
Alternative
form of net
surface radiation

Note that this form is not usable when $\epsilon_i = 1.0$.
The radiation exchange between any two surfaces is:

$$Q_{ij} = A_i F_{ij} J_i - A_j F_{ji} J_j$$

$$= A_i F_{ij} J_i - A_i F_{ij} J_j$$

$$Q_{ij} = A_i F_{ij} (J_i - J_j)$$

E3.14
Radiation between
two surfaces

An important consequence of the developments thus far is that a
gray body radiation conductance may now be defined for the multi-
surface problem:

$$C_{ij} = Q_{ij} / (T_i - T_j)$$

$$C_{ij} = \frac{A_i F_{ij}(J_i - J_j)}{(T_i - T_j)}$$

Presuming that the ϵ_i and ρ_i are known and the F_{ij} calculable for any arbitrary number of N surfaces, the radiosities are determinable by writing E3.11 in matrix form:

$$[A][J] = [B]$$

where:

$$[J] = \begin{matrix} J_1 \\ J_2 \\ \vdots \\ J_N \end{matrix} \qquad [B] = \begin{matrix} \epsilon_1 E_{b1} \\ \epsilon_2 E_{b2} \\ \vdots \\ \epsilon_N E_{bN} \end{matrix}$$

and:

$$[A] = \begin{matrix} (1 - \rho_1 F_{11}) & -\rho_1 F_{12} & \cdots & -\rho_1 F_{1N} \\ -\rho_2 F_{21} & (1 - \rho_2 F_{22}) & \cdots & -\rho_2 F_{2N} \\ \vdots & \vdots & & \vdots \\ -\rho_N F_{N1} & -\rho_N F_{N2} & \cdots & (1 - \rho_N F_{NN}) \end{matrix}$$

The J_i may be found by inverting the matrix $[A]$ and taking the matrix product $[A]^{-1}[B]$:

$$[J] = [A]^{-1}[B]$$

TNETFA has been constructed to take advantage of certain features of this theory. For example, the matrices $[A]$, $[A]^{-1}$ are computed only once because the elements are not temperature-dependent. At the end of each iterative cycle (TNETFA uses a Gauss-Seidel temperature iteration scheme), $[B]$ must be computed and multiplied by $[A]^{-1}$ to obtain $[J]$.

The radiation conductances are computed using E3.15. The $A_i F_{ij}$ product is input in TNETFA via the conductance interconnection scheme. The reciprocity $A_i F_{ij} = A_j F_{ji}$ is consistent with conduction, convection input for which $C_{ij} = C_{ji}$.

3.4.2 Thermal Network Method

Most radiation heat transfer analyses require consideration of the radiation *exchange* problem; i.e., spatial effects and multiple reflections between surfaces must be considered. The algebraic method in Section 3.4.1 is probably unnecessarily complex for some rather classic problems that require consideration of very few surfaces. Many engineering calculations may be made using only a two-surface approximation. The problem of radiation heat transfer from a rectangular U-channel configuration is sufficiently significant that it is in itself probably justification for this chapter (see Section 5.3).

The basis of the thermal network method is developed in the next few paragraphs. Once understood, the resulting rules are rather easily applied to some important cases. Bear in mind, however, that resistance and conductance definitions in the following network theory are not interchangeable with those in the preceding gray body theory discussion.

First, the radiosity equation is rewritten for the ith surface:

$$\rho_i \sum_{j=1}^{N} F_{ij} J_j + \epsilon_i E_{bi} - J_i = 0$$

The term $\epsilon_i J_i$ is added and subtracted:

$$\rho_i \sum_{j=1}^{N} F_{ij} J_j + \epsilon_i E_{bi} - J_i + \epsilon_i J_i - \epsilon_i J_i = 0$$

$$\rho_i \sum_{j=1}^{N} F_{ij} J_j + \epsilon_i E_{bi} - \rho_i J_i - \epsilon_i J_i = 0$$

Each term is then multiplied by A_i:

$$\rho_i A_i \sum_{j=1}^{N} F_{ij} J_j + \epsilon_i A_i E_{bi} - \rho_i A_i J_i - \epsilon_i A_i J_i = 0$$

The third term of the above equation is multiplied by $\sum_{j=1}^{N} F_{ij}$, which is identically one:

$$\rho_i A_i \sum_{j=1}^{N} F_{ij} J_j + \epsilon_i A_i E_{bi} - \rho_i A_i J_i \sum_{j=1}^{N} F_{ij} - \epsilon_i A_i J_i = 0$$

$$\rho_i A_i \sum_{j=1}^{N} F_{ij} (J_j - J_i) - \epsilon_i A_i (J_i - E_{bi}) = 0$$

Finally:

$$\sum_{j=1}^{N} \frac{J_j - J_i}{(A_i F_{ij})^{-1}} = \frac{J_i - E_{bi}}{(\rho_i / \epsilon_i A_i)}$$

The denominators in each of the two terms are identified as resistances:

$$R_{ij} = 1/A_i F_{ij}$$

E3.16
Spatial

$$R_i = (1 - \epsilon_i)/\epsilon_i A_i$$

E3.17
Surface

A thermal radiation equivalent of Kirchhoff's law for electric circuits is recognized:

$$\sum_{j=1}^{N} \frac{J_i - J_j}{R_{ij}} = \frac{E_{bi} - J_i}{R_i}$$

E3.18
Kirchhoff's law

which states that the sum of heat flows into a node (i) equals the sum of the heat flows out. This analogy is aided by Fig. 3-14, a three-surface example, where the radiosity nodes J_i and the black-body emissive powers E_{bi} are the equivalent of potentials, and the heat transfer rates (watts) are the current equivalents.

The net radiative heat loss from the ith surface is:

$$Q_{\text{NET } i} = \frac{E_{bi} - J_i}{R_i}$$

E3.19
Net radiative loss

Relations E3.16, E3.17, E3.18, and E3.19 are used in Section 5.3 to develop a radiation shielding relationship for adjacent fins commonly found on extruded aluminum heat sinks.

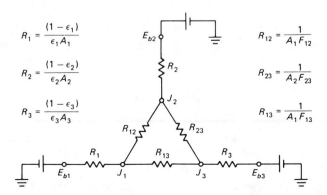

Fig. 3-14. Three-surface thermal radiation network.

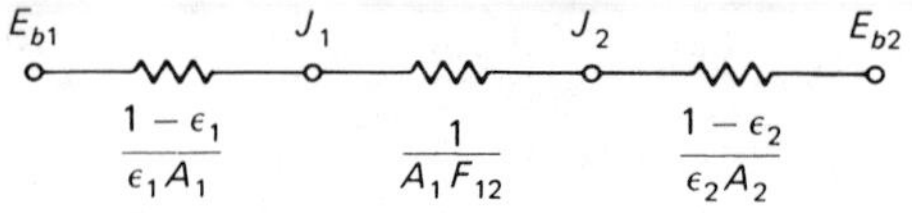

Fig. 3-15. Radiation circuit for two surfaces.

A very important example, useful in proving a commonly used but not always well-understood formula, is the two-surface problem, which is adequate when reflections anywhere other than between A_1 and A_2 are unimportant. The thermal radiation circuit for this problem is illustrated in Fig. 3-15. The net heat transfer between A_1 and A_2 is clearly identical in each of the three elements. The total resistance between E_{b1} and E_{b2} is immediately determined as:

$$R = \frac{1 - \epsilon_1}{\epsilon_1 A_1} + \frac{1}{A_1 F_{12}} + \frac{1 - \epsilon_2}{\epsilon_2 A_2}$$

Then:

$$Q_{NET} = (E_{b1} - E_{b2})/R$$

or:

$$Q_{NET} = \sigma \mathcal{F}_{12} A_1 (T_1'^4 - T_2'^4)$$

E3.20
Two-surface
radiation exchange

and

$$\mathcal{F}_{12} = \frac{1}{\dfrac{1 - \epsilon_1}{\epsilon_1} + \left(\dfrac{1 - \epsilon_2}{\epsilon_2}\right)\left(\dfrac{A_1}{A_2}\right) + \dfrac{1}{F_{12}}}$$

E3.21

Two surface gray body exchange factor

A special case may now be developed from E3.21. Suppose A_1 is totally enclosed by a room (or cabinet, in the case of a component within a system) such that:

$$\frac{\rho_2}{A_2} \ll \frac{\epsilon_2 \rho_1}{\epsilon_1 A_1}$$

which means that either the enclosure is so large that little reflected radiation finds its way back to A_1, or that the surface A_2 is highly absorbing. Also assume that A_1 is not concave. Then:

$$\mathcal{F}_{12} = \cfrac{1}{\cfrac{1 - \epsilon_1}{\epsilon} + 1}$$

$$= \frac{\epsilon_1}{1 - \epsilon_1 + \epsilon_1}$$

$$\boxed{\mathcal{F}_{12} = \epsilon_1}$$

E3.22
Small surface

3.4.3 Radiation Heat Transfer Coefficient

The concept of a radiative heat transfer coefficient was introduced in Section 1.4. In a similar manner, a radiation heat transfer coefficient, h_r, may be defined by:

$$\boxed{Q_r = \mathcal{F} h_r A_s (T_s - T_A)}$$

E3.23
Radiative heat
transfer

where

Q_r = heat transferred from a surface to ambient, watts
A_s = area of surface, cm^2 or in.^{2}
T_s = temperature of surface, °C
T_A = temperature of ambient surface to which heat is being transferred, °C
$\mathcal{F}$ = gray body radiation exchange factor between surface s and ambient
h_r = radiation heat transfer coefficient, watts/cm$^2 \cdot$ °C or watts/in.$^2 \cdot$ °C

The heat transfer coefficient is determined by comparing E3.23 with the Stefan-Boltzmann equation for gray body radiation exchange, E3.20.

$$Q_r = \mathcal{F} A_s \sigma (T_s'^4 - T_A'^4)$$

$$= \mathcal{F} A_s \left[\frac{\sigma (T_s'^4 - T_A'^4)}{T_s - T_A} \right] (T_s - T_A)$$

where we see that:

$$h_r = \frac{\sigma (T_s'^4 - T_A'^4)}{T_s - T_A}$$

or:

$$h_r = 3.657 \times 10^{-11} \left\{ \frac{[(T_s + 273)^4 - (T_A + 273)^4]}{T_s - T_A} \right\}$$ E3.24

Radiation heat transfer coefficient, watts/in.$^2 \cdot °C$

An alternative form may be developed from the expansion

$$a^4 - b^4 = (a - b)(a^3 + a^2b + ab^2 + b^3)$$

to result in:

$$h_r = 3.657 \times 10^{-11}(T_s'^3 + T_s'^2 T_A' + T_s' T_A'^2 + T_A'^3)$$

which may be more difficult to remember than E3.24, but has the advantage of not having a denominator that might be troublesome in iterative problem-solving techniques. Furthermore, if $T_A' \simeq T_s'$, then $h_r \simeq 4\sigma T_A'^3$, or:

$$h_r \simeq 1.463 \times 10^{-10} T_A'^3$$

E3.25
Short form,
$T_A' \simeq T_s'$,
watts/in.$^2 \cdot °C$

The requirement that the ambient and surface temperatures be nearly equal is rather easily met in applications to external cabinet surfaces where often $T_s - T_A < 20°C$, or $T_A' = 25°C + 273°K$ and $T_s' < 318°K$. The short form is accurate to within 10% for $T_A = 0 \rightarrow 100°C$ and $\Delta T < 20°C$. As an example, compute the heat transfer from a 10 in. $\times$ 12 in. cabinet top at an average temperature of 30°C in a room at 20°C. The cabinet surface has a decorative vinyl covering; therefore an emissivity of about 0.8 is appropriate.

$$Q_r = \mathcal{F}A_s h_r (T_s - T_A)$$

For a surface that is small compared to a surface by which it is totally enclosed, $\mathcal{F} = \epsilon$. Using E3.25:

$$h_r \simeq 1.463 \times 10^{-10} T_A'^3$$

$$= 1.463 \times 10^{-10} (20 + 273)^3$$

$$= 3.68 \times 10^{-3} \text{ watt/in.}^2 \cdot °C$$

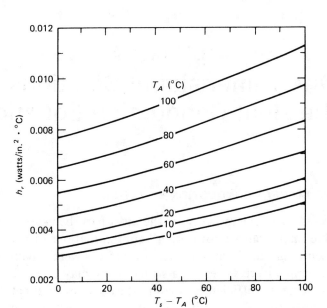

Fig. 3-16. Radiative heat transfer coefficient.

Note that the exact value of h_r from Fig. 3-16 is 3.9×10^{-3}.

$$Q_r = (0.8)(10)(12)(3.68 \times 10^{-3})(10)$$

$$= 3.53 \text{ watts}$$

This is equivalent to a resistance of:

$$R_r = 1/\mathfrak{F}A_s h_r$$

or:

$$R_r = 10°C/3.53 \text{ watts}$$

$$= 2.83°C/\text{watt}$$

Chapter 4
One-Dimensional Solutions to the Heat Conduction Equation

4.1 THE 1-D HEAT CONDUCTION EQUATION

The preceding chapters have, for the most part, been concerned with what one might call "zeroth order" approximations of real problems in the sense that uniform-temperature surfaces were usually assumed. When a temperature varies on a surface, one may break the problem up into elements sufficiently small that each is approximately of uniform temperature. Typically, such problems must be solved using a digital computation technique such as the thermal network method of Chapters 8 and 10.

There are, however, a large number of situations encountered that are soluble by "first order" approximations that not only include surface convection/radiation phenomena, but also account for the effects of temperature gradients within the material. These approximations are addressed in this chapter.

Consider a small element $\Delta x = x_2 - x_1$ of a bar of length L (Fig. 4-1). The temperature gradients through the thickness t and the width w are assumed to be negligible. The temperature gradient in the x-direction causes conductive heat transfer into and out of the element. The heat conducted out at x_2 is less than the heat conducted in at x_1 because of a convective/radiative loss Q_s from an elemental surface area $\Delta A_s = 2(w + t)\Delta x$. The surface losses are quantified by Newtonian cooling, E1.9.

Conservation of energy permits an energy balance on Δx:

$$\text{heat into } \Delta x - \text{heat out of } \Delta x = 0$$

$$\left[-kA_k \left. \frac{dT}{dx} \right|_{x_1} + Q_V \Delta x A_k \right] - \left[-kA_k \left. \frac{dT}{dx} \right|_{x_2} + 2h(w + t)\Delta x T \right] = 0$$

where the average element temperature T is referred to a zero ambient. This means T is the rise above ambient. Q_V is the internal source density in watts per unit volume of the element. Dividing through by $k\Delta x A_k$ and rearranging:

$$\frac{1}{\Delta x}\left[\left. \frac{dT}{dx} \right|_{x_2} - \left. \frac{dT}{dx} \right|_{x_1} \right] - \frac{2h(w + t)}{kA_k} T = \frac{-Q_V}{k}$$

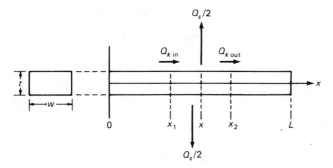

Fig. 4-1. Energy balance on an element $\Delta x = x_2 - x_1$.

Taking the limit as $\Delta x \to 0$:

$$\boxed{\frac{d^2T}{dx^2} - \theta^2 T = \frac{-Q_V}{k}}$$

E4.1
1-D heat conduction
and convection

$$\theta^2 = R_k/L^2 R_s$$

It is interesting to note that the solutions are obtainable in terms of the elementary resistances $R_k = L/kA_k$ and $R_s = 1/hA_s = 1/[2h(w + t)L]$, the "zeroth order" approximations.

The general solution for nonzero θ is:

$$T = c_1 \cosh \theta x + c_2 \sinh \theta x + (\alpha/\theta^2)$$

where $\alpha = Q_V/k$ and cosh, sinh are the standard hyperbolic functions:

$$\cosh \theta x = \frac{e^{\theta x} + e^{-\theta x}}{2}; \quad \sinh \theta x = \frac{e^{\theta x} - e^{-\theta x}}{2}$$

4.2 UNIFORM SOURCE–CONDUCTION

The geometry for this situation is shown in Fig. 4-2. The source heat (watts) is uniformly distributed the length of the device, and surface

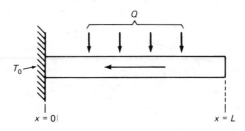

Fig. 4-2. Uniform source–conduction.

losses are negligible. This latter condition requires a modification of the differential equation; i.e., $R_s \rightarrow \infty$ and thus $\theta \rightarrow 0$. Then:

$$\frac{d^2T}{dx^2} = -\frac{Q_V}{k}$$

which is immediately integrable to:

$$T = -Q_V x^2/2k + c_1 x + c_2$$

The constants c_1, c_2 are obtained from the boundary conditions at $x = 0$:

$$T = T_0$$

$$-kA_k \frac{dT}{dx} = -Q_V A_k L, \qquad Q = Q_V A_k L$$

The second boundary condition stems from the condition that all of the source heat $Q_V A_k L$ must pass through $x = 0$. The result is:

$$T(x) = x(Q/kA_k)(1 - x/2L) + T_0$$

The thermal resistance is defined as the hottest temperature above T_0, divided by the total heat dissipation:

$$R = [T(x = L) - T_0]/Q$$

$$= (LQ/2kA_k)/Q$$

$$\boxed{R = \tfrac{1}{2}R_k}$$

E4.2
Uniform
source—conduction

As an application, consider an electronic package where several circuit cards are in a totally sealed box (Fig. 4-3). The designer wishes not to depend on the very restricted space as a heat-convecting

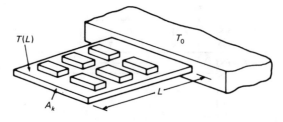

Fig. 4-3. Aluminum core board with negligible air cooling.

medium, and the several cards may act as a set of parallel heat shields, all at nearly the same average temperature. One end of each board is heat sunk to a temperature T_0 such that an additional rise of 20°C to the hottest point on the board is permitted. The problem is to find how many watts can be dissipated on each 6 in. × 6 in. circuit card if a $\frac{1}{16}$ in. thick aluminum (k = 5.0 watts/in. · °C) core is used.

$$T - T_0 = RQ$$

$$Q = (T - T_0)/R$$

$$= (T - T_0)2kwt/L$$

$$= (20)(2)(5)(6)(1/16)/6$$

$$= 12.5 \text{ watts}$$

4.3 UNIFORM SOURCE–CONDUCTION/CONVECTION, ONE END SINKED

Consider the same problem as in Section 4.2 except that surface heat losses are included (Fig. 4-4). The general solution is required:

$$T(x) = c_1 \cosh \theta x + c_2 \sinh \theta x + \alpha/\theta^2$$

The boundary condition:

$$T = T_0 \text{ at } x = 0$$

leads to:

$$c_1 = T_0 - \alpha/\theta^2$$

The boundary condition:

$$-kA_k \frac{dT}{dx} = 0 \text{ at } x = L$$

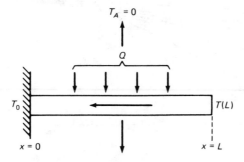

Fig. 4-4. Uniform source–conduction, convection–one end sinked.

leads to:

$$c_2 = -(T_0 - \alpha/\theta^2) \tanh \theta L$$

and since:

$$\frac{\alpha}{\theta^2} = \frac{(Q_V/k)}{(R_k/L^2 R_s)}$$

$$= \frac{(Q/kA_k L)}{(R_k/L^2 R_s)} = QR_s$$

then:

$$T(x) = (T_0 - QR_s)(\cosh \theta x - \tanh \theta L \sinh \theta x) + QR_s$$

Defining:

$$R = \frac{T(x = L)}{Q}$$

$$\boxed{R = \left[\left(\frac{T_0}{Q} - R_s\right)\bigg/ \cosh \sqrt{R_k/R_s}\right] + R_s} \qquad \text{E4.3}$$

Uniform source, one end sinked, conduction/convection

which is plotted in Fig. 4-5 as a computational aid.

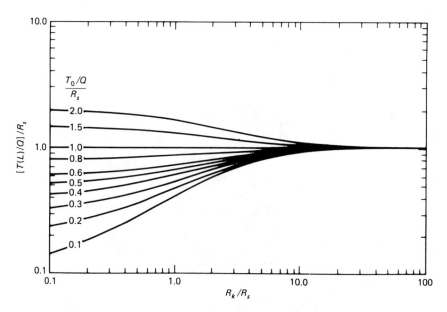

Fig. 4-5. Uniform source–conduction, convection–one end sinked.

The heat Q_0 conducted into the sink wall at temperature T_0 (above ambient) is obtained from:

$$\frac{dT(x)}{dx}\bigg|_{x=0} = (T_0 - QR_s)\theta(\sinh \theta x - \tanh \theta L \cosh \theta x)\bigg|_{x=0}$$

$$= -(T_0 - QR_s)\theta \tanh \theta L$$

$$Q_0 = kA_k(T_0 - QR_s)\theta \tanh \theta L$$

$$\boxed{\frac{Q_0}{Q} = \left(\frac{T_0/Q}{R_s} - 1\right)\frac{\tanh \sqrt{R_k/R_s}}{\sqrt{R_k/R_s}}} \qquad \text{E4.4}$$

Uniform source, one end sinked, conduction/convection

which is plotted in Fig. 4-6.

Suppose the previously discussed 6 in. × 6 in. aluminum core circuit board is reconsidered. An air supply equivalent to a velocity of 300 ft/min. is now available for cooling. The previous $T(L) - T_0 = 20°C$ is still required, and as a worst case $T_0 = 40°C$ above the air ambient.

The surface resistance for forced convection is obtained from E2.24:

$$\bar{h}_L = 0.001092\sqrt{300/6}$$

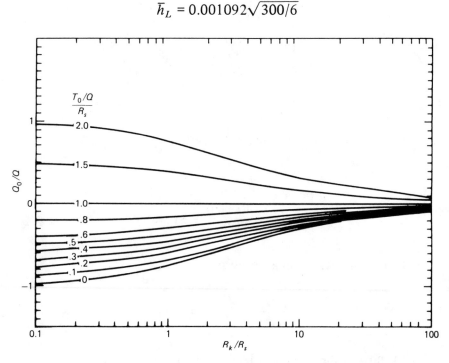

Fig. 4-6. Uniform source—conduction, convection—one end sinked.

and then applying the velocity correction factor $f = 1.5$ from Fig. 2-12 to obtain $\bar{h}_L = 0.0116$ watt/in.^{2} · °C. Then:

$$R_s = 1/\bar{h}_L A_s = 1/(0.0116)(2)(6)(6) = 1.2\text{°C/watt}$$

$$R_k = L/kA_k = 6/(5)(6)(1/16) = 3.2\text{°C/watt}$$

so that:

$$\frac{R_k}{R_s} = \frac{3.2}{1.2} = 2.67$$

An initial estimate of the maximum allowed heat dissipation is obtained from the previous result of $Q = 12.5$ watts when convection was not included. A few iterations are expected to obtain increasingly realistic estimates of Q.

$$\frac{T_0/Q}{R_s} = \frac{40/12.5}{1.2} = 2.67$$

This is out of range of the graphs plotted in Fig. 4-5, so a calculation is required:

$$R = \left[\left(\frac{T_0}{Q} - R_s\right)\Big/\cosh\sqrt{R_k/R_s}\,\right] + R_s$$

$$= \left[\left(\frac{40}{12.5} - 1.2\right)\Big/\cosh\sqrt{2.67}\,\right] + 1.2$$

$$= 1.95$$

A new estimate of Q, called Q', is now obtainable:

$$T = RQ'$$

$$Q' = \frac{T}{R} = \frac{40 + 20}{1.95} = 31 \text{ watts}$$

Clearly a second iteration is necessary.

$$\frac{T_0/Q'}{R_s} = \frac{40/31}{1.2} = 1.08$$

which is now in the range of Fig. 4-5. Then:

$$\frac{R}{R_s} = 1.03, \qquad R = 1.23$$

and a new heat dissipation Q'' is attempted:

$$Q'' = \frac{T}{R} = \frac{60}{1.23} = 49 \text{ watts}$$

A third iteration using $(T_0/Q'')/R_s = 0.68$ gives $R = 1.06$ and:

$$Q''' = \frac{T}{R} = \frac{60}{1.06} = 57 \text{ watts}$$

A fourth and last try using $(T_0/Q''')/R_s = 0.58$ gives $R = 1.01$ and:

$$Q^{iv} = \frac{60}{1.01} = 59 \text{ watts}$$

The heat conducted into the wall is obtained from Fig. 4-6.

$$\frac{Q_0}{Q} = -0.25$$

$$Q_0 = -0.25(59) = -15 \text{ watts}$$

where the negative sign indicates that the direction of heat conduction is the negative x-direction, or into the wall.

4.4 END SOURCE—CONDUCTION/CONVECTION

The pertinent geometry for this very useful case is shown in Fig. 4-7. Note that all heat losses for this system occur via the surface; i.e., neither end is heat sunk.

Since the source is not distributed, it is accounted for through the boundary conditions; thus $Q_V = 0$, and the differential equation is:

$$\frac{d^2 T}{dx^2} - \theta^2 T = 0$$

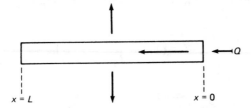

Fig. 4-7. End source—conduction, convection.

which has the solution:

$$T = c_1 \cosh \theta x + c_2 \sinh \theta x$$

The boundary conditions at the ends provide c_1, c_2:

$x = 0$:

$$-kA_k \frac{dT}{dx} = Q, \qquad Q = |Q|$$

$$c_2 = -Q/kA_k\theta$$

$x = L$:

$$-kA_k \frac{dT}{dx} = 0$$

$$c_1 = -c_2 \coth \theta L$$

The temperature function is:

$$T(x) = Q\sqrt{R_k R_s} \left[\coth \sqrt{R_k/R_s} \cosh \sqrt{R_k/R_s} \left(\frac{x}{L}\right) \right.$$

$$\left. - \sinh \sqrt{R_k/R_s} \left(\frac{x}{L}\right) \right]$$

The resistance is defined in a zero ambient:

$$R = T(x = 0)/Q$$

$$\boxed{\frac{R}{R_s} = \sqrt{\frac{R_k}{R_s}} \coth \sqrt{\frac{R_k}{R_s}}}$$

E4.5
End source,
conduction/
convection

which is plotted in Fig. 4-8.

The average temperature, $\bar{T}$, is computed from:

$$\bar{T} = (1/L) \int_0^L T(x)\, dx$$

$$= QR_s$$

One of the most useful examples of a one-dimensional conduc-

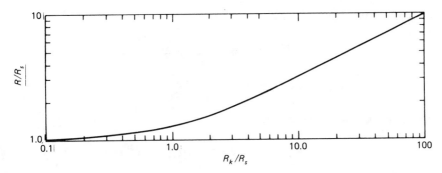

Fig. 4-8. End source–conduction, convection.

tion formula is the finned heat sink shown in Fig. 4-9. The sink is oriented so that the width dimension is actually the sink height, and the transistor heat source is at one end. A centrally located source on an identical sink but with twice the length would have half the resistance computed in this example. The sink length, width, and base thickness are: $L = 4.0$ in., $w = 2.0$ in., $t = 0.2$ in. The fins are spaced so that in either forced or free convection the appropriate flat plate h is used. The total surface area including fins and base is twice the area of the base alone. A maximum average temperature rise of 20°C is permitted.

$$A_s = 2(2wL) = 2(2)(2.0)(4.0) = 32 \text{ in.}^2$$

A worst-case condition will be assumed by an unpainted nonanodized surface, thus permitting neglect of radiation. From Chapter 2:

Natural convection:

$$\bar{h} = 0.0022(\Delta T/w)^{0.35} \text{ from E2.14}$$

$$= 0.0022(20/2)^{0.35}$$

$$= 0.0049 \text{ watt/in.}^2 \cdot °C$$

$$R_s = 1/\bar{h}A_s = 1/(0.0049)(32)$$

$$= 6.38°C/\text{watt}$$

Fig. 4-9. Simple heat sink or chassis element.

Forced convection, $V = 500$ ft/min.:

$$\bar{h} = 0.025 \text{ watt/in.}^2 \cdot \text{°C from Fig. 2-15}$$

$$R_s = 1.25\text{°C/watt}$$

The conduction resistance is:

$$R_k = L/kA_k = 4.0/(5.0)(2.0)(0.2)$$

$$= 2.0\text{°C/watt}$$

The thermal resistance for these two cases may be either com-puted from:

$$\frac{R}{R_s} = \sqrt{\frac{R_k}{R_s}} \coth \sqrt{\frac{R_k}{R_s}}$$

or picked off Fig. 4-8.

Natural convection:

$$\frac{R_k}{R_s} = 0.30, \qquad \frac{R}{R_s} = 1.10, \qquad R = 7.03$$

Forced convection:

$$\frac{R_k}{R_s} = 0.63, \qquad \frac{R}{R_s} = 1.20, \qquad R = 1.50$$

Both the natural and forced convection situations are dominated by the surface resistance, R_s.

The same geometry (Fig. 4-9) and resistance formula is applicable to heat-conducting chassis elements without fins. Consider the ex-ample of a long, thin aluminum bracket with neglected heat losses at contact points with the cabinet.

$$L = 6.0 \text{ in.}, \qquad w = 2.0 \text{ in.}, \qquad t = 0.04 \text{ in.}$$

$$A_s = 2wL = 2(2.0)(6.0) = 24 \text{ in.}^2$$

Natural convection:

$$h = 0.0022(20/2)^{0.35} = 0.0049 \text{ watt/in.}^2 \cdot \text{°C}$$

for a 20°C rise and neglected radiation ($\epsilon = 0.1$) contribution.

$$R_s = 1/hA_s = 8.50°C/watt$$

Forced convection, $V = 500$ ft/min.:

$$\bar{h} = 0.025 \text{ watt/in.}^2 \cdot °C$$

$$R_s = 1/\bar{h}A_s = 1.67°C/watt$$

$$R_k = L/kA_k = 6.0/(5)(2)(0.04) = 15°C/watt$$

The total resistance from Fig. 4-8 is:

Natural convection:

$$\frac{R_k}{R_s} = 1.76, \qquad \frac{R}{R_s} = 1.53, \qquad R = 13.01°C/watt$$

Forced convection:

$$\frac{R_k}{R_s} = 8.98, \qquad \frac{R}{R_s} = 3.01, \qquad R = 5.03°C/watt$$

The effect of a large conduction resistance R_k caused by the small thickness is evidenced by the R/R_s ratios.

R_k, R_s, and R are tabulated in Table 4-1 to permit an easy comparison of three methods that might be used to estimate the total resistance. The series sum $R = R_k + R_s$ overcompensates for R_k in all cases. The next best guess, $R = \frac{1}{2}R_k + R_s$, is much better if R_s is large enough, as in the natural convection problems, but results in very significant errors for small R_s, as in the forced convection cases.

Table 4-1. Summary of thermal resistance
($°C/watt$) predictions for sample problems.

	$\frac{1}{2}$ Sink	Chassis Element
Natural convection:		
R_s	6.38	8.50
R_k	2.00	15.00
$R_k + R_s$	8.38	23.50
$\frac{1}{2}R_k + R_s$	7.38	16.00
R	7.03	13.01
Forced convection:		
R_s	1.25	1.67
R_k	2.00	15.00
$R_k + R_s$	3.25	16.67
$\frac{1}{2}R_k + R_s$	2.25	9.17
R	1.50	5.03

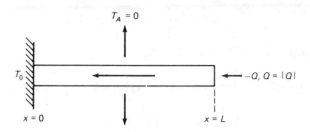

Fig. 4-10. End source–conduction, convection–opposite end sinked.

Clearly the slightly more complex one-dimensional formulae are preferred.

4.5 END SOURCE–CONDUCTION/CONVECTION, ONE END SINKED

The general solution for this problem (Fig. 4-10) is the same as for the preceding, i.e.:

$$T(x) = c_1 \cosh \theta x + c_2 \sinh \theta x$$

but with different boundary conditions.

$x = 0:$ $\qquad\qquad T(x) = T_0$

$x = L:$ $\qquad\qquad -kA_k \dfrac{dT}{dx} = -Q$

Application of the boundary conditions results in a solution:

$$T(x) = T_0(\cosh \theta x - \tanh \theta L \sinh \theta x) + [Q \sinh \theta x/(\theta kA_k \cosh \theta L)]$$

$$\boxed{\dfrac{T(L)/Q}{R_s} = \left(\dfrac{T_0/Q}{R_s} + \sqrt{\dfrac{R_k}{R_s}}\, \sinh \sqrt{\dfrac{R_k}{R_s}}\right) \Big/ \cosh \sqrt{\dfrac{R_k}{R_s}}} \qquad \text{E4.6}$$

End source, sink

which is plotted in Fig. 4-11.

The heat Q_0 conducted into the sink wall at temperature T_0 (above ambient) is found as before:

$$\dfrac{dT(x)}{dx}\bigg|_{x=0} = -\theta \tanh \theta L + \theta Q/\theta kA_k \cosh \theta L$$

$$Q_0 = Q\bigg|_{x=0} = -kA_k \dfrac{dT(x)}{dx}\bigg|_{x=0}$$

$$Q_0 = \theta kA_k (T_0 \tanh \theta L - Q/\theta kA_k \cosh \theta L)$$

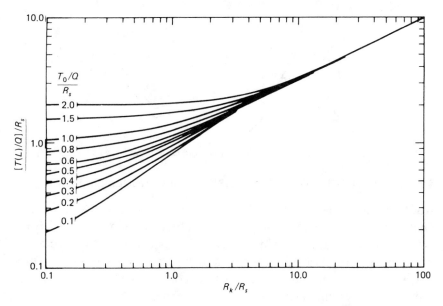

Fig. 4-11.   End source–conduction, convection–one end sinked.

$$\frac{Q_0}{Q} = \frac{\tanh \sqrt{R_k/R_s}}{\sqrt{R_k/R_s}} \left(\frac{T_0/Q}{R_s} - \frac{\sqrt{R_k/R_s}}{\sinh \sqrt{R_k/R_s}} \right)$$ E4.7

End source, sink

E4.7 is plotted in Fig. 4-12.

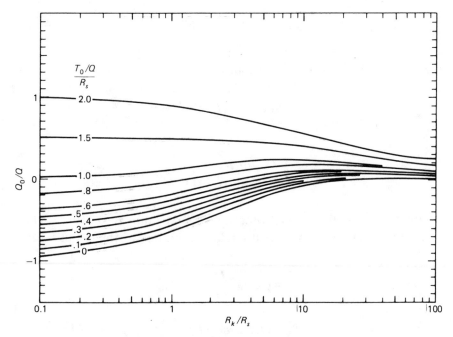

Fig. 4-12.   End source–conduction, convection–one end sinked.

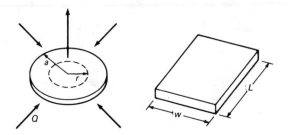

Fig. 4-13. Disk–conduction, convection.

4.6 CONDUCTING/CONVECTING DISK OR RECTANGLE

Another one-dimensional problem is the disk or nearly square rect-angle of thickness t (Fig. 4-13). The actual problem is solved for a circular disk of radius a which may be transformed into a rectangle of equal surface area by requiring a disk radius of $a = \sqrt{wL/\pi}$. Only one side of the surface is assumed to convect heat via h. Heat is input to the system around the circumference.

A one-dimensional differential equation may be derived in polar coordinates in the same manner as for the one-dimensional cartesian coordinate problems:

$$\frac{d^2T}{dr^2} + \frac{1}{r}\frac{dT}{dr} - \frac{h}{kt}T = 0$$

This is reduced to:

$$\frac{d^2T}{dz^2} + \frac{1}{z}\frac{dT}{dz} - T = 0$$

for $z = kr$ and $\alpha^2 = h/kt$ with a solution:

$$T(z) = c_1 I_0(z)$$

or:

$$T(r) = c_1 I_0(\alpha r)$$

where I_0 = modified Bessel function of first kind, zero order. The constant c_1 is found from the boundary condition at the radius $r = a$:

$$-Q = 2\pi a t k \frac{dT}{dr}$$

$$c_1 = Q/[2\pi a k t \alpha I_1(\alpha a)]$$

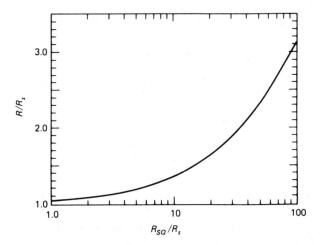

Fig. 4-14. Disk–conduction, convection, peripheral source. From [18], copyright © 1976 by the Institute of Electrical and Electronics Engineers, Inc. Reprinted by permission of the publisher.

or:

$$T(r) = (Q/2\pi akt\alpha)I_0(\alpha r)/I_1(\alpha r)$$

$$\frac{R}{R_s} = \sqrt{\frac{R_{SQ}}{\pi R_s}} \frac{I_0(\sqrt{R_{SQ}/\pi R_s})}{2I_1(\sqrt{R_{SQ}/\pi R_s})}$$

E4.8

Disk, rectangle

where I_1 = modified Bessel function of the first kind, first order, and:

$$R_{SQ} = 1/kt, \qquad R_s = 1/hwL$$

R is from the edge $r = a$, to ambient.

The resistance ratio R/R_s is plotted in Fig. 4-14. For sufficiently small R_{SQ} or large R_s, $R \cong R_s$. Otherwise, the total resistance R is greater than R_s by some factor (R/R_s).

The following problem illustrates application of E4.8. Power transistors may be heat sunk by attachment to chassis side walls or bulkheads. Such a method may be adequate for a totally sealed power supply chamber. The transistors are distributed on the side panels such that some of the heat will be conducted into the aluminum cover, from where it will be radiated and convected to ambient. The cover is 10 in. X 5 in. X 0.03 in. with an $\epsilon = 0.7$ for the external surface. An average surface temperature rise of 10°C above ambient is allowed. The designer wishes to estimate the heat that may be dissipated by the cover.

$$h = h_c + \epsilon h_r$$

$$h_r = 0.0039 \text{ watt/in.}^2 \cdot {}^\circ\text{C from Fig. 3-16}$$

$$h_c = 0.0018\{10/[(5)(10)/2(5 + 10)]\}^{0.33}$$

$$= 0.0033 \text{ watt/in.}^2 \cdot {}^{\circ}\text{C}$$

$$R_s = 1/hA_s = 1/[0.0033 + (0.7)(0.0039)][50]$$

$$= 3.3 {}^{\circ}\text{C/watt}$$

$$R_{SQ} = 1/(4)(0.03) = 8.33 {}^{\circ}\text{C/watt}$$

$$R_{SQ}/R_s = 2.5$$

From Fig. 4-14, $R/R_s = 1.2$, indicating that the distributed conduction resistance adds about 20% to the surface resistance (convection/radiation).

4.7 SPREADING RESISTANCE

The one-dimensional formula in Section 4.6 provides a method of including both conduction and surface losses. A perhaps superior method for some applications follows. A spreading resistance occurs when lines of heat flow diverge from a surface-mounted heat source into a conducting solid. Consider an extruded aluminum heat sink with several TO-3 transistor cases attached. The temperature of a transistor case base to ambient may be written:

$$T_c = R_{sp}Q_i + R_sQ + T_A$$

where

      Q = total power dissipated by sink
      R_s = surface resistance of sink
      Q_i = power dissipated by the one transistor considered
      R_{sp} = spreading resistance at transistor
      T_c = temperature of heat sink directly beneath transistor
      R_s = the combined radiation and convection resistance

    The thermal spreading resistance accounts for the temperature rise through the lines of flow that converge from the plane of the sink into the transistor base. This temperature rise is computed from the product of R_{sp} and the *single* transistor heat dissipation. Thermal spreading effects also occur from a transistor die base to a substrate surface and from a transistor junction to the die base.

    Several one-dimensional formulae are available for computing for square or circular sources. However, the reader must fully appreciate that these are for semi-infinite media, i.e., the material into

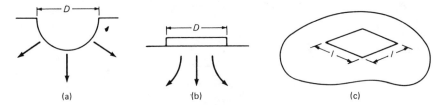

Fig. 4-15. Common geometry for spreading resistance in semi-infinite media.

which the source conducts extends to infinity in a conducting half-space. The three common geometries are illustrated in Fig. 4-15. The following formulae are applicable to the three cases illustrated.

Fig. 4-15	R_{sp}	Remark
(a)	$1/\pi Dk$	
(b)	$1/2Dk$	Ref. 19 uniform T
	$16/3\pi^2 Dk$	Ref. 19 uniform Q, ave. T
(c)	$1.1/2lk$	Ref. 20

Case (a) is included only to caution the reader that it should not be used to simulate planar sources. Case (c) may be extended to non-square sources by consulting the original references.

The thermal spreading resistance for a single square device on a square substrate with finite length, width, and thickness as illustrated in Fig. 4-16 is plotted in Fig. 4-17(a)–(d).* The effects of device dimension l, substrate length L, and thickness t must be included to

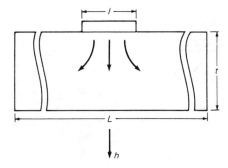

Fig. 4-16. Heat source on plane substrate.

---

*Fig. 4-17 is computed with TAMS (Chapters 7 and 9), using:

$$R_{sp} = \frac{T - T_A}{Q} - \frac{1}{hL^2}, \quad h \simeq 0$$

for the source side and finite for the nonsource side.

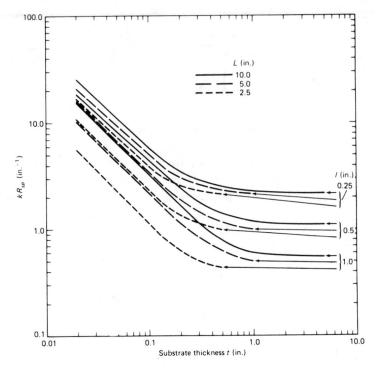

Fig. 4-17(a). Spreading resistance. $h = 0.005$ watt/in.$^2 \cdot °$C.

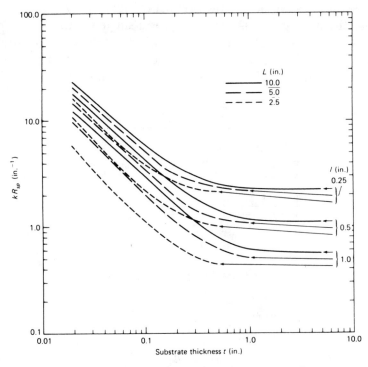

Fig. 4-17(b). Spreading resistance. $h = 0.025$ watt/in.$^2 \cdot °$C.

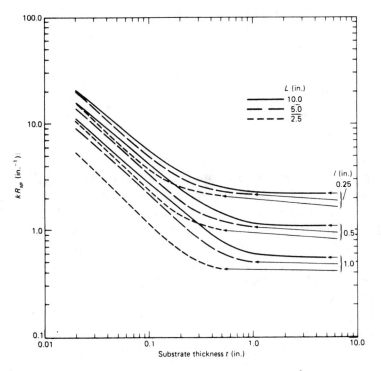

Fig. 4-17(c). Spreading resistance. $h = 0.05$ watt/in.$^2 \cdot °$C.

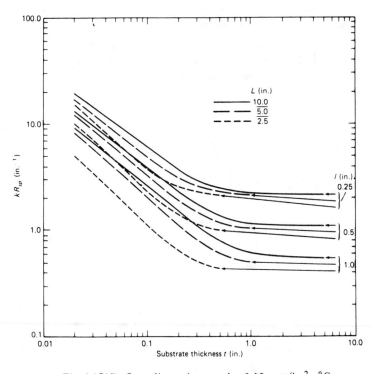

Fig. 4-17(d). Spreading resistance. $h = 0.10$ watt/in.$^2 \cdot °$C.

Table 4-2. Comparison of R_{sp} for $t = \infty$.

	$D, l = 0.25$ in. $k = 5.0$ watts/in. $\cdot\,^\circ$C	$D, l = 0.5$ in. $k = 10.0$ watts/in. $\cdot\,^\circ$C
$1/\pi Dk$	0.26	0.064
$1/2Dk$	0.40	0.100
$16/3\pi^2 Dk$	0.43	0.108
$1.1/2lk$	0.44	0.110

adequately specify the substrate thermal conductivity–spreading resistance product kR_{sp}. The quantity kR_{sp} is not totally independent of k, but the results are within 10% accuracy for $k = 3.5$–5.0 watts/ in. $\cdot\,^\circ$C. Unfortunately, the substrate surface boundary conditions (quantified by the nonsource substrate surface h in this case) also affect kR_{sp}–hence the necessity of providing results for several values of h. The results indicate that $t/l \gtrsim 10$ is equivalent to a semi-infinite medium.

Table 4-2 compares R_{sp} for the various formulae for semi-infinite media. The only unsatisfactory result is $1/\pi Dk$, which is to be expected because of the unrealistic hemispherically shaped source surface.

Unfortunately, most transistor heat sinking applications do not satisfy the $t/l \gtrsim 10$ or even a $t/l \gtrsim 1$ criterion. The only exception is for very small transistor junctions. Even in this case the problem of thermal interaction between small sources may preclude the use of simple formulae. It is highly recommended that the reader use an appropriate digital computation technique such as TAMS and avoid unnecessary and often erroneous guesses.

Chapter 5
Heat Sinks

5.1 INTRODUCTION

Although a great variety of surface configurations are aptly referred to as heat sinks, this chapter is devoted to one of the most common: the extruded aluminum heat sink. The most typical use occurs when the fins are oriented so that a natural convection air draft is permitted to flow upward through the ducts or rectangular U-channels formed by the adjacent fins. Such a system is illustrated in Fig. 5-1. Heat-dissipating transistors are attached either to the flat rear surface or to the front, finned surface where one or more fins are deleted to provide adequate space for the transistors.

The simplest manner in which the relevant temperature rises may be considered is:

$$T_J = \Delta T_{JC} + \Delta T_{CS} + \Delta T_{SA} + T_A$$

where the various temperature differences are:

ΔT_{JC} = transistor junction to case base

ΔT_{CS} = transistor case base to *average* sink surface

ΔT_{SA} = *average* sink surface to ambient

ΔT_{JC} is obtainable from the product of the transistor power dissipation and the component thermal resistance, the latter usually being found in a manufacturer's data book.

ΔT_{CS} is the product of the transistor power dissipation and the sum of the transistor–sink interface and spreading resistance. If two or more transistors are sufficiently close as to interact thermally, this effect is included in the spreading resistance. The interaction effects are sufficiently complicated that digital computer computation capability is required, a subject that is well covered in succeeding chapters.

The temperature rise ΔT_{SA} of the heat sink surface above ambient is the principal subject of this chapter. It requires consideration of sink to ambient radiation, sink to ambient natural convection, and conduction from each fin base to tip. The basics of each of

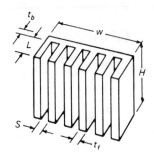

Fig. 5-1. Vertically oriented heat sink.

these phenomena have been discussed in previous chapters and are applied in the following pages to this rather specific but very common geometry.

Heat transfer from the finned side of a heat sink consists of radiation and convection from both the intra-fin passages and the unshielded surface of each of the two outer fins. For the outer fins, the surface conductance is typically:

$$C_{so} = h_o A_o$$

E5.1
Outer surface
conductance

$$A_o = 2H(L + t_b)$$

E5.2
Outer surface area

where h_o is the total heat transfer coefficient $h_o = h_{co} + \epsilon h_r$.

Heat transfer from the interior fin passages is quantified by a conductance that is typically:

$$C_{si} = h_i A_i$$

E5.3
Interior surface
conductance

$$A_i = wH\left[1 + \frac{2(N-1)}{w}L\right]$$

E5.4
Interior surface
area

where $h_i = h_{ci} + \mathcal{F} h_r$. The term $\mathcal{F}$ may be defined as the gray body radiation exchange factor between one complete rectangular U-channel formed by two opposing fin surfaces and the connecting base area. N is the number of fins.

5.2 FIN EFFICIENCY

The majority of finned heat sinks have fins with a length-to-thickness ratio $L/t_f \gg 1$. Thus a temperature gradient, if any, within each fin is usually in the direction of L. The temperature through the thickness t_f is nearly uniform at each cross section.

Since heat transfer is proportional to at least the first power of the fin surface temperature above ambient, any decrease in temperature toward the fin tip will diminish the quantity of heat transfer. Thus the fin should be thick enough to minimize the base-to-tip temperature gradient.

A heat-dissipating fin may be visualized as a one-dimensional bar with a heat source Q at the base. If the small area Ht_b at the tip is neglected, E4.5 may be used for a fin cooled by conduction, convection, and radiation:

$$\frac{R}{R_s} = \sqrt{\frac{R_k}{R_s}} \operatorname{coth} \sqrt{\frac{R_k}{R_s}}$$

In terms of conductances:

$$\frac{C}{C_s} = \sqrt{\frac{C_k}{C_s}} \tanh \sqrt{\frac{C_s}{C_k}}$$

where $C_k = kA_k/L = kHt_f/L$, $C_s = \bar{h}A_s = 2\bar{h}HL$, and $\bar{h}$ is the average total heat transfer coefficient along the fin. The terms R and C are the respective fin resistance and conductance that define the temperature rise of the fin base at T_0 above ambient T_A. The actual heat transfer from the fin surface is:

$$Q_f = C(T_0 - T_A)$$

Using an effective heat transfer coefficient, $h_e = C/A_s$:

$$Q_f = h_e A_s(T_0 - T_A)$$

If the fin were at temperature T_0 from base to tip, the heat transfer would be:

$$Q = \bar{h}A_s(T_0 - T_A)$$

A fin efficiency may be defined as the ratio $\eta = Q_f/Q$:

$$\eta = \frac{C(T_0 - T_A)}{\bar{h}A_s(T_0 - T_A)}$$

$$= C/C_s$$

$$\boxed{\eta = \sqrt{\frac{C_k}{C_s}} \tanh \sqrt{\frac{C_s}{C_k}}}$$

E5.5
Fin efficiency

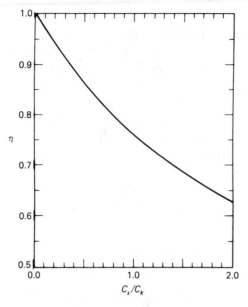

Fig. 5-2. Efficiency of a rectangular fin.

$$C = \eta \bar{h} A_s$$

E5.6
Fin conductance

η is plotted in Fig. 5-2.

It is easy to establish a minimum fin thickness criterion from Fig. 5-2 by requiring $\eta \geqslant 0.95$ (95% efficiency). Thus:

$$\frac{C_s}{C_k} \leqslant 0.2$$

$$\frac{2\bar{h}L^2}{kt_f} \leqslant 0.2$$

$$t_f \geqslant 10\bar{h}L^2/k$$

E5.7
Minimum fin thickness

In natural convection and radiation applications, an $\bar{h} \simeq 0.005$ is typical. If $L = 1.0$ in. and $k = 5$ watts/in. $\cdot$ °C, $t_f \geqslant 10(.005)(1.0)^2/5 = 0.01$ in. Most commercially available, extruded aluminum heat sinks have fins ten times this thickness, so this is not often a severe constraint. However, in forced convection where $\bar{h}$ may be at least ten times greater than for natural convection, fin efficiency should be considered.

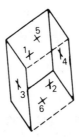

Fig. 5-3. Surface identification of U-channel interior. From [21], copyright © 1979 by the Institute of Electrical and Electronics Engineers, Inc. Reprinted by permission of the publisher.

5.3 GRAY BODY RADIATION EXCHANGE FACTOR

The succeeding treatment of $\mathcal{F}$ follows [21]. All surfaces of the heat sink are assumed to be gray and diffusely reflecting, with the fins and base at a uniform temperature. Admittedly, none of these conditions is satisfied perfectly, but for most engineering calculations, they are more than adequate. A single U-channel with appropriately identified interior surfaces is shown in Fig. 5-3. Numerals 1, 3, and 4 identify the heat sink surfaces, whereas numerals 2, 5, and 6 refer to a nonreflecting ambient.

An equivalent thermal radiation circuit for the U-channel is illustrated in Fig. 5-4. The equivalent surface resistance formula E3.17 is applicable between the radiosity and blackbody potentials, J and E_b:

$$R_1 = R_{19} = (1 - \epsilon)/\epsilon A_3$$

$$R_2 = (1 - \epsilon)/\epsilon A_1$$

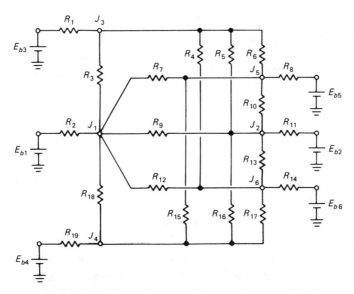

Fig. 5-4. Equivalent circuit for thermal radiation from U-channel interior. From [21], copyright © 1979 by the Institute of Electrical and Electronics Engineers, Inc. Reprinted by permission of the publisher.

Since the ambient (surfaces 2, 5, 6) is taken as nonreflecting ($\epsilon = 1$ or a sink much smaller than the surrounding enclosure):

$$R_8 = R_{11} = R_{14} = 0$$

The spatial resistances interconnecting the radiosity nodes utilize appropriate geometric angle factors and E3.16:

$$R_3 = R_{18} = 1/A_1 F_{13}$$

$$R_4 = R_6 = R_{15} = R_{17} = 1/A_3 F_{35}$$

$$R_7 = R_{12} = 1/A_1 F_{15}$$

$$R_9 = 1/A_1 F_{12}$$

$$R_5 = R_6 = 1/A_3 F_{32}$$

$$= 1/A_3 F_{31}$$

$$= 1/A_1 F_{13}$$

The last result is derived from the reciprocity theorem:

$$A_3 F_{31} = A_1 F_{13}$$

From the assumption of uniform temperature throughout the U-channel:

$$E_{b1} = E_{b3} = E_{b4}$$

and of course:

$$E_{b2} = E_{b5} = E_{b6}$$

thus implying:

$$J_2 = J_5 = J_6$$

which indicates there is no radiation exchange between nodes 2, 5, and 6 so that R_{10} and R_{13} are shorted. One additional simplification is made possible by utilizing the symmetry of the circuit as shown in Fig. 5-5. Additional simplification results in Fig. 5-6. The latter circuit is represented by a system of simultaneous linear alge-

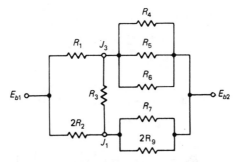

Fig. 5-5. Simplified circuit for U-channel radiation. From [21], copyright © 1979 by the Institute of Electrical and Electronics Engineers, Inc. Reprinted by permission of the publisher.

braic equations:

$$(R_a + R_b + R_e)q_1 - R_e q_2 - R_b q_3 = 0$$

$$-R_e q_1 + (R_c + R_d + R_e)q_2 - R_d q_3 = 0$$

$$-R_b q_1 - R_d q_2 + (R_b + R_d)q_3 = E_{b1} - E_{b2}$$

where:

$$R_a = R_1, \quad R_b = 2R_2, \quad 1/R_c = 1/R_4 + 1/R_5 + 1/R_6,$$

$$R_e = R_3, \quad 1/R_d = 1/R_7 + 1/2R_9$$

The net radiation resistance, R_{NET}, is based on:

$$R_{NET} = (E_{b1} - E_{b2})/q_3$$

Since symmetry was used to solve half the original circuit and thus correspond to half the heat transfer, the complete, net gray body

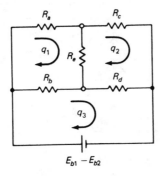

Fig. 5-6. Final circuit for U-channel radiation. From [21], copyright © 1979 by the Institute of Electrical and Electronics Engineers, Inc. Reprinted by permission of the publisher.

shape factor for the U-channel interior is:

$$\mathcal{F} = 2/R_{NET}A_f$$

or:

$$\boxed{\mathcal{F} = 2C_{NET}/[H(S + 2L)]}$$

E5.8
U-channel shielding

where:

$$C_{NET} = [(R_a + R_b + R_e)(R_c + R_d + R_e) - R_e^2]/\{(R_b + R_d)$$
$$\cdot [(R_a + R_b + R_e)(R_c + R_d + R_e) - R_e^2]$$
$$- R_b[R_b(R_c + R_d + R_e) + R_eR_d]$$
$$- R_d[R_d(R_a + R_b + R_e) + R_bR_e]\}$$

and:

$$R_a = (1 - \epsilon)/\epsilon A_3$$
$$R_b = 2(1 - \epsilon)/\epsilon A_1$$
$$R_c = 1/(A_1F_{13} + 2A_3F_{35})$$
$$R_d = 2/(A_1F_{12} + 2A_1F_{15})$$
$$R_e = 1/A_1F_{13}$$

The angle factors F_{12}, F_{13}, F_{15}, and F_{35} are computed using the appropriate parallel and perpendicular plate formulae in Section 3.4.1.

$\mathcal{F}$ has been computed and plotted in Figs. 5-7 through 5-12 for

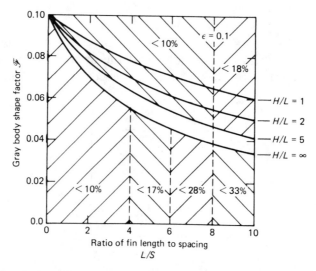

Fig. 5-7. Gray body radiation exchange factors for rectangular U-channel. % indicates error due to three-surface model. From [21], copyright © 1979 by the Institute of Electrical and Electronics Engineers, Inc. Reprinted by permission of the publisher.

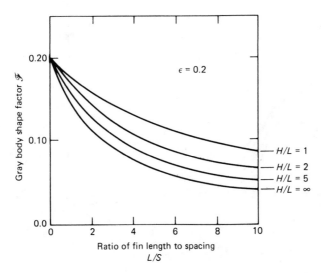

Fig. 5-8. Gray body radiation exchange factors for rectangular U-channel. From [21], copyright © 1979 by the Institute of Electrical and Electronics Engineers, Inc. Reprinted by permission of the publisher.

L/S of 0 to 10 and H/L from 1 to ∞ for $\epsilon = 0.1$, 0.2, 0.4, 0.6, 0.8, and 1.0. An H/L ratio of about 10 is sufficiently close to ∞. Note that the maximum $\mathcal{F}$ equals ϵ, as it should. As S increases for fixed L, shielding effects diminish, but that does not necessarily decrease the radiation resistance because the total finned surface area also decreases. This is easily illustrated by an example.

Consider a heat sink with all dimensions except S fixed where $H = w = 6$ in., $L = 1$ in., and $t_f = 0.1$ in. The interior area A_i is de-

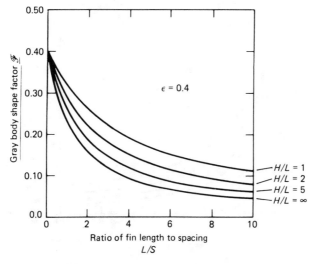

Fig. 5-9. Gray body radiation exchange factors for rectangular U-channel. From [21], copyright © 1979 by the Institute of Electrical and Electronics Engineers, Inc. Reprinted by permission of the publisher.

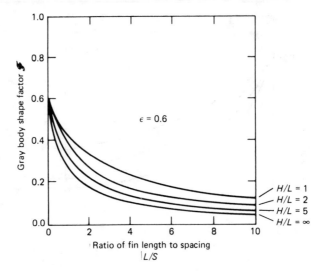

Fig. 5-10. Gray body radiation exchange factors for rectangular U-channel. From [21], copyright © 1979 by the Institute of Electrical and Electronics Engineers, Inc. Reprinted by permission of the publisher.

termined by the number of fins, N:

$$A_i = wH \left[1 + \frac{2(N-1)}{w} L \right]$$

The spacing, S, is found from:

$$w = N t_f + (N-1) S$$

$$S = (w - N t_f)/(N-1)$$

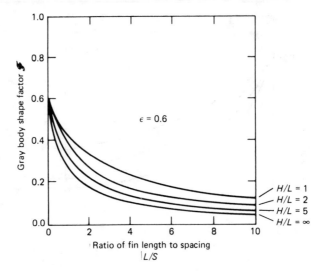

Fig. 5-11. Gray body radiation exchange factors for rectangular U-channel. Error is $\leqslant 6\%$ for three-surface model. From [21], copyright © 1979 by the Institute of Electrical and Electronics Engineers, Inc. Reprinted by permission of the publisher.

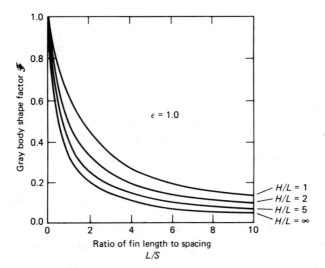

Fig. 5-12. Gray body radiation exchange factors for rectangular U-channel. From [21], copyright © 1979 by the Institute of Electrical and Electronics Engineers, Inc. Reprinted by permission of the publisher.

The product $\mathfrak{F}A_i$ is then computed from the appropriate $\mathfrak{F}$ vs. L/S, H/L, ϵ curve. The results are plotted in Fig. 5-13 for $\epsilon = 0.1$ and 0.8 for $N = 2$ (one U-channel) to 20. The fin spacing, $S = 0.21$, for $N = 20$ is about as small as or smaller than usually found in commercially available, off-the-shelf extrusion stock. Within this range of N, the case $\epsilon = 0.8$ indicates a definite optimum number of fins

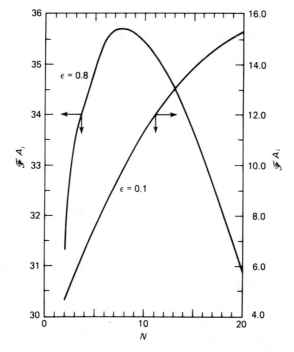

Fig. 5-13. Radiation properties for a heat sink with $H = W = 6.0$ in., $L = 1.0$ in., $t_f = 0.1$ in.

of $N = 8$ ($S = 0.7425$ in.). The optimum for $\epsilon \doteq 0.1$ is in excess of $N = 20$. It must be emphasized that these optima are for radiation only. The effects of natural convection will be considered later in this chapter.

The error resulting from the use of only three surfaces may be estimated by studying [22], a summary of shielding effect computations for $\epsilon = 0.1$ and 0.8 using a many-surface model of greater complexity than considered here. The three-surface model errors are indicated in Figs. 5-7 and 5-11.

5.4 NATURAL CONVECTION FROM PARALLEL VERTICAL PLATES

One of the most frequently quoted references applicable to finned heat sinks is Elenbass's study, [23], of flow between parallel plates. Elenbass's work indicates an optimum spacing, S_{opt}:

$$\frac{S_{opt}}{H} (Gr_s Pr) \simeq 50$$

$$\frac{S_{opt}}{H} = \frac{50}{(Gr_s Pr)}$$

$$= 50 \bigg/ \left[\left(\frac{g\beta\rho^2}{\mu^2} \right) \Delta T S_{opt}^3 Pr \right]$$

$$\frac{S_{opt}^4}{H} = 50 \bigg/ \left[\left(\frac{g\rho^2}{\mu^2} \right) \beta \Delta T Pr \right]$$

Evaluating $(g\rho^2/\mu^2)$ and Pr at T_s and β at T_A, as was done by Elenbaas, and selecting $T_s = 75°C$, $T_A = 25°C$:

$$\frac{S_{opt}^4}{H} = 50/[(4.03 \times 10^5)(3.35 \times 10^{-3})(50)(0.72)]$$

$$S_{opt} = 0.18 H^{1/4}$$

On this basis, a sink 6 in. high would have an $S_{opt} = 0.28$ in.

5.5 NATURAL CONVECTION FROM VERTICAL U-CHANNELS

An upper limit to a natural convection heat transfer coefficient for vertical heat sinks is given by the vertical plate formula E2.13:

$$h_H = 0.0024 (\Delta T/H)^{0.25}$$

For reasons to be discussed later in this section, the above is used rather than the empirical relationship E2.14.

In practice, convection from rectangular U-channels is well approximated by flat plate theory for large fin spacings. Van de Pol and Tierney, [24], have developed a curve fit for the U-channel Nusselt number from a variety of related but limited studies. The resulting formula is applicable to fin dimensions encountered in most applications. Although [24] indicates applicability to laminar flow only, this author has not found this a significant limitation. Empirical studies have shown realistic predictions obtainable for heights up to 12 in. and sink temperature rises up to 80°C above a room ambient of 25°C.

Van de Pol and Tierney's result is:

$$Nu_r = \frac{Ra^*}{\psi}\left\{1 - \exp\left[-\psi\left(\frac{0.5}{Ra^*}\right)^{3/4}\right]\right\}$$

$$\psi = \frac{24(1 - 0.483\,e^{-0.17/a})}{\left\{\left[1 + \frac{a}{2}\right][1 + (1 - e^{-0.83a})(9.14a^{1/2}e^{VS} - 0.61)]\right\}^3}$$

where

$$Ra^* = (r/H)\,Gr_r Pr$$

$$r = 2LS/(2L + S)$$

$$a = S/L$$

$$V = -11.8\ (\text{in.}^{-1})$$

All physical properties except β are evaluated at the surface temperature; β is evaluated at the fluid temperature, i.e., ambient. Van de Pol and Tierney claim that for $L/S \to 0$, Nu_r becomes the classical vertical flat plate formula, E3.13. Thus it is convenient to consider the ratio h_c/h_H where h_c is the convective heat transfer coefficient for the U-channel interior. The flat plate condition $L/S \to 0$ results in $h_c/h_H = 1.0$.

To facilitate computation, h_c/h_H is plotted vs. H/S, L/S, and S in Figs. 5-14 through 5-17 for average surface temperature rises of $\Delta T = 10, 25, 50,$ and $100°C$ for an ambient of 25°C. If extended completely to $L/S = 0$, all of the curves h_c/h_H would, to within a few percent, pass through $h_c/h_H = 1.0$.

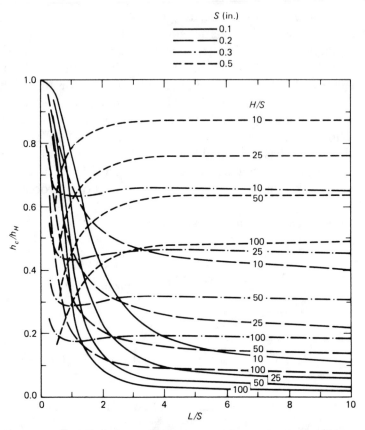

Fig. 5-14. Correction factor to vertical plate heat transfer coefficient for finned heat sink. $\Delta T = 10°C$.

A paradox appears if some of the h_c/h_H curves are studied as $L/S \to 0$. For example, for $S = 0.3$ in. in Fig. 5-17, a dip occurs in h_c/h_H for $L/S \simeq 0.8$. The dip is even more pronounced for larger values of S. This is somewhat misleading because for any fixed H/S, H must also increase as S is increased. There is probably a limit to the maximum allowed H for satisfactory application of the Van de Pol/Tierney results. The reader should therefore view with suspicion any computations requiring use of a curve from Figs. 5-14 through 5-17 in the vicinity of this dip.

The heat sink considered in Section 5.3 as a radiation shielding application is now considered for natural convection properties. Assuming a 50°C rise of the average surface temperature above a 25°C ambient, the convection conductance for the interfin area has been computed and plotted vs. the number of fins N in Fig. 5-18. The h_c was computed from the Van de Pol/Tierney correlation.

The radiation conductance C_r is also plotted using the Section 5.3 results for $\mathcal{F}_i A_i$ and h_r for $T_A = 25°C$, $T_s = 75°C$. The total interior conductance is $C = C_c + C_r$. The maximum radiation and convection conductances occur at $N = 8$ and 15 fins, respectively. The

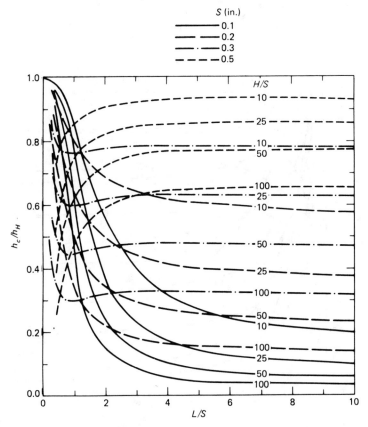

Fig. 5-15. Correction factor to vertical plate heat transfer coefficient for finned heat sink. $\Delta T = 25°C$.

relatively small contribution of C_r is dominated by C_c; thus the optimum fin spacing for a maximum total conductance is determined by the convection properties of the U-channel array. The $S_{opt} = 0.33$ in. is not very different from the value 0.28 in. previously calculated from the less exact relationship of Elenbass.

5.6 COMPLETE ANALYSIS OF AN EXTRUDED HEAT SINK

A cross-sectional view of a commercially produced aluminum heat sink is shown in Fig. 5-19. The quoted thermal resistance is based on the manufacturer's measurements of temperature rise per unit power dissipation. The temperature is presumed to be in the vicinity of a centrally located transistor(s). The resistance is quoted for a 75°C rise, a 3 in. extrusion length, and a black anodized surface finish ($\epsilon = 0.8$). The heat sink is oriented so that the 3 in. extrusion length becomes the fin height H. A theoretical calculation of the thermal resistance is detailed in the following paragraphs.

The heat sink may be considered as consisting of two regions for

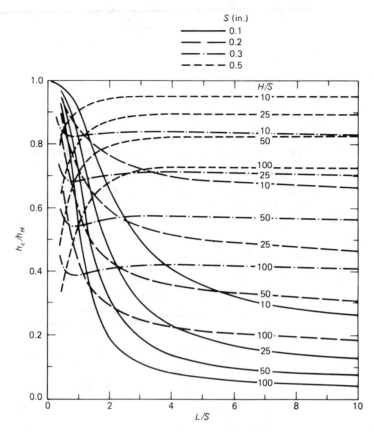

Fig. 5-16. Correction factor to vertical plate heat transfer coefficient for finned heat sink. $\Delta T = 50°C$.

calculating convection and radiation properties: a region of small fin spacing and a region of large fin spacing.

Small spaced fin region:

$$S = 0.273 \text{ in.}^*$$

$$L/S = [(1.5 - 0.188)/2]/0.273 = 2.40$$

$$H/S = 3.0/0.273 = 10.99$$

$$H/L = 3.0/0.66 = 4.55$$

From Fig. 5-11:

$$\mathcal{F} = 0.21$$

---

*From personal communication with Thermalloy, Inc.

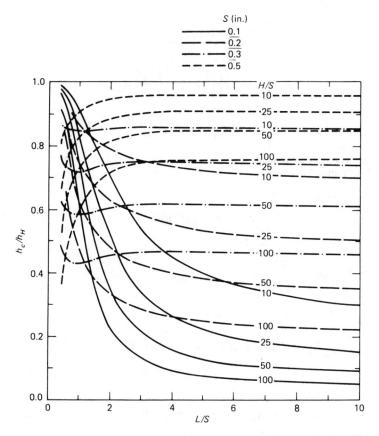

Fig. 5-17. Correction factor to vertical plate heat transfer coefficient for finned heat sink. $\Delta T = 100°C$.

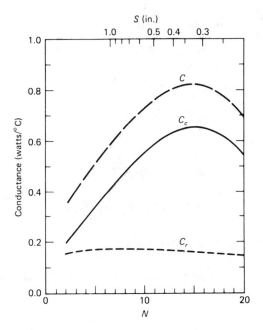

Fig. 5-18. Heat transfer properties of a heat sink with $H = W = 6.0$ in., $L = 1.0$ in., $t_f = 0.1$ in., $\epsilon = 0.8$, $T_A = 25°C$.

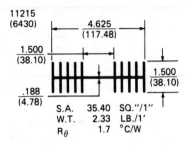

Fig. 5-19. Extruded aluminum heat sink. Reprinted from [25] with permission of Therm-alloy, Inc., Dallas, Texas.

From Fig. 3-16:

$$h_r = 0.0056$$

From Figs. 5-16 and 5-17:

$h_c/h_H = 0.84, 0.86$, respectively, for $S \simeq 0.3$ in., $H/S = 10$, $L/S = 2.4$

Use $h_c/h_H = 0.85$ for $\Delta T = 75°C$.

$$h_H = 0.0024(\Delta T/H)^{0.25} = 0.0024(75/3)^{0.25}$$

$$= 0.0054 \text{ watt/in.}^2 \cdot °C$$

$$h_c = (h_c/h_H)h_H = (0.85)(0.0054)$$

$$= 0.0046 \text{ watt/in.}^2 \cdot °C$$

The fin efficiency may be estimated using:

$$h = h_c + \mathcal{F}h_r$$

$$= 0.0046 + (0.21)(0.0056)$$

$$= 0.0058 \text{ watt/in.}^2 \cdot °C$$

For a single fin:

$$C_k = kA_k/L = kHt_f/L$$

$$= (5.0)(3.0)(0.094)/0.66$$

$$= 2.14 \text{ watts/}°C$$

$$C_s = h(2HL) = (0.0058)(2)(3.0)(0.66)$$

$$= 0.023 \text{ watt/}°C$$

$$C_s/C_k = 0.023/2.14 = 0.011$$

From Fig. 5-2, $\eta = 1.0$. Thus the consequences of a non-unity fin efficiency may be eliminated for this heat sink.

The surface area consists of four sections of five fins each, and each section has a width $W = (4.625 - 1.50)/2 = 1.56$ in. The total interior surface area is A_I:

$$A_I = 4A_i = 4\left\{WH\left[1 + \frac{2(N-1)}{W}L\right]\right\}$$

$$= 4\left\{(1.56)(3)\left[1 + \frac{2(4)}{1.56}(0.66)\right]\right\}$$

$$= 82.1 \text{ in.}^2$$

The total convection and radiation conductance for the interior, small fin spaced region is:

$$C_I = (h_c + \mathcal{F}h_r)A_I$$

$$= [0.0046 + (0.21)(0.0056)](82.1)$$

$$= 0.47 \text{ watt/}^\circ\text{C}$$

Large spaced fin region:

$$S = 1.50 \text{ in.}$$

$$L/S = 0.66/1.50 = 0.44$$

$$H/L = 3.0/0.66 = 4.55$$

$$H/S = 3.0/1.5 = 2.0$$

h_c/h_H is not plotted in Figs. 5-14 through 5-17 for $H/S = 2$, but from the results for the smaller spacing, it is sufficient to use $h_c/h_H \simeq 1.0$:

$$h_c = (h_c/h_H)h_H = (1.0)(0.0054) = 0.0054$$

From Fig. 5-11:

$$\mathcal{F} = 0.52$$

The total interior surface area for the large spacing is:

$$A_I = 2A_i = 2\left\{WH\left[1 + \frac{2(N-1)}{W}L\right]\right\}$$

$$= 2\left\{(1.50)(3)\left[1 + \frac{2(1)(0.66)}{1.50}\right]\right\}$$

$$= 16.92 \text{ in.}^2$$

The total convection and radiation conductance for the interior, large spaced fin region is:

$$C_{sI} = h_i A_i = (h_c + \mathcal{F} h_r) A_I$$

$$= [0.0054 + (0.52)(0.0056)](16.92)$$

$$= 0.14 \text{ watt}/°C$$

The two outer unshielded surfaces contribute also:

$$A_0 = 2[2H(L + t_b/2)]$$

$$= 2[2(3)(0.66 + 0.188/2)]$$

$$= 9.05 \text{ in.}^2$$

The contributing conductance is:

$$C_0 = h_H A_0 = (h_H + \epsilon h_r) A_0$$

$$= [0.0054 + (0.8)(0.0056)](9.05)$$

$$= 0.089 \text{ watt}/°C$$

The total surface conductance for this entire sink is the sum of the individual conductances, i.e.:

$$C_s = 0.47 + 0.14 + 0.089 = 0.70$$

$$R_s = 1/C_s = 1/0.70 = 1.4°C/\text{watt}$$

The resistance to heat conduction from the center mounted transistor to the various fins is estimated by two methods. The first is to use the solution to the one-dimensional problem of an end source on a bar with conduction and surface losses. The solution for the 1/2 sink is E4.5:

$$\frac{R_{1/2}}{R_{s2}} = \sqrt{\frac{R_k}{R_{s2}}} \coth \sqrt{\frac{R_k}{R_{s2}}}$$

where $R_k = (W/2)/kA_k$ for a full sink width W, and R_{s2} is a 1/2 sink surface resistance:

$$R_k = \frac{(4.625/2)}{(5)(3)(0.188)} = 0.82°C/\text{watt}$$

$$R_{s2} = 2(1.4) = 2.8°C/\text{watt}$$

$$\frac{R_k}{R_{s2}} = 0.29$$

Therefore:

$$\frac{R_{1/2}}{R_{s2}} = 1.09$$

$$R_{1/2} = (1.09)(2.8) = 3.05°C/watt$$

The total sink has a resistance:

$$R = \tfrac{1}{2}R_{1/2} = 1.5°C/watt$$

using the one-dimensional bar method.

The second method is to use the spreading results from Fig. 4-17 for $l = 1.0$ in. (assume a TO-3 package) at $t = 0.19$ in. Recalling that Fig. 4-17 was computed for a nonzero h on the nonsource surface only, the equivalent single-sided plate h that results in a conductance equal to the total heat sink conductance is computed from:

$$WHh = C_s$$

$$h = C_s/WH = 0.70/(4.625)(3.0)$$

$$= 0.05 \text{ watt/in.}^2 \cdot °C$$

From Fig. 4-17(c) for $t = 0.19$ in., $l = 1.0$ in., $L = 2.5$–5.0 in.:

$$kR_{sp} = 0.7 \rightarrow 1.2$$

$$R_{sp} \simeq 1.0/5.0$$

$$= 0.2°C/watt$$

$$R = R_{sp} + R_s$$

$$= 0.2 + 1.4$$

$$= 1.6°C/watt$$

$R = 1.6°C/watt$ is within 6% of the manufacturer's test result of $1.7°C/watt$.

5.7 ANALYSIS OF AN EXTRUDED HEAT SINK WITH AN OFFSET TRANSISTOR MOUNTING FLANGE

The device considered here is illustrated in Fig. 5-20. The dimensions are $H = 6.0$ in., $S = 0.391$ in., $L_1 = 1.50$ in., $L_2 = 6.0$ in., $w_1 = 2.5$ in., and $w_2 = 8.4$ in. The fin length L is less than L_1 by the base

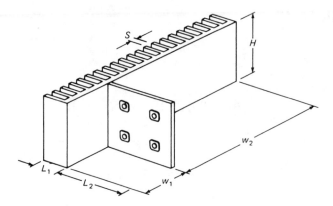

Fig. 5-20. Extruded heat sink with flange for transistor mounting.

thickness $t_b = 0.25$ in.; thus $L = 1.25$ in. The surface finish is non-anodized aluminum such that $\epsilon = 0.1$. The fin efficiency is $\eta = 1.0$.

The objective is to estimate the total power that may be nearly uniformly distributed on the flange such that the maximum temperature on the flange (i.e., transistor case temperature neglecting an interface rise) does not exceed about 75°C in a 25°C ambient. Both front and rear sink surfaces and the flange are included in the analysis.

The finned section has a total U-channel area for 23 fins of:

$$A_f = (w_1 + w_2)H \left[1 + \frac{2(N-1)L}{(w_1 + w_2)}\right]$$

$$= (2.5 + 8.4)(6)\left[1 + \frac{2(22)(1.25)}{(2.5 + 8.4)}\right] = 395 \text{ in.}^2$$

The outer, unshielded fins and flat rear surface contribute an area:

$$A_0 = 2H(L + t_b) + (w_1 + w_2)H$$

$$= 2(6)(1.25 + 0.25) + (2.5 + 8.4)(6) = 83.4 \text{ in.}^2$$

The total surface conductance, excluding the flange, is then:

$$C_s = (h_c + \mathcal{F}h_r)(395) + (h_H + \epsilon h_r)(83)$$

The contributions by the lengths w_1 and w_2 are proportionately:

$$C_{1s} = \frac{w_1}{(w_1 + w_2)} C_s$$

$$C_{2s} = \frac{w_2}{(w_1 + w_2)} C_s$$

It is not expected that the sink and flange will both be at a uniform 75°C temperature. The finned sections should be at a rise somewhat lower than 50°C; thus an initial guess is required to make a first iteration of the temperature-dependent quantities h_c, h_H, and h_r. A first trial of $\Delta T = 30°C$ is used. The pertinent ratios for h_c/h_H are $L/S = 3.2$ and $H/S = 15$. Figure 5-15 for h_c/h_H at $\Delta T = 25°C$ is sufficient. Thus:

$$h_c/h_H = 0.84$$

$$h_H = 0.0024(\Delta T/H)^{0.25} = 0.0036 \text{ watt/in.}^2 \cdot °C$$

$$h_c = (0.84)(0.0036) = 0.0030 \text{ watt/in.}^2 \cdot °C$$

$$h_r = 0.0045 \text{ watt/in.}^2 \cdot °C \text{ from Fig. 3-16}$$

The radiation shielding factor from Fig. 5-7 using $L/S = 3.2$ and $H/L = 4.8$ is $\mathcal{F} = 0.068$ at $\epsilon = 0.1$. Mutual shielding between the flange and rear sink surface is neglected.
Then:

$$C_s = [0.0030 + (0.068)(0.0045)](395)$$

$$+ [0.0036 + (0.1)(0.0045)](83)$$

$$= 1.31 + 0.34 = 1.64 \text{ watts/°C}$$

$$C_{1s} = (2.5/10.9)(1.64) = 0.84 \text{ watt/°C}$$

$$C_{2s} = (8.4/10.9)(1.64) = 1.26 \text{ watts/°C}$$

The effect of conduction resistance on the w_1 and w_2 portions is computed using the one-dimensional solution plotted in Fig. 4-8 for R_k/R_s.

For section 1:

$$R_k = l/kA_k = w_1/kHt_b$$

$$= (2.5)/(5.0)(6.0)(0.25) = 0.33 \text{ watt/°C}$$

and $R_s = 1/C_{1s} = 1/0.38$. Thus $(R_k/R_s)_1 = 0.13$ and $(R/R_s)_1 = 1.04$ (Fig. 4-8), so that $R_1 = 2.74°C/watt$.

For section 2:

$$R_k = w_2/kHt_b$$

$$= (8.4)/(5.0)(6.0)(0.25) = 1.12°C/watt$$

and $R_s = 1/C_{2s} = 1/1.25 = 0.79°C/watt$. Thus $(R_k/R_s)_2 = 1.41$ and $(R/R_s)_2 = 1.43$ (Fig. 4-8), so that $R_2 = 1.13°C/watt$.

The net resistance, excluding the flange, is the parallel sum of R_1 and R_2:

$$R = R_1 R_2/(R_1 + R_2) = (2.74)(1.13)/(2.74 + 1.13)$$

$$= 0.80 \text{ watt}/°C$$

The flange is treated as another one-dimensional problem (uniform source, one end sinked at T_0 above ambient, convection and radiation losses). The average flange temperature rise will be midway between the first finned surface estimate of 30°C and maximum flange rise of 50°C; i.e., the average flange ΔT is estimated at 40°C, for which:

$$h_H = 0.0024(40/6)^{0.25} = 0.0039 \text{ watt/in.}^2 \cdot °C$$

$$h_r = 0.0047 \text{ watt/in.}^2 \cdot °C$$

The flange surface conductance is:

$$C_s = (h_H + \epsilon h_r)2(L_2 H)$$

$$= [0.0039 + (0.1)(0.0047)]2(6)(6)$$

$$= 0.31 \text{ watt}/°C$$

and $R_k = L_2/kHt = (6)/(5)(6)(0.25) = 0.80°C/watt$. Thus $(R_k/R_s) = (0.8)/(1/0.31) = 0.25$.

The total heat dissipation Q is first estimated from the finned sink resistance, i.e., $Q \simeq \Delta T/R = (30°C)/(0.8°C/watt) = 38$ watts. Using $Q \simeq 40$ watts, $(T_0/Q)/R_s = (50/40)(1/0.31) = 0.39$. From Fig. 4-5, $T(L) = (0.45)R_s Q = (0.45)(1/0.31)(40) = 58°C$, which is 8°C greater than the desired 50°C. An improved estimate of the total power dissipation is simply $Q \simeq 40(50/58) = 35$ watts.

Chapter 6
Airflow

6.1 AIRFLOW BASICS

In earlier chapters the reader saw that both air and surface temperature predictions require an estimate of airflow. A surface temperature is the sum of the temperature rises from ambient to local air
and local air to surface. Air temperature is computed using a volumetric flow rate. Surface temperature in forced convection applications is computed using an air velocity.

Two basic equations from fluid mechanics are required. The
first, based on conservation of mass for uniform density, relates air
velocity and cross-sectional area. Referring to Fig. 6-1 for air flowing
through a duct that necks down from a cross-sectional area A_1 to A_2:

$$A_1 V_1 = A_2 V_2$$

E6.1
Conservation of mass

where V_1, V_2 are the air velocities corresponding to A_1, A_2, respectively. Any system of units may be used as long as it is remembered
the V_1, V_2 are in linear distance per unit time. It is also assumed that
the velocities are uniform over the cross section of the duct. The
product AV is the volumetric flow rate G:

$$G = VA/144$$

E6.2
Volumetric flow rate

where

G = volumetric airflow, ft^3/min.
V = air velocity, ft/min.
A = duct cross-sectional area, in.^{2}

The second basic equation is Bernoulli's equation. Again referring to Fig. 6-1:

$$\frac{p_1}{\rho g} + \frac{V_1^2}{2g} + z_1 = \frac{p_2}{\rho g} + \frac{V_2^2}{2g} + z_2$$

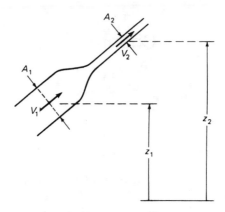

Fig. 6-1. Geometry relevant to Bernoulli's equation.

for an ideal frictionless fluid in streamline flow (vortex-free) where:

p_1 = pressure, in units of force/area

ρ = air density, in units of mass/volume

g = acceleration due to gravity, in units of linear distance/time^2

z = elevation above some arbitrary reference point, in units of distance

The condition of ideal frictionless flow is very seldom encountered in practice because effects such as frictional energy loss from the fluid to the duct wall result in a sum of terms on the right side of Bernoulli's equation not equal to the sum of those on the left side. Thus, practical applications require an additional term to compensate for losses between any two points 1, 2.

An additional modification is made due to an unfortunate choice of units for density. The preceding representation of Bernoulli's equation is in an absolute system of units. For airflow applications in electronic equipment, this would suggest that density ρ would be in slugs/volume and pressure in pounds force/area. However, nearly all expositions in the United States concerning fans and airflow use ρ in pounds mass/volume. The conversion is:

$$\rho \text{[slugs/volume]} = \rho \text{[pound mass/volume]}/32.174$$

$$\rho_s = \rho_{\text{lb}_m}/g$$

Inserting this last result into Bernoulli's equation, dropping the lb_m

from ρ_{lb_m}, and adding the loss term:

$$p_1 + \frac{\rho V_1^2}{2g} + \rho z_1 = p_2 + \frac{\rho V_2^2}{2g} + \rho z_2 + \Delta p_{1\text{-}2}$$

E6.3

Bernoulli's equation for a real fluid

where $\Delta p_{1\text{-}2}$ is the loss term between points 1 and 2, and ρ is in units of pound mass/volume. Applications to electronic equipment are such that $z_1 \simeq z_2$. Terms p_1 and p_2 are referred to as static pressures, whereas $\rho V_1^2/2g$ and $\rho V_2^2/2g$ are velocity pressures.

Another convention that must be considered is that pressure measurements and calculations are made in terms relative to the pressure that would support a column of water one inch in height. Bernoulli's equation then becomes:

$$h_{s1} + h_{V1} = h_{s2} + h_{V2} + h_L$$

E6.4
Pressure
head equation

where

h_{s1}, h_{s2} = static pressure "head"
h_{V1}, h_{V2} = velocity pressure "head"
h_L = total pressure loss

The velocity pressure head is:

$$h_V = \frac{\rho_{air} V^2}{2g(\rho \Delta z)_{H_2O}}$$

for Δz of 1 in. or 1/12 ft. For water at 62.4 lb_m/ft^3 and standard air at 0.075 lb_m/ft^3:

$$h_V = \frac{0.075 V^2}{2(32.2)(62.4/12)}$$

for V ft/sec. The use of G ft^3/min. and cross-sectional area A in.^{2} requires time and area conversion factors $(1/60)^2$, $(144)^2$, respectively.

$$h_V = \frac{(144)^2(0.025)}{(60)^2(2)(32.2)(62.4/12)} \frac{G^2}{A^2}$$

$$h_V = 1.290 \times 10^{-3} \frac{G^2}{A^2}$$

E6.5
Velocity head
for standard air

where

h_V = velocity head, in. H$_2$O
G = airflow, ft^3/min.
A = cross-sectional area, in.^{2}

The total pressure head is the sum of the velocity and static pressure heads:

$$h_T = h_V + h_s$$

6.2 FORCED AIRFLOW IN AN ENCLOSURE

A simple forced air cooled cabinet for an electronic system is shown in Fig. 6-2. Typical of many electronic packaging schemes, air enters an inlet in the bottom and side panels and is drawn through the chassis into the fan mounted on the rear panel and out into the room. The air inlet and internal mechanical structure resist the airflow. The net resistance is represented by a system loss curve such as indicated in Fig. 6-3.

Most fan manufacturers supply fan static pressure vs. airflow curves as part of the performance specifications. When both the system resistance and fan performance curves are plotted together, the actual total system airflow is determined by the perpendicular dropped from the intersection point to the airflow axis.

The single fan performance curve may be used to indicate the effects of using two identical fans in parallel or series (push–pull). Consider the fan curve in Fig. 6-4. The parallel combination is constructed by following several horizontal constant pressure lines from zero airflow out to the fan curve. The corresponding point on the two-fan curve is at this constant pressure, but twice the airflow. If this is done for several points, a complete, two-parallel-fan curve is established. A note of caution is indicated here: This example shows a common, monotonically decreasing pressure curve. Some air-

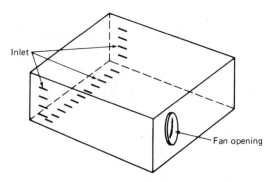

Fig. 6-2. Simple forced air cooled enclosure.

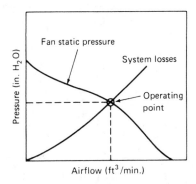

Fig. 6-3. Fan and system curves for forced air cooled enclosure.

moving systems such as centrifugal blowers do not always lend themselves to such a simple technique. The interested reader is encouraged to look at a detailed fan treatise such as [26].

An equivalent curve may also be estimated for two identical fans in series. The method (Fig. 6-5) is similar to the one just described except that a constant airflow line is followed vertically up to the single fan curve. A point for the two-fan curve at this constant airflow is at twice the pressure for one fan. The complete curve is constructed in this manner.

Some interesting qualitative comments are in order concerning fan applications. Consider a cabinet that has small airflow resistance. Reference to Figs. 6-4 and 6-5 indicates that a considerable airflow increase may be expected by using two parallel fans, but very little improvement for series fans. The opposite tends to be true for high resistance systems, where only series addition of fans may be expected to increase airflow in a nontrivial amount. Measurements in real systems indicate, however, that the predicted airflow increase for the addition of a second fan is usually not fully realized.

Before we proceed with the general description of three systems typically encountered, a few definitions are required. The reader is encouraged to obtain a set of appropriate ASHRAE standards that

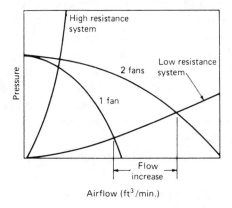

Fig. 6-4. Two identical fans in parallel.

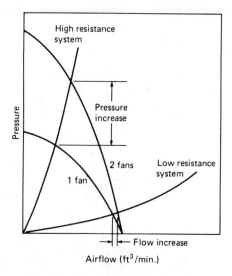

Fig. 6-5. Two identical fans in series.

will be useful in setting up an airflow and fan characterization facility, [27]. Using:

$$h_{td} = \text{total pressure head at fan discharge}$$

$$h_{ti} = \text{total pressure head at fan inlet}$$

$$h_{Vd} = \text{velocity pressure head at fan discharge}$$

the following two definitions are given:

$$h_T = \text{fan total pressure} \equiv h_{td} - h_{ti}$$

$$h_{fs} = \text{fan static pressure} \equiv h_T - h_{Vd}$$

It is important to note that h_T is the difference in inlet and discharge total pressures. However, h_{fs} is obtained by the difference of $(h_{td} - h_{ti})$ and h_{Vd}, not $(h_{Vd} - h_{Vi})$. A plot of h_{fs} (in. H_2O) vs. G (ft^3/min.) is the fan characterization provided by most manufacturers.

Consider an electronic system with a fan used as a blower (Fig. 6-6). The losses between points 2 (fan discharge) and 3 include friction, expansion, and contraction losses. Methods of quantifying these losses are given in Section 6.3, but for now:

$$H_L \equiv \text{sum of all losses between 2 and 3}$$

Flow at the fan inlet is presumed loss-less or is included in the fan characteristic curve h_{fs} vs. G.

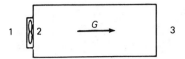

Fig. 6-6. Blower fan system.

Bernoulli's equation with losses, E6.4, is applied between 2 and 3:

$$h_{V2} + h_{s2} = h_{V3} + h_{s3} + H_L$$

At 3, external to the cabinet, the air velocity is zero, and the static pressure is set at zero as a reference point:

$$h_{V3} + h_{s3} = 0$$

$$h_{V2} + h_{s2} = H_L$$

$$h_{t2} = H_L$$

At 1, the air velocity is zero, and the static pressure is the zero reference value. Assuming loss-less conditions at the fan inlet:

$$h_T = h_{t2} - h_{t1}$$

$$= h_{t2} - (h_{V1} + h_{s1})$$

$$= h_{t2}$$

$$h_{fs} = h_T - h_{V2}$$

$$= h_{t2} - h_{V2}$$

The velocity head h_{V2} is the velocity pressure h_{Vd} at the fan discharge, and as shown above, $h_{t2} = H_L$. Thus:

$$\boxed{h_{fs} = H_L - h_{Vd}}$$

E6.6
Blower fan

Note that if E6.6 is rewritten, $h_{fs} + h_{Vd} = H_L$, implying the system operating point is determined by the intersection of the system resistance curve H_L and a fan static pressure curve h_{fs} corrected by adding h_{Vd}.

Now consider the exhaust fan system of Fig. 6-7. Applying

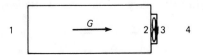

Fig. 6-7. Exhaust fan system.

Bernoulli's equation with losses H_L between 1 and 2:

$$h_{s1} + h_{V1} = h_{s2} + h_{V2} + H_L$$

$$h_{s1} + h_{V1} = 0$$

$$h_{s2} + h_{V2} = -H_L$$

Similarly, between 3 and 4:

$$h_{s3} + h_{V3} = (h_{s4} + h_{V4}) + H_{L3-4}$$

$$= (0 + 0) + H_{L3-4}$$

where H_{L3-4} is the loss for an infinite expansion. Comparing E6.5 and the loss term for an infinite expansion from Fig. 6-9, it is seen that:

$$H_{L3-4} = h_{V3}$$

and:

$$h_{s3} = 0$$

Using:

$$h_T = h_{t3} - h_{t2}$$

$$= (h_{s3} + h_{V3}) - (h_{s2} + h_{V2})$$

$$= (0 + h_{V3}) - (h_{s2} + h_{V2})$$

$$= h_{V3} - (-H_L)$$

$$= h_{V3} + H_L$$

then:

$$h_{fs} = h_T - h_{Vd}$$

$$= (h_{V3} + H_L) - h_{Vd}$$

Note that 3 is the fan discharge location. Thus $h_{V3} = h_{Vd}$, and:

$$\boxed{h_{fs} = H_L}$$

E6.7
Exhaust fan

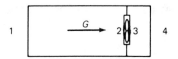

Fig. 6-8. Intermediate fan system.

H_L is the system curve indicated in Fig. 6-3 for an exhaust fan application. In an exhaust fan application, the system resistance curve
H_L intersects an uncorrected (by h_{Vd}) fan static pressure curve h_{fs}
at the system operating point.

The third and last system is one in which the fan is at an intermediate location. The loss between 3 and 4 in Fig. 6-8 may be a
low resistance finger guard or a high resistance EMI screen. The
sum of losses between the cabinet inlet and fan inlet is H_{L1-2}:

$$h_{V1} + h_{s1} = h_{V2} + h_{s2} + H_{L1-2}$$

$$h_{V1} + h_{s1} = 0$$

$$h_{t2} = h_{V2} + h_{s2}$$

$$= h_{V1} + h_{s1} - H_{L1-2}$$

$$= -H_{L1-2}$$

The sum of losses between 3 and 4 is H_{L3-4}:

$$h_{V3} + h_{s3} = h_{V4} + h_{s4} + H_{L3-4}$$

$$h_{V4} + h_{s4} = 0$$

$$h_{t3} = h_{V3} + h_{s3}$$

$$= h_{V4} + h_{s4} + H_{L3-4}$$

$$= H_{L3-4}$$

The definitions of fan total and static pressures are used next:

$$h_T = h_{t3} - h_{t2}$$

$$= H_{L3-4} - (-H_{L1-2})$$

$$= H_{L1-2} + H_{L3-4}$$

$$= H_L$$

H_L is clearly the sum of all system losses excluding the fan for this system.

$$h_{fs} = h_T - h_{V3}$$

Since h_{V3} is the velocity pressure at discharge:

$$h_{V3} = h_{Vd}$$

$$\boxed{h_{fs} = H_L - h_{Vd}}$$

E6.8
Intermediate fan

In summary, note that a system curve for a blower or intermediate fan requires either (1) the subtraction of the velocity discharge pressure from the sum of the losses or (2) the addition of the losses to the fan static pressure. It is important to emphasize that E6.6 and E6.8 require downstream losses between the ambient and a location precisely at the fan discharge. E6.7 and E6.8 require upstream system losses between the external ambient air and a fan inlet location that is just short of the fan inlet (excluding fan inlet losses).

If the fan test conditions are such that the fan pressures are measured sufficiently remote from the fan inlet and exit that inlet and exit losses are both included as part of h_{fs}, then the fan-system operating point is determined by $h_{fs} = H_L$ for all three fan location possibilities. In this case H_L does not include fan inlet and exit losses.

6.3 ESTIMATING AIRFLOW RESISTANCE ELEMENTS

The losses in the simple representation of cabinets in Figs. 6-6, 6-7, and 6-8 are determined by summing resistance elements for air inlets, circuit cards, and even contractions and expansions. For a total system flow G, H_L is computed from a sum of individual resistances R_i.

$$\boxed{\frac{H_L}{G^2} = \text{sum of system resistances}}$$

E6.9

Total system loss

Formulae for the most common situations encountered in air-cooled electronic equipment are listed in Fig. 6-9, which should be self-explanatory except for case (b), the perforated or slotted plate. The area A in either of these situations is the total open area through which the air passes.

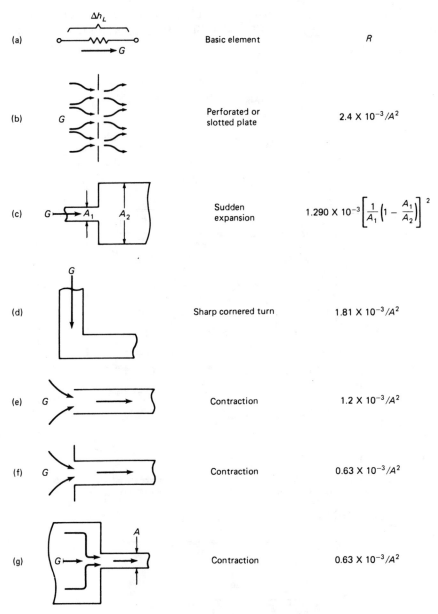

(a)		Basic element	R
(b)	G	Perforated or slotted plate	$2.4 \times 10^{-3}/A^2$
(c)	$G \to A_1 \quad A_2$	Sudden expansion	$1.290 \times 10^{-3}\left[\frac{1}{A_1}\left(1 - \frac{A_1}{A_2}\right)\right]^2$
(d)	G	Sharp cornered turn	$1.81 \times 10^{-3}/A^2$
(e)	G	Contraction	$1.2 \times 10^{-3}/A^2$
(f)	G	Contraction	$0.63 \times 10^{-3}/A^2$
(g)	G	Contraction	$0.63 \times 10^{-3}/A^2$

Fig. 6-9. Airflow resistance formulae.

Very little general information is available concerning pressure loss for flow over circuit cards. The reprinted data in Fig. 6-10 are among the best, and are used by selecting the illustrated card style that best resembles the real hardware and using the appropriate resistance formula for ΔH_L from Table 6-1.

In practice, most airflow systems consist of more than one resistance element: often several flow paths in a series/parallel structure. The rules for combining airflow resistances in series and parallel differ somewhat from those for heat flow circuits.

The representation of two series elements R_1, R_2 by an equivalent element R is shown in Fig. 6-11. The total pressure losses across R_1 and R_2 are:

$$\Delta H_{L1} = R_1 G_1^2$$

$$\Delta H_{L2} = R_2 G_2^2$$

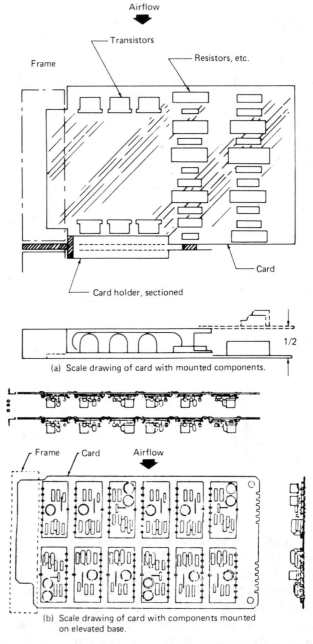

(a) Scale drawing of card with mounted components.

(b) Scale drawing of card with components mounted on elevated base.

Fig. 6-10. Circuit card geometry referring to Table 6-1. Reprinted from [28]. Author: Donald Hay, McLean Engineering Division of Zero Corporation, Princeton Junction, N.J. 08550.

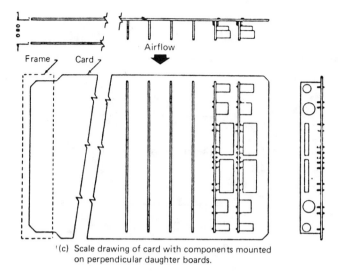

'(c) Scale drawing of card with components mounted
on perpendicular daughter boards.

Fig. 6-10. (*Continued*)

The total pressure loss across both elements is:

$$\Delta H_L = RG^2$$

The determination of R is based on:

$$G = G_1 = G_2$$

$$\Delta H_L = \Delta H_{L1} + \Delta H_{L2}$$

$$= R_1 G_1^2 + R_2 G_2^2$$

$$= (R_1 + R_2)G^2$$

**Table 6-1. Circuit card airflow resistance formulae referred to Fig. 6-10.
Reprinted from [28]. Author: Donald Hay, McLean Engineering Division
of Zero Corporation, Princeton Junction, N.J. 08550.**

Card Geometry	Reference Figure 6-10	Free Passage	Card Spacing (in.)	R_L Formulae
Childless	a	62%	½	$1.35nL\,10^{-3}(1/A)^{(2.00-0.03n)}$
Childless	a	81%	1	$3.08nL\,10^{-4}(1/A)^{(2.00-0.01n)}$
Childless	a	70%*	½	$1.93nL\,10^{-3}(1/A)^{(2.00-0.03n)}$
// daughter	b	74%	0.80	$1.95nL\,10^{-3}(1/A)^2$
// daughter	b	87%	1.60	$1.43nL\,10^{-3}(1/A)^2$
⊥ daughter	c	58%	0.80	$5.18nL\,10^{-4}(1/A)^2$
⊥ daughter	c	79%	1.60	$3.24nL\,10^{-4}(1/A)^2$

*This formulae includes the pressure drop caused by the card holder while this is omitted in those above.
n = no. of card rows through which air flows.
L = card dimension (in.) parallel to flow.
A = total cross-sectional area (in.^{2}) at entrance including card edges.

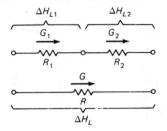

Fig. 6-11. Series addition of airflow resistances.

Hence:

$$R = R_1 + R_2$$

E6.10
Series airflow elements

Two parallel elements are depicted in Fig. 6-12. In this case, the total pressure loss across each element is identical, but the airflow is split into two components, G_1 and G_2.

$$\Delta H_L = \Delta H_{L1} = \Delta H_{L2}$$

$$G = G_1 + G_2$$

$$\sqrt{\frac{\Delta H_L}{R}} = \sqrt{\frac{\Delta H_{L1}}{R_1}} + \sqrt{\frac{\Delta H_{L2}}{R_2}}$$

$$\frac{1}{\sqrt{R}} = \frac{1}{\sqrt{R_1}} + \frac{1}{\sqrt{R_2}}$$

E6.11
Parallel airflow
elements

Note that the presence of a square root distinguishes the parallel sum of airflow elements from heat flow elements.

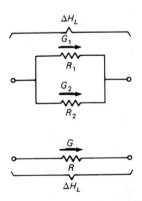

Fig. 6-12. Parallel addition of airflow elements.

The airflow in each element is easily determined:

$$G_2 = G - G_1$$

$$\frac{G_2}{G} = \frac{G - G_1}{G} = \frac{\sqrt{\dfrac{\Delta H_L}{R}} - \sqrt{\dfrac{\Delta H_L}{R_1}}}{\sqrt{\dfrac{\Delta H_L}{R}}}$$

$$= \frac{\dfrac{1}{\sqrt{R}} - \dfrac{1}{\sqrt{R_1}}}{\dfrac{1}{\sqrt{R}}} = \frac{\dfrac{1}{\sqrt{R_1}} + \dfrac{1}{\sqrt{R_2}} - \dfrac{1}{\sqrt{R_1}}}{\dfrac{1}{\sqrt{R_1}} + \dfrac{1}{\sqrt{R_2}}}$$

$$= \frac{\dfrac{1}{\sqrt{R_2}}}{\dfrac{1}{\sqrt{R_1}} + \dfrac{1}{\sqrt{R_2}}}$$

$$\frac{G_2}{G} = \left\{ \frac{1}{1 + \sqrt{\dfrac{R_2}{R_1}}} = \sqrt{\dfrac{R}{R_2}} \right.$$

E6.12
Branch airflow in
parallel circuit

Many complex circuits may be subdivided into series and parallel elements. Once this has been accomplished, the airflow and associated air temperature in each stream may be estimated.

The circuit formulae in Table 6-1 and Fig. 6-10 are applicable with E6.10, E6.11, and E6.12 only if the exponent of $(1/A)$ is very nearly 2. When this is not the case, graphical methods may be used.

A simple example is used to illustrate some of the concepts developed in this section. Consider a section of an enclosure with two adjacent chambers. One chamber contains nine circuit cards and is 5 in. wide and 8 in. high. Each card dissipates 5 watts and is 9 in. long and 8 in. high. The free passage area is estimated at 75% of the total card cage cross section. From Fig. 6-10, the most appropriate resistance formula is estimated to be:

$$R_1 = \frac{1.95nL \times 10^{-3}}{A^2}$$

$$= \frac{1.95(1.0)(9) \times 10^{-3}}{[(8)(5)]^2} = 1.1 \times 10^{-5}$$

The chamber adjacent to the card cage dissipates 8 watts and is rather open except for four inlet slots, each measuring 3 in. × 0.25 in. Denoting the combined slot resistance as R_2:

$$R_2 = \frac{2.4 \times 10^{-3}}{A^2}$$

$$= \frac{2.4 \times 10^{-3}}{[(4)(3)(0.25)]^2} = 2.7 \times 10^{-4}$$

The net resistance R is found from E6.11:

$$\frac{1}{\sqrt{R}} = \frac{1}{\sqrt{R_1}} + \frac{1}{\sqrt{R_2}}$$

$$= \frac{1}{\sqrt{1.1 \times 10^{-5}}} + \frac{1}{\sqrt{2.7 \times 10^{-4}}}$$

$$R = 7.6 \times 10^{-6}$$

The total airflow in this section of the system is known from measurements to be 10 ft³/min. Therefore, we can determine the component airflows:

$$\frac{G_2}{G} = \sqrt{\frac{R}{R_2}} = \sqrt{\frac{7.6 \times 10^{-6}}{2.7 \times 10^{-4}}} = 0.17$$

$$G_2 = (0.17)(10) = 1.7 \text{ ft}^3/\text{min.}$$

$$G_1 = G - G_2$$

$$= 10 - 1.7 = 8.3 \text{ ft}^3/\text{min.}$$

An air temperature of 50°C at the inlet to each chamber is expected. E2.30 is used to compute the air temperature rises:

$$\Delta T = \frac{2(T_I + 273)}{\left(333.5 \dfrac{G}{Q} - 1\right)}$$

$$\Delta T_1 = \frac{2(50 + 273)}{\left[333.5 \left(\dfrac{8.3}{45}\right) - 1\right]} = 11°C$$

$$\Delta T_2 = \frac{2(50 + 273)}{\left[333.5 \left(\dfrac{1.7}{8}\right) - 1\right]} = 9°C$$

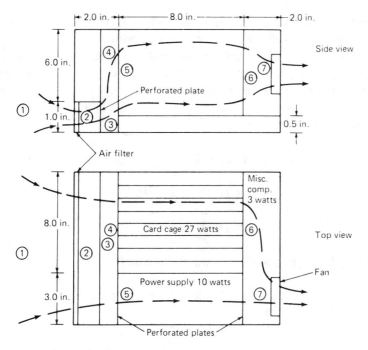

Fig. 6-13. Sample problem. Forced air cooled enclosure.

The average air velocity in the card cage is:

$$V = \frac{144G}{A} = \frac{144(8.3)}{(0.75)(8)(5)}$$

$$= 40 \text{ ft/min}.$$

A more complex example, an airflow analysis of a complete system, is in order. Side and top views of an enclosure with length, width, and height dimensions of 12, 11, and 7 inches respectively are depicted in Fig. 6-13.

The airflow enters through a filter located across the lower one inch of the front panel and through a perforated plate with the same dimensions as the filter panel. The flow then divides into two parallel paths: one through a seven-card card cage, the other through a power supply chamber with a rather open air passage. The inlet and exit to the power supply consist of perforated plates. Both flow paths come together at the entrance to the fan.

The airflow resistance circuit for the system is shown in Fig. 6-14. The numbered points identify the corresponding locations in

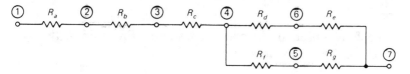

Fig. 6-14. Forced air flow circuit for cabinet illustrated in Fig. 6-13.

the instrument, Fig. 6-13. Since the fan is drawing air from the cabinet, all interior locations are negatively pressurized.

The resistance elements are computed using the formulae from Section 6.3. These values are tabulated in Table 6-2. The following calculations determine the total system resistance R.

From ① to ④:

$$R(①, ④) = R_a + R_b + R_c$$

$$= 1.0 \times 10^{-3} + 1.6 \times 10^{-4} + 6.6 \times 10^{-6}$$

$$= 1.2 \times 10^{-3}$$

Two parallel paths connect ④ and ⑦, in which:

$$R_f + R_g = 5.1 \times 10^{-5} + 5.1 \times 10^{-5}$$

$$= 1.0 \times 10^{-4}$$

$$R_d + R_e = 1.1 \times 10^{-6} + 4.4 \times 10^{-6}$$

$$= 5.5 \times 10^{-6}$$

Thus:

$$\frac{1}{\sqrt{R(④, ⑦)}} = \frac{1}{\sqrt{R_d + R_e}} + \frac{1}{\sqrt{R_f + R_g}}$$

$$= \frac{1}{\sqrt{5.5 \times 10^{-6}}} + \frac{1}{\sqrt{1.0 \times 10^{-4}}}$$

$$R(④, ⑦) = 3.6 \times 10^{-6}$$

The total resistance R is:

$$R = R(①, ⑦) = R(①, ④) + R(④, ⑦)$$

$$= 1.2 \times 10^{-3} + 3.6 \times 10^{-6}$$

$$= 1.2 \times 10^{-3}$$

The fan is used in an exhaust fashion. From E6.9, the system curve is the sum of the losses:

$$H_L = RG^2$$

Table 6-2. Summary of airflow elements for sample problem illustrated in Fig. 6-13.

Node I to Node J		Function	Value
1	2	Filter	$R_a = 1 \times 10^{-3}$, mfgs. data
2	3	Perforated plate	$R_b = 2.4 \times 10^{-3}/[(11)(1)(0.35)]^2 = 1.6 \times 10^{-4}$
3	4	Expansion	$R_c = 1.29 \times 10^{-3}\left\{\dfrac{1}{(1)(11)}\left[1 - \dfrac{(1)(11)}{(6.5)(8)}\right]\right\}^2 = 6.6 \times 10^{-6}$
4	6	Card cage	$R_d = \dfrac{3.08(1)(8)(10^{-4})}{[(8)(6)]^2} = 1.1 \times 10^{-6}$
4	5	Perforated plate	$R_f = 2.4 \times 10^{-3}/[(6.5)(3.0)(0.35)]^2 = 5.1 \times 10^{-5}$
5	7	Perforated plate	$R_g = 5.1 \times 10^{-5}$
6	7	Contraction	$R_e = 0.63 \times 10^{-3}/[(2)(6)]^2 = 4.4 \times 10^{-6}$

A manufacturer-provided fan curve is shown in Fig. 6-15. It is seen that the sum of the various losses reduces the unloaded fan flow from about 14 to 6 ft^3/min.

Flow in the card cage path is computed from:

$$G_{cc} = \frac{G}{1 + \sqrt{\dfrac{R_d + R_e}{R_f + R_g}}}$$

$$= \frac{6}{1 + \sqrt{\dfrac{5.5 \times 10^{-6}}{1.0 \times 10^{-4}}}}$$

$$= (0.81)(6) = 4.9 \text{ ft}^3/\text{min}.$$

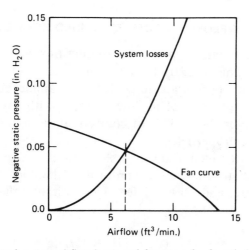

Fig. 6-15. Computed system airflow losses and fan curve for forced air flow system illustrated in Fig. 6-13.

Flow in the power supply is, of course:

$$G_{ps} = G - G_{cc} = 6.0 - 4.9 = 1.1 \text{ ft}^3/\text{min.}$$

It now remains to compute the appropriate air temperature rises. This application example presumes an inlet air temperature of $T_I = 55°C$. Using:

$$\Delta T = \frac{2(T_I + 273)}{\left(333.5 \dfrac{G}{Q} - 1\right)}$$

the temperature rises are:

$$\Delta T_{cc} = \frac{2(55 + 273)}{\left[\dfrac{(333.5)(4.9)}{27} - 1\right]} = 11°C$$

$$\Delta T_{ps} = \frac{2(55 + 273)}{\left[\dfrac{(333.5)(1.1)}{10} - 1\right]} = 18°C$$

The air temperature rise at the fan inlet is computed using the total upstream heat dissipation of 40 watts:

$$\Delta T_{FI} = \frac{2(55 + 273)}{\left[\dfrac{(333.5)(6)}{40} - 1\right]} = 13°C$$

6.4 FORCED AIRFLOW IN DUCTS: EXTRUDED HEAT SINKS

Convective heat transfer properties of duct flow with emphasis upon forced air cooling of extruded heat sinks was dealt with in Section 2.5. It is now appropriate to consider the applicable technique of predicting the quantity of airflow obtained with any given fan and heat sink combination. This of course requires a knowledge of airflow resistance properties.

A typical extruded heat sink is shown in Fig. 6-16. A cover plate is indicated so that airflow is confined to the ducts formed by the fins, base plate, and cover plate. The hydraulic diameter for this geometry is:

$$D = 4SH/(2S + 2H) = 2SH/(S + H)$$

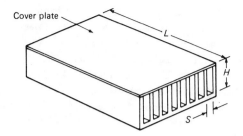

Fig. 6-16. Geometry for forced convection cooled extruded heat sink.

From E2.15:

$$Re_D = VD(\rho/\mu)(1/5) = (VD/\nu)(1/5)$$

$$= \frac{144}{5} \frac{(G/A_c)D}{\nu}$$

where G (one-channel flow), A_c, D, and ν have dimensions of ft³/min., in.², in., and in.²/sec., respectively.

The loss, H_L, in total pressure head is the sum of three contributions: a contraction at the entrance; friction down the duct length; an expansion at the exit. The total resistance for standard air conditions and laminar flow is:

$$R = \frac{1.29 \times 10^{-3}}{N_p^2 A_c^2} \left[K_c + K_e + 4\bar{f}\, \frac{L}{D} \right] \qquad \text{E6.13}$$

Extruded sink laminar airflow resistance

where

N_p = number of channels
A_c = cross-sectional area of each channel, in.²
K_c = contraction coefficient
K_e = expansion coefficient $\simeq 1.0$
$\bar{f}$ = average friction coefficient for length L

and R has dimensions of in. H$_2$O/(ft³/min.)² such that:

$$H_L = RG^2$$

Han, [29], computed friction coefficients for rectangular channels. Parallel plate results from [29] were recomputed and tabulated (Table 6-3) consistent with the form of the friction coefficient re-

Table 6-3. Mean friction coefficients
for laminar flow between parallel plates.
From *Handbook of Heat Transfer* by
W. M. Rohsenow and J. P. Hartnett, [3].
Copyright © 1973 by McGraw-Hill, Inc.
Used with the permission of McGraw-Hill
Book Company.

$L/(DRe_D)$	$\bar{f}Re_D$
0.0_4431	168.4
0.0_3209	88.89
0.0_3354	73.14
0.0_3686	57.60
0.00159	42.63
0.00260	35.73
0.00338	32.60
0.00448	29.78
0.00529	28.76
0.00567	27.83
0.00644	26.97
0.00733	26.21
0.00845	25.56
0.00910	25.27
0.00983	25.01
0.01067	24.77
0.01114	24.67
0.01165	24.57
0.01221	24.47
0.01283	24.40
0.01352	24.32
0.01427	24.25
0.01518	24.20
0.01611	24.14
0.01695	24.10
0.02059	24.06
∞	24.00

quired by E6.13. Laminar flow contraction coefficients computed by Lundgren et al., [30], are plotted in Fig. 6-17.

The resistance for turbulent flow uses an "apparent" friction factor $\bar{f}_{app}$:

$$R = \frac{1.29 \times 10^{-3}}{N_p^2 A_c^2} \left[K_c + K_e + 4\bar{f}_{app} \frac{L}{D} \right] \qquad \text{E6.14}$$

Extruded sink turbulent airflow resistance

An abstract of a study by Deissler, [31], was published as Fig. 4, p. 7-7 in [3]. The $\bar{f}_{app}$ from this abstraction is reprinted here as Fig. 6-18. Recommended K_c and K_e are 0.4 and 1.0, respectively.

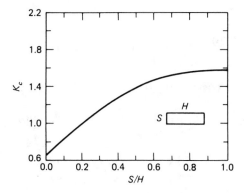

Fig. 6-17. Contraction coefficients for laminar duct flow. Reprinted from [30] with permission of The American Society of Mechanical Engineers.

A sample problem is now in order. Consider the two back-to-back heat sinks in Fig. 2-20, for which the following data are applicable.

$$N_{p.} = 24 \text{ channels}$$

$$A_c = (0.23 \text{ in.})(1.0 \text{ in.}) = 0.23 \text{ in.}^2$$

$$D = 2SH/(S + H)$$

$$= 2(0.23)(1.0)/(0.23 + 1.0)$$

$$= 0.37 \text{ in.} \simeq 0.4 \text{ in.}$$

$$K_c = 1.05 \text{ from Fig. 6-17}$$

$$K_e = 1.0$$

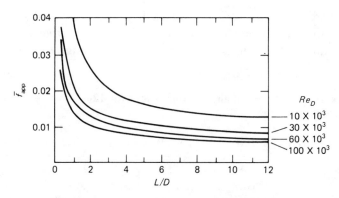

Fig. 6-18. Friction coefficient for turbulent flow in the hydrodynamic entry length of a circular tube. From *Handbook of Heat Transfer*, Editors, W. M. Rohsenow and J. P. Hartnett, Fig. 4, p. 7-7, copyright © 1973 by McGraw-Hill, Inc. Used with permission of McGraw-Hill Book Co.

In the first case, $L = 12$ in. Then

$$R = \frac{1.29 \times 10^{-3}}{(24)^2 (0.23)^2} \left[2.05 + 4\bar{f} \left(\frac{12}{0.4} \right) \right]$$

$$= 4.23 \times 10^{-5} (2.05 + 120\bar{f}) \text{ for } Re_D < 2000$$

Re_D is computed for an assumed average bulk temperature of 50°C. From Chapter 2:

$$\nu = 0.0213 \exp [(0.0045\bar{T}_B)] = 0.027$$

and:

$$Re_D = \frac{144}{5} \frac{G_T}{N_p A_c \nu} D$$

$$= \frac{144}{5} \frac{G_T (0.4)}{(24)(0.23)(0.027)}$$

$$= 77 G_T$$

where $N_p A_c$ is the total cross-sectional area for a total flow G_T. The resistance R and resultant pressure loss H_L are listed in Table 6-4 for several values of G_T. The static pressure head $H_L = R G^2$ is plotted in Fig. 6-19 for $L = 12$ in. A comparison of the theoretical computations with a few measurements indicates that a conservative airflow prediction would be obtained.

The same heat sink, but cut down to $L = 6$ in., is analyzed in a similar fashion:

$$\frac{L}{D} = \frac{6.0}{0.4} = 15$$

so that:

$$R = 4.23 \times 10^{-5} (2.05 + 60\bar{f}) \text{ for } Re_D < 2000$$

The pressure calculations are tabulated in Table 6-5 and plotted in Fig. 6-20 along with some measurements. In this case, the theoretical–experimental correlation is rather good for the range of flow values examined.

Table 6-4. Summary of airflow calculations for 12-in.-long heat sink.

G_T(ft³/min.)	Re_D	$L/(D Re_D)$	$\bar{f} Re_D$ (Fig. 6-17)	$\bar{f}$	R (in. H₂O/ (ft³/min.)²)	H_L(in. H₂O)
1	77	0.39	24	0.31	1.67×10^{-3}	0.0017
5	385	0.078	24	0.062	4.0×10^{-4}	0.010
10	770	0.039	24	0.031	2.19×10^{-4}	0.022
20	1540	0.019	24	0.016	1.68×10^{-4}	0.067
25	1925	0.016	24	0.012	1.50×10^{-4}	0.094

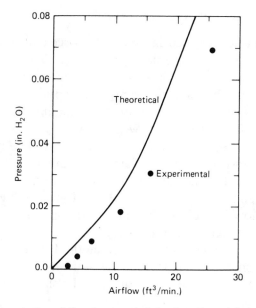

Fig. 6-19. Pressure loss vs. airflow for extruded aluminum heat sink, length = 12.0 in.

Table 6-5. Summary of airflow calculations for 6-in.-long heat sink.

G_T(ft^3/min.)	Re_D	$L/(DRe_D)$	$\bar{f}Re_D$	$\bar{f}$	R (in. H$_2$O/ (ft^3/min.)^2)	H_L (in. H$_2$O)
1	77	0.196	24	0.31	8.7 × 10^{-4}	8.7 × 10^{-4}
2.5	193	0.078	24	0.12	3.9 × 10^{-4}	0.0024
5	385	0.039	24	0.062	2.4 × 10^{-4}	0.0060
10	770	0.019	24	0.031	1.65 × 10^{-4}	0.0165
15	1155	0.013	24	0.021	1.4 × 10^{-4}	0.032

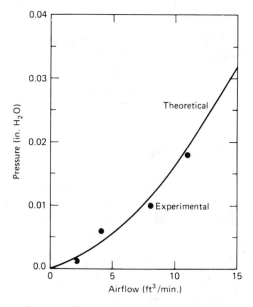

Fig. 6-20. Pressure loss vs. airflow for extruded aluminum heat sink, length = 6.0 in.

6.5 NATURAL CONVECTION AIR DRAFT

Many of the principles used in predicting forced air flow are applicable to naturally induced ventilating air drafts. The major difference is that whereas fan-cooled instruments utilize a fan as the driving force for the air, the naturally ventilated cabinet flow is forced by the pressure difference between the cabinet interior and exterior. This pressure difference occurs when heat dissipation into the cabinet interior air causes a decrease in air density. The net difference in interior–exterior air density from the cabinet bottom to top results in a force that pushes the air upward and out.

The inlet and exit ventilation grills and intervening obstructions resist the airflow in the same manner that was described for forced air cooling. Resistances in the same flow path are added according to the series resistance formula. A method of compensation for parallel flows in a manner that is theoretically consistent with the basic physics of flow either has not been developed or has not been widely published.

The buoyancy pressure for natural ventilation is obtained as follows. Over an infinitesimal increment of height, the buoyancy pressure is found from the difference between the internal cabinet air density ρ and the external air density ρ_0:

$$dp_B = (\rho_0 - \rho)\, dz$$

From the ideal gas law:

$$\rho = \rho_0 \left(\frac{273.15 + T_0}{273.15 + T} \right)$$

where T_0 and T are temperatures of air with densities ρ_0 and ρ, respectively. Combining the pressure and temperature formulae:

$$dp_B = \rho_0 \left(1 - \frac{273.15 + T_0}{273.15 + T} \right) dz$$

which must be integrated over the height for which T increases from T_0.

Referring to Fig. 6-21 for a simple enclosure, a linear temperature rise is postulated. The air is presumed to increase in temperature starting from the lower edge of card-mounted, heat-dissipating components, up to the top of the cards. This overall distance is the dissipation height d, and:

$$T = \frac{\Delta T}{d} (z - z_0) + T_0$$

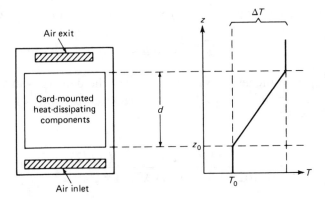

Fig. 6-21. Temperature gradient model for natural ventilation air draft.

where ΔT is the increase in internal air temperature. This of course assumes well-mixed air, a condition not necessarily met in practice.

The incremental pressure equation becomes:

$$dp_B = \rho_0 \left[1 - \frac{T_0 + 273.15}{T_0 + 273.15 + \left(\frac{\Delta T}{d} \right)(z - z_0)} \right] dz$$

$$= \rho_0 \, dz - \rho_0 (T_0 + 273.15) \frac{dz}{T_0 + 273.15 + \left(\frac{\Delta T}{d} \right)(z - z_0)}$$

The total change Δp_B in pressure is found by integrating the above expression:

$$\Delta p_B = \rho_0 d - \rho_0 (T_0 + 273.15) \int_{z_0}^{z_0 + d} \frac{dz}{T_0 + 273.15 + \frac{\Delta T}{d}(z - z_0)}$$

$$\Delta p_B = \rho_0 d \left[1 - \frac{T_0 + 273.15}{\Delta T} \ln \left(1 + \frac{\Delta T}{T_0 + 273.15} \right) \right]$$

An approximation to this is suggested in [32], Section 504.4, as:

$$\Delta p_B \simeq \rho_0 d \left[\frac{(\Delta T / 2)}{\left(\frac{\Delta T}{2} \right) + T_0 + 273.15} \right]$$

which, when compared to the logarithmic version, is accurate to within 5% for $T_0 = -40°C$ to $100°C$ and $\Delta T = 0°C$ to $100°C$.

It is desirable to calculate Δp_B in inches of water, consistent with the units in which system resistance losses are estimated. The buoyancy pressure head relative to a 1-in.-high water column is then:

$$H_B = \Delta p_B / (\rho \Delta z)_{H_2O}$$

where H_B is in inches of water and $\rho \Delta z$ is the density, water column product for Δz (in inches). For standard air at $\rho = 0.075 \ lb_m / ft^3$ and water at $\rho = 62.4 \ lb_m / ft^3$:

$$H_B = \frac{(0.075)(1/12)d}{(62.4)(1/12)} \left[\frac{(\Delta T/2)}{(\Delta T/2) + T_0 + 273.15} \right]$$

$$= 0.0012d \left[\frac{\Delta T/2}{(\Delta T/2) + T_0 + 273.15} \right]$$

for d in inches and ΔT in °C. The air temperature rise is easily related to an airflow by either E2.30 or E2.28, i.e.:

$$\Delta T = \frac{2(T_0 + 273)}{\left(333.5 \ \dfrac{G}{Q} - 1 \right)}, \text{ or } 5.997 \times 10^{-3} \ \frac{Q}{G} \ (\bar{T}_B + 273)$$

T_0 is the inlet air temperature.

As an illustration, consider a simple cabinet with free inlet area, exit area, and dissipation height of 7.5 in.^{2}, 7.5 in.^{2}, and 14 in., respectively. Airflow resistances other than inlet and exit grills are considered negligible for this example. The total resistance for the two elements in series is:

$$R = 2(2.4 \times 10^{-3})/A^2$$

$$= 2(2.4 \times 10^{-3})/(7.5)^2$$

$$= 8.5 \times 10^{-5}$$

Using E2.30 for an inlet temperature of $T_0 = 50°C$, the buoyancy pressures are plotted in Fig. 6-22 for $Q = 5$, 10, and 20 watts and G up to 5 ft^3/min. The actual air drafts are obtained from the intersection of the resistance and buoyancy curves, i.e., 1.4, 1.8, and 2.3 ft^3/min. for the three wattages. The respective well-mixed air temperature rises are computed to be 7, 11, and 18°C.

The dependence of H_B on several variables (Q, G, d) makes the graphical matching of H_B and H_L rather tedious. An iterative nongraphical technqiue is next described. This method is easily adapted to programmable calculators.

At the point of intersection of the buoyancy and resistance

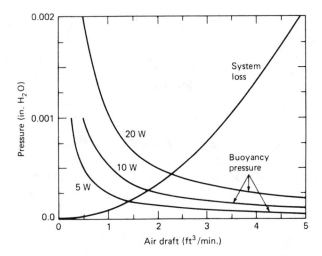

Fig. 6-22. Air draft characteristics for a simple cabinet with 7.5 in.^{2} inlet, exit free areas and a 14 in. dissipation height.

curves:

$$H_B = H_L$$

where:

$$H_L = RG^2$$

for a total resistance R to flow G. After a little algebra, one obtains:

$$(T_0 + 273)RG^3 + 3 \times 10^{-3}(\bar{T}_B + 273)QRG^2$$

$$- 3.6 \times 10^{-6}d(\bar{T}_B + 273)Q = 0$$

G is computed using Newton's method to solve for the real root x of a function $f(x)$. Using a first guess or preceding iteration x, the next iteration x' is estimated from:

$$x' = x - f/(df/dx)$$

Applying Newton's method to the air draft relation:

$$f = (T_0 + 273)RG^3 + 3 \times 10^{-3}(\bar{T}_B + 273)QRG^2$$

$$- 3.6 \times 10^{-6}d(\bar{T}_B + 273)Q = 0$$

It is clear that for $G > 0$, f has only one real root. Finally:

$$G' = G - (\{[(T_0 + 273)RG + 3 \times 10^{-3}(\bar{T}_B + 273)QR]G\}G$$

$$- 3.6 \times 10^{-6}d(\bar{T}_B + 273)Q)/[3(T_0 + 273)RG$$

$$+ 6 \times 10^{-3}(\bar{T}_B + 273)QR]G$$

which can be simplified using $T_0 = 20°C$ and $\overline{T}_B = 30°C$:

$$G' = G - \frac{(293RG + 0.909RQ)G^2 - 1.09 \times 10^{-3}dQ}{(909RG + 1.82RQ)G} \qquad \text{E6.15}$$

Air draft for $T_0 \simeq 20°C$, $\overline{T}_B \simeq 30°C$

The following example illustrates the iterative air draft computation. A cabinet is constructed such that the inlet and exit are the dominant airflow resistances. The free inlet and exit areas are each 1 in.^{2}, and the dissipation height is 8 in. Five watts of heat are dissipated into the airstream.

$$R = 2(2.4 \times 10^{-3})/A^2$$

$$= 2(2.4 \times 10^{-3})/(1)^2$$

$$= 0.0048$$

The air draft equation becomes:

$$G' = G - \{ [(293)(0.0048) G + (0.909)(0.0048)(5)] G^2$$

$$- 1.09 \times 10^{-3}(8)(5)\}/[(909)(0.0048) G$$

$$+ 1.82(0.0048)(5)] G$$

$$= G - \frac{(1.41G + 0.0218)G^2 - 0.0436}{(4.36G + 0.0437)G}$$

Several iterative results are listed in Table 6-6. The designated zeroth iteration, is actually a first guess, 0.1 ft^3/min. in this case, and is substituted for G. The first iteration results in $G' = 0.975$, which is in turn substituted for G in the iteration formula. The sixth and seventh iterations produce identical answers to the third decimal place.

Table 6-6. Iteration of ventilative
air draft for cabinet with 8 in.
dissipation height.

Iteration No.	G (ft^3/min.)
0	0.1
1	0.975
2	0.668
3	0.473
4	0.362
5	0.317
6	0.309
7	0.309

6.6 AIR DRAFT COMBINED WITH CABINET LOSSES BY NATURAL CONVECTION AND RADIATION

The air draft prediction method discussed in Section 6.5 is a very useful technique with accuracy adequate for most practical problems. The air temperature rise for a well-ventilated cabinet is usually predicted with only a few iterations. Many packaging designs, however, are such that heat dissipation into the air draft is only part of the problem—external cabinet losses via radiation and natural convection may be equally significant.

The thermal circuit in Fig. 6-23 is a realistic representation of the dominant heat transfer mechanisms in simple electronic enclosures. The elements R_{CE}, R_r represent the thermal resistances of the external natural convection and radiation from the cabinet at an average surface temperature T_s to an ambient T_A. It is assumed that the ambient air and surrounding room walls are at identical temperatures. The element R_{CI} represents natural convection between the internal air at T_I and the cabinet walls at T_s. The temperature drop across the walls may be neglected for metal cabinets and most plastic systems also.

The major effect of radiation exchange within the cabinet is assumed to be one of minimizing surface temperature differences. Shielding effects by circuit cards and other internal obstructions tend also to allow the omission of internal radiation elements.

Heat loss by the ventilating air draft is quantified by the element R_f.

The simplicity of the circuit in Fig. 6-23 is deceiving, principally because the elements R_{CI}, R_{CE}, and R_r are temperature-dependent, and the element R_f must be computed using a somewhat lengthy process. Unfortunately, iterative techniques must be used. This problem is sufficiently important in practical applications to justify a detailed exposition in the following paragraphs. With a little patience, it will be shown that these iterations may be accomplished without the aid of a digital computer, although a modest programmable calculator somewhat reduces the effort.

The formulae for the appropriate temperature rises are derived first:

$$T_I - T_A = (R_I + R_E)Q_s$$

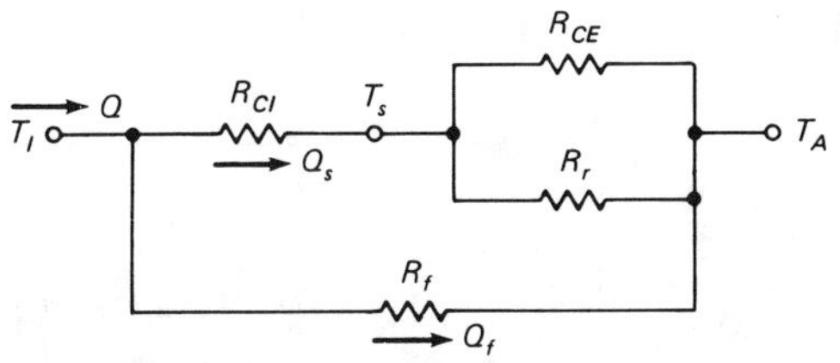

Fig. 6-23. Thermal circuit for cabinet cooled by radiation, natural convection, and ventilation air draft.

and:

$$T_I - T_A = R_f Q_f$$

where R_I and R_E are the net internal and external resistances:

$$R_I = R_{CI}$$

$$R_E = R_{CE} R_r / (R_{CE} + R_r)$$

Then:

$$Q_f = Q_s (R_I + R_E)/R_f$$

Noting that $Q = Q_f + Q_s$, Q_f and Q_s are determined:

$$Q_s = Q \frac{R_f}{R_f + R_I + R_E}$$

$$Q_f = Q \frac{R_I + R_E}{R_f + R_I + R_E}$$

The surface temperature rise is:

$$T_s - T_A = R_E Q_s$$

$$= \frac{R_{CE} R_r R_f Q}{(R_{CE} + R_r)(R_f + R_{CI} + R_E)}$$

$$\boxed{T_s - T_A = \frac{R_{CE} R_r Q}{(R_{CE} + R_r)\left[1 + \dfrac{1}{R_f}\left(R_{CI} + \dfrac{R_{CE} R_r}{R_{CE} + R_r}\right)\right]}} \quad \text{E6.16}$$

Cabinet surface temp. rise

The internal air temperature rise above the surface is:

$$\boxed{T_I - T_s = \frac{R_{CI} Q}{\left[1 + \dfrac{1}{R_f}\left(R_{CI} + \dfrac{R_{CE} R_r}{R_{CE} + R_r}\right)\right]}} \quad \text{E6.17}$$

Internal air temp. rise above surface

and of course:

$$T_I - T_A = (T_I - T_s) + (T_s - T_A) \qquad \text{E6.18}$$

Internal air temp. rise above surface

Next the formulae for the resistance elements are obtained:

$$R_{CE} = 1 \Big/ \Big\{ 0.0022 \left[\frac{T_s - T_A}{0.5WL/(W + L)} \right]^{0.25} WL$$

$$+ 0.0011 \left[\frac{T_s - T_A}{0.5WL/(W + L)} \right]^{0.25} WL$$

$$+ 2(0.0024) \left[\frac{T_s - T_A}{H} \right]^{0.25} WH$$

$$+ 2(0.0024) \left[\frac{T_s - T_A}{H} \right]^{0.25} LH \Big\}$$

for cabinet height, length, and width or H, L, and W, respectively. Simplifying:

$$R_{CE} = 1 \Big/ \Big\{ 0.0033 \left[\frac{T_s - T_A}{0.5WL/(W + L)} \right]^{0.25} WL$$

$$+ 0.0048 \left[\frac{T_s - T_A}{H} \right]^{0.25} (L + W)H \Big\} \qquad \text{E6.19}$$

Convection resistance external to cabinet

Similarly:

$$R_{CI} = 1 \Big/ \Big\{ 0.0033 \left[\frac{T_I - T_s}{0.5WL/(W + L)} \right]^{0.25} WL$$

$$+ 0.0048 \left[\frac{T_I - T_s}{H} \right]^{0.25} (L + W)H \Big\} \qquad \text{E6.20}$$

Convection resistance internal to cabinet

The element R_f is easily obtained from the temperature rise:

$$T_I - T_A = 5.997 \times 10^{-3} \frac{Q_f}{G} (\bar{T}_B + 273)$$

$$= 1.82 \frac{Q_f}{G} \text{ for } T_A = 20°C, \quad \bar{T}_B \simeq 30°C$$

Using the definition:

$$R_f = \frac{T_I - T_A}{Q_f}$$

$$R_f \simeq \frac{1.82}{G}$$

E6.21
Thermal resistance
for air draft

It is worthwhile to rewrite the iterative formula for the air draft using the variable definitions in this section.

$$G' = G - [(293 R_a G + 0.909 R_a Q_f) G^2$$

$$- 1.09 \times 10^{-3} dQ_f]/(909 R_a G + 1.82 R_a Q_f) G$$

E6.22

Air draft for cabinet

Note that the heat Q_f is the quantity dissipated into the airstream. R_a is the airflow resistance in units of in. $H_2O/(ft^3/min.)^2$.

$$R_a = 2.4 \times 10^{-3} \left(\frac{1}{A_{IN}^2} + \frac{1}{A_{EX}^2} \right)$$

for an airflow resistance circuit dominated by the inlet and exit areas A_{IN}, A_{EX}, respectively.

The following outline is a technique that may be used to obtain an iterative solution:

1. Start with computing the nonventilated solution, i.e., $R_f = \infty$.
2. Guess the starting temperatures, e.g., $T_I - T_s = T_s - T_A = 10°C$.
3. Compute R_{CI}, R_{CE}, R_r; use the most recent estimate of $T_I - T_s$, $T_s - T_A$.
4. Compute $T_I - T_s$, $T_s - T_A$.
5. Compare step 4 results with the previous estimate (first guess or preceding iteration). If agreement is satisfactory, go to step 6. Otherwise, return to step 3.

6. Compute $T_I - T_A = (T_I - T_s) + (T_s - T_A)$.
7. Guess the starting G, e.g., 1.0 ft^3/min.
8. Compute $Q_f = (T_I - T_A)G/1.82$.
9. Compute G'.
10. Compare G' with G. If agreement is satisfactory, go to step 11. Otherwise reset $G = G'$ and return to 8.
11. Compute R_f using the latest G'.
12. Compute new R_{CI}, R_{CE}, R_r using the most recent values of $T_I - T_s$, $T_s - T_A$.
13. Compute new $T_I - T_s$, $T_s - T_A$.
14. Compare step 13 results with the previous estimate. If agreement is satisfactory, proceed to step 15. Otherwise return to step 12.
15. Compute $T_I - T_A$ and compare with previous $T_I - T_A$. If there is satisfactory agreement, go to step 16. Otherwise return to step 8.
16. Set G = most recent result.
17. Compute Q_f.
18. Compute G'.

An illustrative example may be easier to follow than the preceding procedural description. Suppose an enclosure has dimensions $W = 10$ in., $L = 10$ in., $H = 8$ in., and a dissipation height $d = 8$ in. The exterior surface emissivity is $\epsilon = 0.8$, and the internal power dissipation is $Q = 40$ watts. The inlet and exit areas are each 4 in.^{2}. The basic formulae for this problem may be somewhat simplified.

$$R_a = 2.4 \times 10^{-3}/[1.0/(4)^2 + 1.0/(4)^2]$$

$$= 3.0 \times 10^{-4}$$

$$G' = G - \{[293(3.0 \times 10^{-4})G + 0.909(3.0 \times 10^{-4})Q_f]G^2$$

$$- 1.09 \times 10^{-3}(8)Q_f\}/[909(3.0 \times 10^{-4})G$$

$$+ 1.82(3.0 \times 10^{-4})Q_f]G$$

$$G' = G - \frac{[0.0879G + 2.73 \times 10^{-4}Q_f]G^2 - 8.72 \times 10^{-3}Q_f}{(0.273G + 5.46 \times 10^{-4}Q_f)G}$$

$$R_{CI} = 1 \bigg/ \left[(0.0033)\left(\frac{T_I - T_s}{2.5}\right)^{0.25}(100) + (0.0048)\left(\frac{T_I - T_s}{8}\right)^{0.25}(160)\right]$$

$$= 1.39/(T_I - T_s)^{0.25}$$

Similarly:

$$R_{CE} = 1.39/(T_s - T_A)^{0.25}$$

For modest ambient temperatures and surface temperature rises, $h_r \simeq 0.004$; thus:

$$R_r \simeq 1/2 \epsilon h_r (WL + WH + LH)$$

$$= 1/2(0.8)(0.004)(260)$$

$$= 0.6°C/watt$$

The results of each step in the iterative procedure are tabulated in Table 6-7. The first set of temperature rises is iterated for $R_f = \infty$. This iteration loop is terminated when $T_I - T_s$ and $T_s - T_A$ result in equal values of 23.5 and 13.1°C, respectively, for two successive iterations. The sum, $T_I - T_A = 36.6°C$, is the nonventilated cabinet solution.

The air draft effect is initiated by making a first guess of $G = 1.0$ ft³/min. The air draft is iterated until two successive results agree.

A new set of temperature rise iterations is initiated with $R_f = 1.82/G$. R_{CI} and R_{CE} are computed using $T_I - T_s = 23.5°C$ and $T_s - T_A = 13.1°C$, respectively. The remaining calculations are self-explanatory.

The problem may be terminated at the last calculation of $T_I - T_A = 25.1°C$, but it is probably worthwhile to compute a final Q_f (15.9 watts) and make a last iteration for G. The results show an internal cabinet air temperature rise of 25.1°C, a ventilation draft of 1.15 ft³/min., and a loss of 15.9 watts into the air draft. A heat loss of 40 - 15.9 = 24.1 watts is transferred by radiation and natural convection by the cabinet exterior. It is very important to note that neither the ventilation nor the surface losses (convection/radiation) could be neglected to simplify the calculations and still give sufficiently accurate results.

The following two examples illustrate that some latitude exists in the placement of ventilation holes. A portable enclosure is illustrated in Fig. 6-24 with two different ventilation arrangements for the top panel: 11 in.² or 20 in.². Heat dissipation was provided by uniformly distributed, low-profile components on ten vertically oriented circuit cards. The cabinet exterior was covered with a thin opaque vinyl material with an emissivity of about 0.8. Air temperature measurements were obtained by averaging the readings from thermocouples distributed the length of the upper edge of several cards.

Calculations were made using the four-element model iteration scheme just described. A dissipation height $d = 4$ in. and an airflow

Table 6-7. Summary of iterative results.

R_f (°C/watt)	R_{CI} (°C/watt)	R_{CE} (°C/watt)	$T_I - T_s$ (°C)	$T_s - T_A$ (°C)	$T_I - T_A$ (°C)	Q_f (watts)	G, G' (ft³/min.)
			Guess 10	Guess 10		0	0
	0.78	0.78	31.3	13.6			
	0.59	0.72	23.5	13.1			
	0.63	0.73	23.5	13.1			
					36.6		
							Guess 1.0
						20.1	1.29
						25.9	1.34
						27.0	1.36
						27.4	1.37
1.33	0.63	0.73	14.7	7.7			
	0.71	0.83	15.8	7.8			
	0.70	0.83	15.6	7.8			
	0.70	0.83	15.6	7.8			
					23.4		
						17.6	1.22
						15.6	1.15
						14.8	1.12
						14.4	1.11
1.64	0.70	0.83	17.1	8.5			
	0.68	0.81	16.8	8.5			
	0.69	0.81	16.9	8.5			
	0.69	0.81	16.8	8.5			
	0.69	0.81	16.9	8.5			
					25.4		
						15.5	1.14
						15.9	1.15
1.58	0.69	0.81	16.6	8.4			
	0.69	0.82	16.7	8.4			
					25.1		1.15
						15.9	1.15

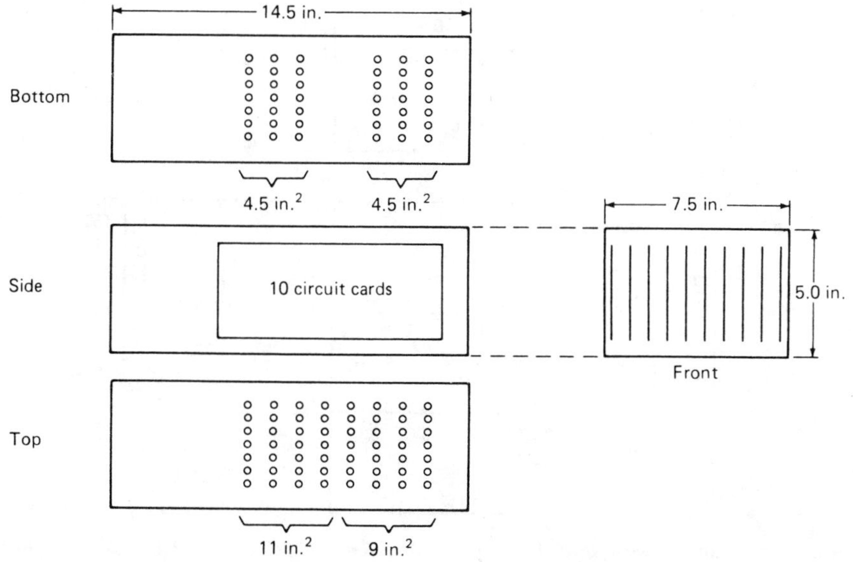

Fig. 6-24. Cabinet perforated in top and bottom panels. Surface emissivity $\epsilon = 0.8$. Ambient air $T_A = 20°C$.

Table 6-8. Air temperature rise for a vented cabinet in Fig. 6-24.

Vent Area (in.^{2})		Power Dissipation (watts)					
		25		50		100	
Inlet	Exit	Calc.	Exper.	Calc.	Exper.	Calc.	Exper.
0	0	28°C	27 ± 3°C	51°C	62 ± 6°C	91°C	100 ± 2°C
11	11	15	21 ± 4	25	35 ± 6	42	55 ± 9
11	20	14		23	27 ± 4	38	41 ± 5

resistance of:

$$R_a = 2.4 \times 10^{-3}/[(1/11)^2 + (1/11)^2]$$

or:

$$R_a = 2.4 \times 10^{-3}/[(1/11)^2 + (1/20)^2]$$

was used. No adjustment of the cabinet surface area was made to compensate for the perforations. The measured and computed results are summarized in Table 6-8.

The second application is a larger cabinet, a benchtop system illustrated in Fig. 6-25. The cabinet surface was unpainted and presumed to have a surface emissivity of about 0.1. The ventilation was a mix of slots and perforations in the front, rear, and side panels. The slots were 35% and 15% of the actual respective inlet and exit areas. Heat dissipation was provided by several wire-wound resistors, although a subsequent replacement by nine, 8 in. high circuit cards

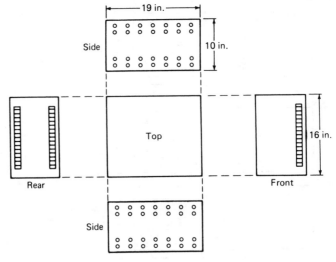

Fig. 6-25. Cabinet with slotted and perforated vents. Surface emissivity $\epsilon = 0.1$. Inlet area = 34.1 in.^{2}, exit area = 24.6 in.^{2}, ambient air $T_A = 20°C$.

Table 6-9. Air temperature rise for vented cabinet in Fig. 6-25.

Power Dissipation (watts)	Calculated			Exper.
	Q_f (watts)	G (ft³/min.)	ΔT (°C)	ΔT (°C)
25	16.0	4.1	7.0	7.8
50	33.7	5.2	11.8	12.5
100	69.3	6.6	19.1	19.8

did not drastically alter the results. The measurements and computations are summarized in Table 6-9.

The two preceding examples verify the validity of the theoretical method. However, one point in particular must be emphasized. Measurements of the systems just described, and several others not discussed here, indicate that systems analyses using flat plate convection models produce temperatures that usually correlate best with average maximum measurements, i.e., data recorded for the warmest air regions of a cabinet. This is usually in the vicinity of the top cover panel. A note of caution: the measured and computed air temperature rises agree very well, but such is not always the case. It is hoped that the reader will make some of his/her own measurements and draw relevant conclusions.

6.7 TWO-DIMENSIONAL ANALYSIS OF AN IDEAL FLUID*

Most forced airflow patterns in electronic enclosures are too complex to be analyzed by any method other than a lumped resistance technique as described in Sections 6.2 and 6.3. Additional physical insight is sometimes provided when the flow is expected to follow a two-dimensional pattern such as between two widely spaced circuit cards with a well-defined boundary, entrance, and exit.

When the fluid is assumed to be nonviscous and incompressible, the flow characteristics may be described by a set of orthogonal curves: velocity potentials and streamlines. The flow follows the direction taken by the streamlines, the latter intersecting the potentials at right angles.

Both the potential ϕ and the stream function ψ are solutions of Laplace's equation:†

$$\nabla^2\phi = 0, \qquad \nabla^2\psi = 0$$

The x and y velocity components, u and v, respectively, are obtain-

---

*Reference 33 is a good general text for basic fluid mechanics.

† $\nabla^2\phi = \dfrac{\partial^2\phi}{\partial x^2} + \dfrac{\partial^2\phi}{\partial y^2}$.

able from ϕ or ψ:

$$u = -\frac{\partial \phi}{\partial x}, \qquad v = -\frac{\partial \phi}{\partial y}$$

$$u = -\frac{\partial \psi}{\partial y}, \qquad v = \frac{\partial \psi}{\partial x}$$

Numerical solutions of $\nabla^2 \phi = 0$ and $\nabla^2 \psi = 0$ require finite difference approximations:

$$\nabla^2 \phi = \frac{\phi(x + \Delta x, y) - \phi(x, y)}{(\Delta x)^2} + \frac{\phi(x - \Delta x, y) - \phi(x, y)}{(\Delta x)^2}$$

$$+ \frac{\phi(x, y + \Delta y) - \phi(x, y)}{(\Delta y)^2} + \frac{\phi(x, y - \Delta y) - \phi(x, y)}{(\Delta y)^2}$$

$$= 0$$

where Δx and Δy define the size of the subdivision into which the two-dimensional space is divided. The equation for ψ is of identical form, but ψ is substituted for ϕ.

If ϕ_i is used to represent $\phi(x, y)$ and ϕ_j for any other ϕ, e.g., $\phi(x - \Delta x, y)$, $\phi(x + \Delta x, y)$, then the finite difference representation of $\nabla^2 \phi = 0$ may be written as:

$$\sum_{j \neq i} C_{ij}(\phi_i - \phi_j) = 0$$

$$C_{ij} = 1/(\Delta x)^2 \quad \text{in } x\text{-direction}$$

$$C_{ij} = 1/(\Delta y)^2 \quad \text{in } y\text{-direction}$$

The summation $\Sigma_{j \neq i}$ is over all nearest neighbors j connected to i. The exclusion $j \neq i$ excludes connection of i to itself. The velocity components in finite difference form are:

$$u = -\frac{\phi(x + \Delta x, y) - \phi(x, y)}{\Delta x} = -\sqrt{C_{ij}} \,(\phi_i - \phi_j)$$

$$v = -\frac{\phi(x, y + \Delta y) - \phi(x, y)}{\Delta y} = -\sqrt{C_{ij}} \,(\phi_i - \phi_j)$$

As in all field problems, boundary conditions must be applied. A boundary across which no flow is permitted is trivially established by the grid boundaries.

A flow sink is established by an equipotential $\phi = 0$ on the appropriate boundary defining the air exit.

Inlet air flow is quantified by imagining a set of fictitious boundary nodes i' immediately outside the entrance nodes i. For inlet flow u_0 in the x-direction:

$$u_0 = -(\phi_i - \phi_{i'})/\Delta x$$

$$(\phi_i - \phi_{i'}) = -\Delta x u_0$$

Now suppose that the summation over the fictitious nodes i' is separated out from the remainder of $\nabla^2 \phi = 0$:

$$\sum_{j \neq i, i'} C_{ij}(\phi_i - \phi_j) + \frac{1}{(\Delta x)^2}(\phi_i - \phi_{i'}) = 0$$

Since $(\phi_i - \phi_{i'}) = -\Delta x u_0$:

$$\sum_{j \neq i, i'} C_{ij}(\phi_i - \phi_j) - \frac{u_0}{\Delta x} = 0$$

which is readily solved for ϕ_i in a form suitable for iteration:

$$\phi_i = \frac{\sum_{j \neq i, i'} C_{ij}\phi_j + (u_0/\Delta x)_i}{\sum_{j \neq i, i'} C_{ij}}$$

The $j \neq i'$ indicates that the "fictitious" nodes just outside the air inlet are to be excluded from the grid solution. The subscript on $(u_0/\Delta x)$ indicates this source term is to be included only at nodes i along the inlet.

The stream function equation in finite form, ready for iterative solution, is developed in precisely the same manner as the potential function except that a source or air inlet term is not required:

$$\psi_i = \frac{\sum_{j \neq i} C_{ij}\psi_j}{\sum_{j \neq i} C_{ij}}$$

The stream function boundary conditions are provided by the boundaries constraining the flow. This is illustrated in Fig. 6-26.

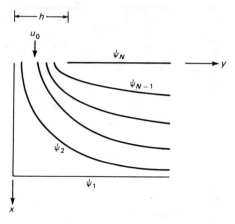

Fig. 6-26. Two-dimensional stream function.

Considering ψ_1 and ψ_N, the inlet flow velocity in the x-direction is:

$$u_0 = -\left.\frac{\partial \psi}{\partial y}\right|_{x=0,\, y=0 \to y=h}$$

$$\simeq -(\psi_{j+1} - \psi_j)/h$$

which applied across the inlet is:

$$\psi_N - \psi_1 = -u_0 h$$

$$\boxed{\psi_1 = \psi_N + u_0 h}$$

E6.23
Stream function
boundary conditions

ψ_N may be set at any value, e.g., $\psi_N = 0$. Then $\psi_1 = u_0 h$.

A sample problem in Section 8.10 illustrates details of digital computation of potential and stream function calculations and plotted results.

Chapter 7

Application of Digital Computer Methods I—A Continuous Solution*

7.1 INTRODUCTION

Chapters 1–6 are intended to give the reader a solid foundation and provide the self confidence that is associated with independence from a large computing facility and its associated paraphernalia such as punched cards, computer printouts, magnetic tapes, etc.

Unfortunately the time comes when the arithmetic chores are exceedingly formidable, and satisfactorily accurate temperature predictions are impossible without digital computation aids. In the absence of reasonable digital computation skills and capability, an engineer cannot be really competent in thermal analysis and design.

The inclusion of actual program listings and instructions in a basic text is rather unusual. These programs are not short training routines to be discarded after a cursory look or two, but are rather complete and should answer the needs of most equipment designers. Neither are the programs an afterthought. They are an integral part of the text, and the reader should attempt to implement their usage.

This chapter is intended to demonstrate most of the features of TAMS (see Chapter 9) by a group of selected examples. The problems are not intended to serve as examples of good package design, but are chosen for their illustrative value. The decision as to the desirability of at least a cursory reading of Chapters 9 and 10 prior to this is left to the reader. Actually learning how to use the programs probably is best accomplished by studying each problem input and comparing it with the program rules in Chapters 9 and 10.

Confidence in program features not covered by the sample problems is obtained by computing solutions to *very simple* problems and comparing program solutions with desk calculator answers. This particularly applies to the thermal network method of Chapters 8 and 10.

---

*Portions of this chapter are abstracted from "TAMS–Thermal Analyzer for Multilayer Structures," courtesy Tektronix, Inc.

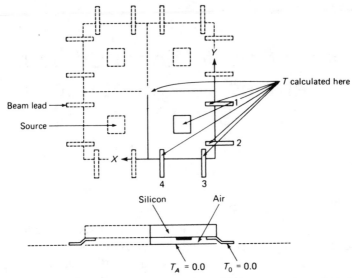

Fig. 7-1. Sample problem. One quadrant of a 16 lead, four-source beam lead chip. $k = 2.123$ watts/in. $\cdot$ °C for silicon, 0.0007 watt/in. $\cdot$ °C for air gap. $A = B = 0.025$ in., $t = 0.005$ in. for silicon and $t = 0.001$ in. air gap. Each beam lead resistance = 450°C/watt. Reprinted from the Proceedings of the 1978 International Microelectronics Symposium, [37], by permission of the International Society for Hybrid Microelectronics, P.O. Box 3255, Montgomery, AL 36109.

7.2 PROBLEM—BEAM LEAD CHIP

This example may not be representative of most equipment designers' applications, but it is an ideal vehicle for illustration of some TAMS features. The device under consideration has 16 leads and four heat sources distributed in the fashion shown in Fig. 7-1. The complete chip measures 0.05 in. × 0.05 in. and is 0.005 in. thick. A 0.001 in. air gap offsets the source plane from the substrate. Note in Fig. 7-1 that the TAMS z-axis must be into the plane of the page.

The TAMS model requires two different thermal conductivities: $k = 2.123$ watts/in. $\cdot$ °C and $k = 0.0007$ watts/in. $\cdot$ °C for silicon and air, respectively. The identification of four layers for the TAMS input is accomplished by using one silicon layer and three air layers.

Symmetry of the chip in the x–y plane permits calculations for only one quadrant. This is so because there is no heat transfer across a line of symmetry, and this also satisfies the TAMS edge boundary condition of $\partial T/\partial x = \partial T/\partial y = 0$.

The choice of a substrate source plane at the silicon–air interface, plus the approximation of the substrate as an isothermal infinite heat sink ($h_2 = \infty$), is accommodated by using ITEST = 6 on input. Note that a literal value of $h_2 = 1.0$ is input, but this value is not used for ITEST = 2 or 6; thus any value may be input.

Beam lead heat conduction is simulated by specifying 450°C/watt resistance for each lead attached via a 0.003 in. × 0.003 in. bond pad. The beam leads terminate at the same isothermal substrate as seen directly below the chip face, and since an ambient $T_A = 0.0$, all $T_0 = 0.0$ for the beam leads.

DATA SET

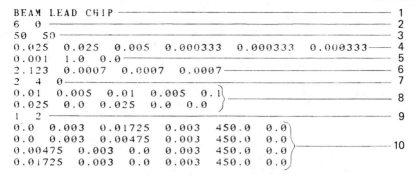

BEAM LEAD CHIP						1
6 0						2
50 50						3
0.025	0.025	0.005	0.000333	0.000333	0.000333	4
0.001	1.0	0.0				5
2.123	0.0007	0.0007	0.0007			6
2 4	0					7
0.01	0.005	0.01	0.005	0.1		8
0.025	0.0	0.025	0.0	0.0		
1 2						9
0.0	0.003	0.01725	0.003	450.0	0.0	
0.0	0.003	0.00475	0.003	450.0	0.0	10
0.00475	0.003	0.0	0.003	450.0	0.0	
0.01725	0.003	0.0	0.003	450.0	0.0	

Fig. 7-2. TAMS input for beam lead chip.

The program outputs temperatures at all source centers. In this problem, the temperature is also desired at the chip center. This is accomplished by defining a zero power source with zero edge dimensions at the appropriate quadrant corner.

The standard TAMS input defining this multi-source (2), multi-resistance (4) problem is listed in Fig. 7-2. The output from a line printer is shown in Fig. 7-3. The first column of source temperatures does not include the effect of conduction through the beam leads, a rather irrelevant result in this problem. The second column of source temperatures includes resistance conduction effects as indicated. Since the ambient (substrate) $T_A = 0.0$, all results are actually rises above the substrate surface temperature.

Note that the heat transfer through each of the four beam leads as well as the temperature at the center of each bond point is also printed. The difference between the source power of 0.1 watt and total of the F(I) is 0.0060 watt and is conducted through the air space to the substrate.

The last item relevant to setting up this problem is actually the first computer run required of TAMS just prior to the multi-source run. This is the Fourier-series convergence test. The input is similar to Fig. 7-2 with the following exceptions:

1. The 6 in line 2 is increased by 10.
2. Line 3 is made 70   70 (maximum number of series terms in each of the x, y components).
3. Line 7 is set for a single source, no resistance problem, i.e., 1   0   0.
4. Line 8 is set for the smallest source plane area, in this case the first resistance which has a bond pad of 0.003 in. × 0.003 in. The power is arbitrary.
5. The second source is deleted.
6. Data set 9 is set for one source, i.e., 1.
7. Data set 10 is deleted.

The line printer output in Fig. 7-4 shows very little change in the

```
BEAM LEAD CHIP

TAMSIV-THERMAL ANALYZER FOR MULTILAYER SYSTEMS
NOV. 11, 1978 VERSION

THERMAL ANALYSIS FOR NEWTON'S LAW COOLING AT Z=0 AND AN ISOTHERMAL SURFACE TA AT Z=C4
SOURCES AND LEADS AT Z=C1

SUBSTRATE DIMENSIONS AND PHYSICAL CONSTANTS
A= .02500  B= .02500  T1= .00500    T2= .00033    T3= .00033    T4= .00033
H1= .1000E-02  H2= .1000E+01
K1= .2123E+01  K2= .7000E-03  K3= .7000E-03  K4= .7000E-03

NUMBER OF SOURCES= 2  NUMBER OF RES.= 4  NR1= 4  NR2= 0

LMAX= 50  MMAX= 50

TA= 0.0

SOURCE DATA
SOURCE NO. I   XS(I)    DXS(I)    YS(I)    DYS(I)    Q(I)
1              .0100    .0050     .0100    .0050     .100
2              .0250    0.0000    .0250    0.0000    0.000

RES. DATA
RES. NO. I     XR(I)    DXR(I)    YR(I)    DYR(I)    R(I)        T0(I)
1              0.0000   .0030     .0173    .0030     .450E+03    0.0
2              0.0000   .0030     .0048    .0030     .450E+03    0.0
3              .0048    .0030     0.0000   .0030     .450E+03    0.0
4              .0173    .0030     0.0000   .0030     .450E+03    0.0

THERMAL FLUX IN RES. AND TEMPERATURES CALCULATED AT RES. PAD CENTERS
RES. NO. I     F(I)         TR(I)
1              .244E-01     11.0
2              .226E-01     10.2
3              .226E-01     10.2
4              .244E-01     11.0

TEMPERATURES CALCULATED AT SOURCE CENTERS
SOURCE NO. I   TS(I) WITH SOURCES ONLY    TS(I) WITH SOURCES AND RES.
1              232.4                      18.6
2              227.4                      14.7
```

Fig. 7-3. Standard TAMS output for beam lead chip.

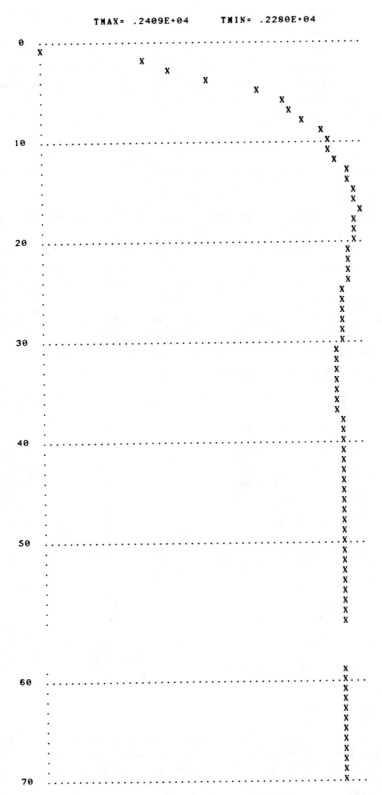

Fig. 7-4. Convergence test for beam lead chip.

single source temperature computation after LMAX = MMAX = 20. Values of 50 and 50 used in the multi-source run are actually much larger than necessary.

It should be pointed out here that although the convergence test plot may vary not only with the source dimensions but also with the x, y location, it is not necessary to run this plot for every source and resistor. It is sufficient to select the smallest device and then use an LMAX, MMAX slightly in excess of the minimum indicated by the plot.

7.3 PROBLEM—SINGLE CHIP IC PACKAGE

This example demonstrates the use of the multi-layer features of TAMS for a single chip package with forced air cooling (Fig. 7-5). A small extruded heat sink with fin height, substrate length, and width of 0.24, 0.62 and 1.32 in., respectively, and N/W = 4.5 fins/in. width and a fin spacing S = 0.182 in. is epoxied to the ceramic package. A 0.16 in. square chip with uniform power distribution is used for a heat source.

The TAMS model presumes flat surfaces and does not therefore automatically take the fin surface area into account. The fin area is compensated for by increasing the true heat transfer coefficient in proportion to the additional area. If:

N = number of fins

L = substrate length

W = substrate width

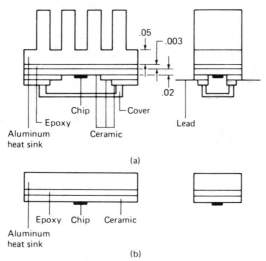

Fig. 7-5. Single chip IC package with an aluminum heat sink epoxied to ceramic base. (a) Actual package. (b) TAMS model. Dimensions in inches. From [10], copyright © 1979 by the Institute of Electrical and Electronics Engineers, Inc. Reprinted by permission of the publisher.

$\bar{h}$ = true heat transfer coefficient

h_2 = TAMS heat transfer coefficient for finned side

H = fin height

then:

$$\frac{h_2}{\bar{h}} = \frac{\left(1 + 2\frac{N}{W}H\right)WL}{WL}$$

$$h_2 = \left(1 + 2\frac{N}{W}H\right)\bar{h}$$

As an example, at an air velocity of 1000 ft/min., we find from Fig. 2-14 that $\bar{h} = 0.085$ watt/in.$^2 \cdot$ °C. Then:

$$h_2 = [1 + (2)(4.50)(0.24)](0.085)$$

$$= 0.269 \text{ watt/in.}^2 \cdot \text{°C}$$

A very small heat loss from the package into the board is quantified by an $h_1 \simeq 0.005$ watt/in.$^2 \cdot$ °C.

Note in Fig. 7-5 that the TAMS model ignores all ceramic super-structure other than the 0.02 in. base. This is justified because significant ceramic heat conduction takes place in the vicinity of the chip, whence it passes through the epoxy and into the aluminum, so the superstructure does not play much of a role in the heat problem.

Standard multi-source input/output is listed in Figs. 7-6 and 7-7. The single source convergence test follows in Fig. 7-8, from which it is clear that LMAX = MMAX = 40 is more than sufficient.

The same single chip package resistance (temperature rise for 1 watt) is computed without the heat sink at an air velocity of 1000 ft/min. Figures 7-9, 7-10, and 7-11 are the relevant input/output for a 0.02 in. base thickness.

DATA SET

```
SINGLE CHIP PACKAGE  -  SINKED  -  V=1000 —— 1
 1    0 ——————————————————————————————————————— 2
40    40 ———————————————————————————————————————— 3
1.32   0.62    0.02    0.003    0.025    0.025—— 4
0.005  0.269   0.0 ——————————————————————————————— 5
1.0    0.03    5.0    5.0 ———————————————————————— 6
 2    0    0 ———————————————————————————————————— 7
0.58   0.16    0.23    0.16    1.0⎫
0.724  0.0     0.31    0.0     0.0⎭ ———————————————— 8
 1    2 ——————————————————————————————————————————— 9
```

Fig. 7-6. TAMS input for single chip package, sinked.

SINGLE CHIP PACKAGE - SINKED - V=1000

TAMSIV-THERMAL ANALYZER FOR MULTILAYER SYSTEMS
NOV. 11, 1978 VERSION

THERMAL ANALYSIS FOR NEWTON'S LAW COOLING AT Z=0 AND C4
SOURCES AND LEADS AT Z=0

SUBSTRATE DIMENSIONS AND PHYSICAL CONSTANTS
A= 1.32000 B= .62000 T1= .02000 T2= .00300 T3= .02500 T4= .02500
H1= .5000E-02 H2= .2690E+00
K1= .1000E+01 K2= .3000E-01 K3= .5000E+01 K4= .5000E+01

NUMBER OF SOURCES= 2 NUMBER OF RES. = 0 NR1= 0 NR2= 0

LMAX= 40 MMAX= 40

TA= 0.0

SOURCE DATA

SOURCE NO. I	XS(I)	DXS(I)	YS(I)	DYS(I)	Q(I)
1	.5800	.1600	.2300	.1600	1.000
2	.7240	0.0000	.3100	0.0000	0.000

TEMPERATURES CALCULATED AT SOURCE CENTERS

SOURCE NO. I	TS(I) WITH SOURCES ONLY	TS(I) WITH SOURCES AND RES.
1	8.5	
2	7.6	

Fig. 7-7. Standard TAMS output for single chip package, sinked.

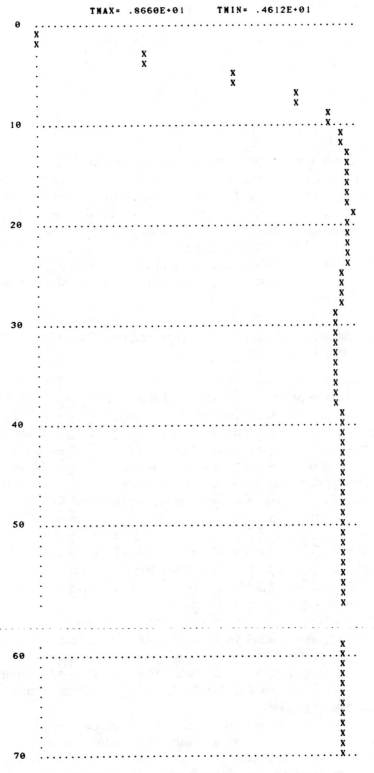

Fig. 7-8. TAMS convergence test for single chip package, sinked.

DATA SET

```
                                                          1
SINGLE CHIP PACKAGE  -  UNSINKED  -  V=1000 ——— 1
1   0  ———————————————————————————————————— 2
20    20 ———————————————————————————————————— 3
1.32   0.62   0.005   0.005   0.005   0.005——— 4
0.005   0.085   0.0  ——————————————————————— 5
1.0   1.0   1.0   1.0  ——————————————————————— 6
2   0   0  ———————————————————————————————— 7
0.58   0.16   0.23   0.16   1.0⎫
                               ⎬———————————— 8
0.724   0.0   0.31   0.0   0.0⎭
1   2 ——————————————————————————————————————— 9
```

Fig. 7-9. TAMS input for single chip package, unsinked.

Computations at other air velocities are plotted in Fig. 7-12 along with a few measurements for an actual device with a rather uniform distribution of heat over the source. The plotted theory does not include the temperature gradients due to the chip thickness. The unsinked package calculations are obtained from an average of results for 0.02 and 0.04 in. thick substrates. Averaging in the 0.04 in. base results tends to compensate for the additional ceramic superstructure in the nonsinked package. The theoretical–experimental correlation is satisfactory.

7.4 PROBLEM—NATURAL CONVECTION COOLED POWER SUPPLY HEAT SINK

This example illustrates one of the most common applications of TAMS: rather precise computation of hot spots on an extruded heat sink such as is illustrated in Fig. 7-13. Some of the significant features are the nonuniformly distributed fins, different ambient temperatures for each side of the sink, and nonuniformly distributed heat sources representing transistors dissipating a total of 40 watts.

The appropriate TAMS coordinate system in Fig. 7-14 illustrates the placement of four transistors as rectangular heat sources on the flat side of the heat sink, i.e., the $z = 0$ plane. The coordinates of the source corner nearest the x, y origin for each transistor, as well as the device x, y dimensions and heat dissipation, are used as TAMS input. Heat convection and radiation for the source side are specified by a combined (radiation + convection) heat transfer coefficient h_1 for an average surface temperature.

Convection and radiation from the finned side should be accounted for in a manner that considers the nonuniform fin distribution. This may not be done by specifying a heat transfer coefficient h_2 for the total 9 in. × 5.0 in. plate. Therefore the TAMS input for h_2 is set at a very small value, say 1.0×10^{-10}, so as to make this mechanism negligible.

Six lumped parameter resistances, $R_1 \cdots R_6$, are used to remove heat from the $z = t$ surface. The subdivision into six regions is determined by convenience according to the geometry and the anticipated thermal gradient in the heat sink base. Three equal sections across

SINGLE CHIP PACKAGE - UNSINKED - V=1000

TAMSIV-THERMAL ANALYZER FOR MULTILAYER SYSTEMS
NOV. 11, 1978 VERSION

THERMAL ANALYSIS FOR NEWTON'S LAW COOLING AT Z=0 AND C4
SOURCES AND LEADS AT Z=0

SUBSTRATE DIMENSIONS AND PHYSICAL CONSTANTS
A= 1.32000   B= .62000   T1= .00500   T2= .00500   T3= .00500   T4= .00500
H1= .5000E-02 H2= .8500E-01
K1= .1000E+01 K2= .1000E+01 K3= .1000E+01 K4= .1000E+01

NUMBER OF SOURCES= 2   NUMBER OF RES.= 0   NR1= 0   NR2= 0

LMAX= 20   MMAX= 20

TA= 0.0

SOURCE DATA
SOURCE NO. I   XS(I)   DXS(I)   YS(I)   DYS(I)   Q(I)

1   .5800   .1600   .2300   .1600   1.000

2   .7240   0.0000   .3100   0.0000   0.000

TEMPERATURES CALCULATED AT SOURCE CENTERS
SOURCE NO. I   TS(I) WITH SOURCES ONLY   TS(I) WITH SOURCES AND RES.

1   26.4

2   24.0

Fig. 7-10.   Standard TAMS output for single chip package, unsinked.

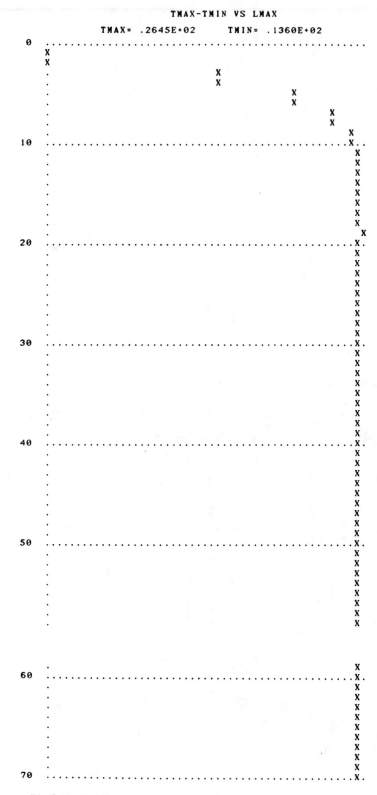

Fig. 7-11. TAMS convergence test for single chip package, unsinked.

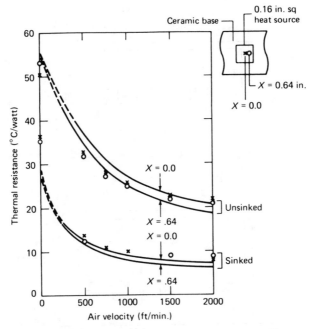

Fig. 7-12. Theoretical and experimental results for air-cooled alumina ceramic package with and without a heat sink. Reprinted from the Proceedings of the 1978 International Microelectronics Symposium, [37], by permission of the International Society for Hybrid Microelectronics, P.O. Box 3255, Montgomery, AL 36109.

the width are reasonable according to the dimensions of the finned section. The temperature gradients, except in the immediate vicinity of the transistors, should not be particularly large; therefore the rectangle edge dimensions are maintained approximately equal in the x and y directions.

It is important to be aware that the TAMS model will transfer heat in proportion to the surface temperature rise above ambient. Where h_1 or h_2 is significant, the error in convection/radiation is

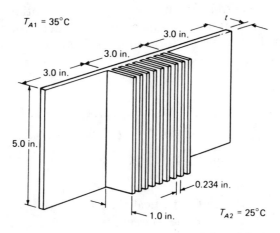

Fig. 7-13. Nonuniformly finned power supply heat sink.

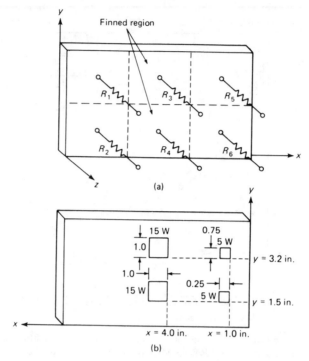

Fig. 7-14. TAMS model of power supply heat sink. (a) Finned side. (b) Source side.

due to using a uniform h over the surface, whereas h should ideally vary from point to point owing to the dependence of h_c and h_r on T_s.

When lumped resistances are used to simulate surface losses, as in this problem, the loss will depend on $T_s - T_A$, but the T_s for a complete section, such as for R_1 in Fig. 7-14(a), will be at the center of the defined rectangle. Thus if this single T_s is not a good average for the rectangle defined by R_1 coordinates, the computed final results from TAMS may be somewhat in error. This is usually not much of a problem, but the program user should be aware of this consideration.

Nearly all TAMS input is geometric description; h_1, $R_1 \cdots R_6$ are the exceptions. The temperature dependence of these quantities is estimated by a straightforward consideration of the total conductance of the complete system:

Source side:

$$C_s = (h_H + \epsilon h_r)(9 \text{ in.})(5 \text{ in.})$$

$$= (h_H + \epsilon h_r)(45)$$

where h_H and h_r are the vertical plate convection and radiation heat transfer coefficients for the source side.

Finned side, finned section:

Interior of finned area:

$$A_I = (3 \text{ in.})(5 \text{ in.}) \left[1 + \frac{2(N-1)}{W} L \right]$$

for

$$N = 9 \text{ fins}$$

$$L = \text{fin length} = 1.0 \text{ in.}$$

$$W = \text{width of finned section} = 3.0 \text{ in.}$$

$$A_I = (3)(5) \left[1 + \frac{2(9-1)}{3} (1) \right] = 95 \text{ in.}^2$$

Exterior of the finned area, neglecting shielding and boundary layer interference at the two right angles:

$$A_0 = 2(1 \text{ in.})(5 \text{ in.}) = 10 \text{ in.}^2$$

$$C_3 + C_4 = 2C_3$$

$$= \left[\left(\frac{h_c}{h_H} \right) h_H + \mathcal{F} h_r \right] A_I + (h_H + \epsilon h_r) A_0$$

$$= \left[\left(\frac{h_c}{h_H} \right) A_I + A_0 \right] h_H + (\mathcal{F} A_I + \epsilon A_0) h_r$$

$$C_1 + C_2 + C_5 + C_6 = 4C_1$$

$$= (h_H + \epsilon h_r)(4)(3.0 \text{ in.})(2.5 \text{ in.})$$

$$= (h_H + \epsilon h_r)(30)$$

Using an average front and rear surface temperature $\bar{T}_s$, the total heat convected and radiated is the sum of the heat Q_s transferred to T_{A1} from the source side and Q_f transferred to T_{A2} from the finned side:

$$Q = Q_s + Q_f$$

$$Q_s = C_s(\bar{T}_s - T_{A1})$$

$$Q_f = (2C_3 + 4C_1)(\bar{T}_s - T_{A2})$$

The coefficients h_H and h_r are computed using E2.13 and E3.24, respectively. The ratio (h_c/h_H) for the fin channels is estimated from Figs. 5-14 through 5-17 for:

$$(L/S) = 1/0.234 = 4.3, \quad (H/S) = 5.0/0.234 = 21.4, \quad S = 0.23$$

$\mathfrak{F} = 0.14$ is obtained from Fig. 5-11 for $(L/S) = 4.3$ and $(H/L) = 5.0/1.0 = 5.0$. Naturally, the h's and the resultant conductances are temperature-dependent. Table 7-1 is a tabulation of some of the relevant quantities, including Q_s and Q_f, the heat transferred from the source and finned sides of the sink, respectively.

The quantities Q_s, Q_f and $Q = Q_s + Q_f$ are plotted in Fig. 7-15 against the average surface temperature $\bar{T}_s$. A total of $Q = Q_s + Q_f = 12 + 28 = 40$ watts is seen to produce a $\bar{T}_s = 69°C$. The various temperature-dependent input variables for TAMS may now be computed with sufficient accuracy.

At the source side:

$$h_1 = Q_s/[WH(\bar{T}_s - T_{A1})]$$

$$= 12.0/[(9)(5)(69 - 35)]$$

$$= 0.0078 \text{ watt/in.}^2 \cdot °C$$

Using Eiii.1 to provide for the elevated $T_{A1} = 35°C$ ambient:

$$h_1' = h_1[1 - (T_{A1} - T_{A2})/(\bar{T}_s - T_{A2})]$$

$$= 0.0078(1 - 10/44)$$

$$= 0.0060 \text{ watt/in.}^2 \cdot °C$$

Convection and radiation from the finned side are provided for, not by h_2 (which is set at 1.0×10^{-10}), but rather $R_1 \cdots R_6$:

$$R_1 = R_2 = R_5 = R_6 = 1/C_1$$

$$= 4/[(h_H + \epsilon h_r)(30)]$$

For $\bar{T}_s = 69°C$ and $T_{A2} = 25°C$, $h_H = 0.0041$, $h_r = 0.0048$.

$$R_1 = 4/\{[0.0041 + (0.8)(0.0048)](30)\}$$

$$= 16.6°C/\text{watt}$$

Table 7-1. Intermediate results for computation of sample problem from Section 7.4.

		Source Side					Finned Side					
$\bar{T}_s$ (°C)	$\bar{T}_s - T_{A1}$ (°C)	h_H (watts/ in.$^2 \cdot$°C)	h_r (watts/ in.$^2 \cdot$°C)	C_s (watts/ °C)	Q_s (watts)	$\bar{T}_s - T_{A2}$ (°C)	h_H (watts/ in.$^2 \cdot$°C)	$\dfrac{h_c}{h_H}$	h_r (watts/ in.$^2 \cdot$°C)	$2C_3$ (watts/ °C)	$4C_1$ (watts/ °C)	Q_f (watts)
50	15	0.0032	0.0046	0.31	4.65	25	0.0036	0.5	0.0044	0.30	0.21	12.75
75	40	0.0040	0.0052	0.37	14.80	50	0.0043	0.6	0.0050	0.40	0.25	32.50
100	65	0.0046	0.0058	0.42	27.30	75	0.0047	0.62	0.0056	0.44	0.28	54.00

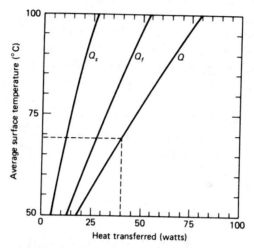

Fig. 7-15. Heat distribution computed for average surface temperature.

The (h_c/h_H) for the finned section is estimated to be 0.6.

$$R_3 = R_4 = 1/C_3$$

$$= 2/\{[(h_c/h_H)A_I + A_0]h_H + (\mathcal{F}A_I + \epsilon A_0)h_r\}$$

$$= 5.40°C/\text{watt}$$

The standard TAMS input is listed in Fig. 7-16, for which a few comments are in order. The fifth line of input has an ambient of

```
POWER SUPPLY HEAT SINK
1   0
50   50
9.0   5.0   0.03125   0.03125   0.03125   0.03125
0.0060   1.0E-10   25.0
3.5   3.5   3.5   3.5
4   0   6
1.0   0.25   1.5   0.75   5.0
1.0   0.25   3.2   0.75   5.0
4.0   1.0   1.5   1.0   15.0
4.0   1.0   3.2   1.0   15.0
1   0   3   4
0.0   3.0   2.5   2.5   16.6   25.0
0.0   3.0   0.0   2.5   16.6   25.0
3.0   3.0   2.5   2.5   5.4   25.0
3.0   3.0   0.0   2.5   5.4   25.0
6.0   3.0   2.5   2.5   16.6   25.0
6.0   3.0   0.0   2.5   16.6   25.0
```

Fig. 7-16. Standard TAMS input for power supply heat sink.

POWER SUPPLY HEAT SINK

TAMSIV-THERMAL ANALYZER FOR MULTILAYER SYSTEMS
NOV. 11, 1978 VERSION

THERMAL ANALYSIS FOR NEWTON'S LAW COOLING AT Z=0 AND C4
SOURCES AND LEADS AT Z=0

SUBSTRATE DIMENSIONS AND PHYSICAL CONSTANTS
A= 9.00000   B= 5.00000   T1= .03125   T2= .03125   T3= .03125   T4= .03125
H1= .6000E-02   H2= .1000E-09
K1= .3500E+01   K2= .3500E+01   K3= .3500E+01   K4= .3500E+01

NUMBER OF SOURCES= 4   NUMBER OF RES. = 6   NR1= 0   NR2= 6

LMAX= 50   MMAX= 50

TA= 25.0

SOURCE DATA

SOURCE NO. I	XS(I)	DXS(I)	YS(I)	DYS(I)	Q(I)
1	1.0000	.2500	1.5000	.7500	5.000
2	1.0000	.2500	3.2000	.7500	5.000
3	4.0000	1.0000	1.5000	1.0000	15.000
4	4.0000	1.0000	3.2000	1.0000	15.000

RES. DATA

RES. NO. I	XR(I)	DXR(I)	YR(I)	DYR(I)	R(I)	T0(I)
1	0.0000	3.0000	2.5000	2.5000	.166E+02	25.0
2	0.0000	3.0000	0.0000	2.5000	.166E+02	25.0

Fig. 7-17.   Standard TAMS output for power supply heat sink.

3	3.0000	3.0000	2.5000	2.5000	.540E+01	25.0
4	3.0000	3.0000	0.0000	2.5000	.540E+01	25.0
5	6.0000	3.0000	2.5000	2.5000	.166E+02	25.0
6	6.0000	3.0000	0.0000	2.5000	.166E+02	25.0

THERMAL FLUX IN RES. AND TEMPERATURES CALCULATED AT RES. PAD CENTERS

RES. NO. I	F(I)	TR(I)
1	.296E+01	74.1
2	.282E+01	71.8
3	.962E+01	77.0
4	.845E+01	70.6
5	.224E+01	62.2
6	.219E+01	61.4

TEMPERATURES CALCULATED AT SOURCE CENTERS

SOURCE NO. I	TS(I) WITH SOURCES ONLY	TS(I) WITH SOURCES AND RES.
1	179.4	75.4
3	183.2	76.2
4	184.7	77.3

Fig. 7-17. *(Continued)*

25°C consistent with the previously described method of calculating h_1'. The computed temperatures shown in the output of Fig. 7-17 are surface temperatures (not rises) in a 25°C room ambient on the finned side and a 35°C ambient on the source side (presumably the cabinet interior). The convergence plot in Fig. 7-18 was computed for the first 0.25 in. × 0.75 in. source on a 0.125 in. sink base. An LMAX = MMAX = 50 is seen to be sufficient.

Inputs identical to Fig. 7-16 with the exception of the fourth line were used for a second and third TAMS run, respectively. The variations in the fourth input line were from 0.03125 to 0.125 for the last four variables to simulate heat sink base thicknesses from 0.125 in. to 0.5 in. The maximum temperature (source 4) is plotted vs. sink thickness in Fig. 7-19.

7.5 PROBLEM—CERAMIC SUBSTRATE WITH FINNED HEAT SINK AND CONDUCTING LEADS

An alumina substrate epoxied to a small extruded aluminum heat sink is shown in Fig. 7-20. The substrate extends beyond the heat sink width to provide additional space for very low power dissipating circuitry. The significant heat-dissipating devices consist of two 1-watt, 0.2 in. square chips, a 2-watt, 0.1 in. square chip, and a 2-watt resistor network measuring 0.6 in. × 0.25 in. An array of edge leads is represented by three thermal resistances of 1150°C/watt each. The heat sink and substrate surfaces are cooled by natural convection. Radiation is neglected because of the presumed presence of surrounding high-temperature surfaces.

The average surface temperature rise is estimated as follows:

$$\frac{1}{R} = \frac{2}{R_E} + \frac{1}{R_s} + \frac{1}{R_B}$$

where

R = average total convection resistance of assembly
R_E = convection resistance of each substrate end extending beyond heat sink
R_s = average convection resistance of heat sink
R_B = average convection resistance of 1.8 in. wide section of substrate source side

Using h_H = natural convection h for a vertical plate of height H:

$$R_E = 1/2(0.8 \text{ in.})(1.0 \text{ in.})h_H$$

$$R_B = 1/(1.8 \text{ in.})(1.0 \text{ in.})h_H$$

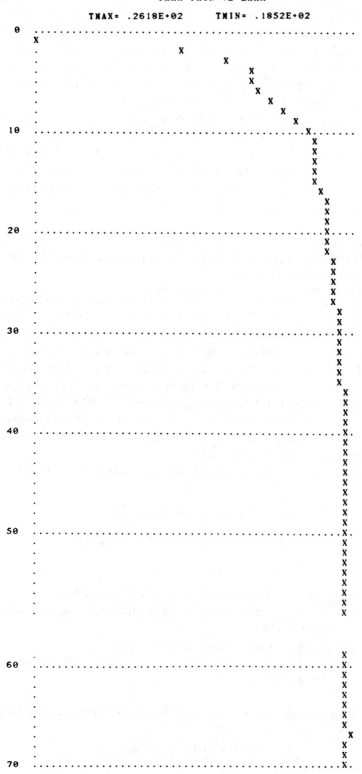

Fig. 7-18. TAMS convergence test for power supply heat sink.

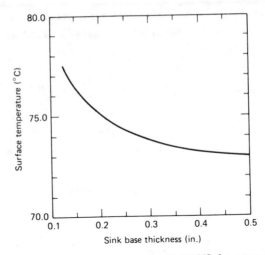

Fig. 7-19. Source no. 4 temperature computed with TAMS for power supply heat sink from Section 7.4.

The finned heat sink has an interior, between fin area A_I and an exterior area A_0:

$$A_I = WH \left[1 + \frac{2(N-1)}{W} L \right]$$

$$= (1.8)(1.0) \left[1 + \frac{2(7)}{(1.8)} (1.25) \right] = 19.3 \text{ in.}^2$$

$$A_0 = 2H(L + t_b)$$

$$= 2(1.0)(1.25 + 0.125) = 2.75 \text{ in.}^2$$

The convection conductance C_s for the sink is:

$$C_s = 1/R_s = h_H A_0 + (h_c/h_H)h_H A_I$$

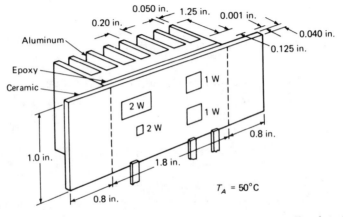

Fig. 7-20. Hybridized circuits on extruded heat sink cooled ceramic substrate.

Table 7-2. Computed characteristics of heat sunk hybrid.

$\overline{\Delta T}_s$ (°C)	h_c/h_H	h_H (watts/in.$^2\cdot$°C)	R_E (°C/watt)	R_s (°C/watt)	R_B (°C/watt)	R (°C/watt)	Q (watts)	h_2 (watts/in.$^2\cdot$°C)
10	0.58	0.0043	145.4	16.7	129.2	12.3	0.81	0.0333
25	0.73	0.0054	115.7	11.0	102.9	8.5	2.94	0.0505
50	0.79	0.0064	97.7	8.7	86.8	6.8	7.35	0.0639

where:

$$h_H = 0.0024 \left(\frac{\overline{T}_s - T_A}{H} \right)^{0.25}$$

and (h_c/h_H) is determined from Figs. 5-14, 5-15, 5-16 for $\overline{T}_s - T_A =$ 10, 25, 50°C, respectively. The effect of a 50°C ambient (opposed to 25°C as used for Fig. 5-14, etc.) is to introduce less than a 15% error in h_c/h_H. The various resistances are tabulated in Table 7-2.

The total convected heat Q is obtained from:

$$Q = (\overline{T}_s - T_A)/R$$

An "effective" heat transfer coefficient h_2 to simulate the fins for TAMS is computed from:

$$h_2 = 1/WHR_s$$

$$= 1/(1.8 \text{ in.})(1.0 \text{ in.})R_s$$

Q and h_2 are also listed in Table 7-2 for $\overline{T}_s - T_A =$ 10, 25, 50°C.

Convection characteristics at all temperature rises between 0 and 50°C are estimated by interpolating a graph of Q and h_2, Fig. 7-21. At 6 watts, the actual hybrid dissipation, Fig. 7-21 indicates:

$$\overline{T}_s - T_A = 43°C$$

$$h_2 = 0.062 \text{ watt/in.}^2 \cdot °C$$

It is now possible to compute R_E and h_1 for the 43°C rise:

$$h_H = 0.0024 \left(\tfrac{43}{1}\right)^{0.25}$$

$$= 0.0062 \text{ watt/in.}^2 \cdot °C$$

$$R_E = 1/(1.6)(0.0062)$$

$$= 100.8°C/watt$$

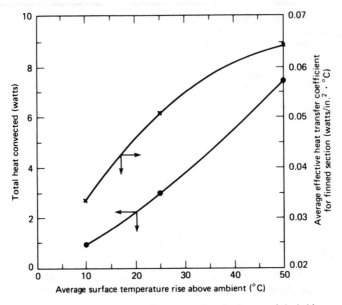

Fig. 7-21. Pertinent convection properties for heat sunk hybrid.

The source surface heat transfer coefficient $h_1 = h_H = 0.0062$.

The coordinate system for the TAMS model is shown in Fig. 7-22, where $R_1 = R_2 = R_E$ and $R_3 = R_4 = R_5 = 1150°C/watt$. The first, second, third, and fourth TAMS layers are alumina ($t = 0.04$ in., $k = 0.8$ watt/in. $\cdot$ °C), epoxy ($t = 0.001$ in., $k = 0.007$ watt/in. $\cdot$ °C), and two aluminum layers (each $t = 0.0625$ in., $k = 4.0$ watts/in. $\cdot$ °C), respectively.

The TAMS input file is listed in Fig. 7-23. The corresponding TAMS single source convergence plot in Fig. 7-24 indicates that LMAX = MMAX = 40 terms is an adequate series truncation.

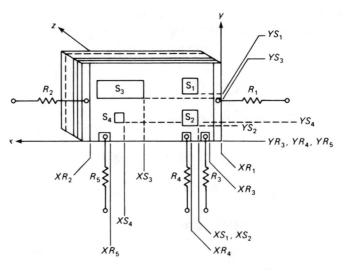

Fig. 7-22. TAMS model of heat sunk hybrid.

DATA SET

```
HEAT  SUNK  HYBRID                                    1
 1    0                                               2
40     40                                             3
 1.8    1.0    0.04    0.001    0.0625    0.0625       4
 0.0062    0.062    0.0                                5
 0.8    0.007    4.0    4.0                            6
 4    5    0                                           7
 0.30    0.20    0.60    0.20    1.0
 0.30    0.20    0.20    0.20    1.0
 1.00    0.60    0.55    0.25    2.0                   8
 1.25    0.10    0.25    0.10    2.0
 1    2    3    4                                      9
 0.00    0.10    0.0    1.00    100.80    0.0
 1.70    0.10    0.0    1.00    100.80    0.0
 0.15    0.10    0.0    0.10    1150.0    0.0          10
 0.40    0.10    0.0    0.10    1150.0    0.0
 1.45    0.10    0.0    0.10    1150.0    0.0
```

Fig. 7-23. TAMS standard input for heat sunk hybrid.

Standard TAMS output for the multi-source computer analysis is listed in Fig. 7-25, which shows the 2-watt, 0.1 in. × 0.1 in. source to be 56.6°C above the 50°C ambient. Note that the substrate overhang convects $0.397 + 0.409 = 0.806$ watt. The leads conduct a rather trivial $0.0350 + 0.0350 + 0.0362 = 0.106$ watt.

7.6 PROBLEM—COPPER-CLAD EPOXY-GLASS CIRCUIT CARD

This example is used to illustrate vividly an extreme case of thermal spreading resistance that leads to a rather interesting temperature profile. A method of determining temperatures on the nonsource surface is also demonstrated.

The basic geometry is illustrated in Fig. 7-26. The dimensions are $A = 6.0$ in., $B = 3.0$ in., $t_1 = t_4 = 0.0014$ in. and $t_2 = t_3 = 0.031$ in. The two middle layers represent the epoxy-glass material with a thermal conductivity estimated to 0.00737 watt/in. · °C. The two outer layers are solid copper sheets (with the exception of transistor mounting holes) with a $k = 10.0$ watts/in. · °C. Sources 1, 3, 5, and 7 are nondissipating fictitious sources used to obtain more complete profile data.

The total average heat transfer coefficients h_1, h_2 are estimated in a manner similar to that illustrated in Section 7.4. Using a 25°C ambient:

$$h_1 = h_2 = 3.657 \times 10^{-11} \epsilon \frac{[(\bar{T}_s + 273.14)^4 - (298.14)^4]}{\bar{T}_s - 25}$$

$$+ 0.0024 \left(\frac{\bar{T}_s - 25}{3}\right)^{0.25}$$

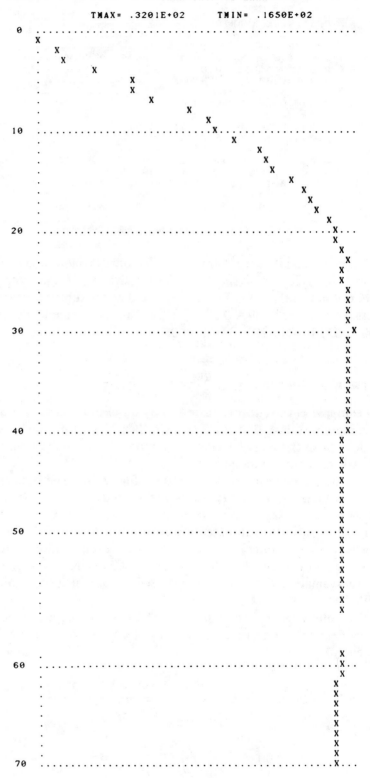

Fig. 7-24.   TAMS convergence plot for heat sunk hybrid.

HEAT SUNK HYBRID

TAMSIV-THERMAL ANALYZER FOR MULTILAYER SYSTEMS
NOV. 11, 1978 VERSION

THERMAL ANALYSIS FOR NEWTON'S LAW COOLING AT Z=0 AND C4
SOURCES AND LEADS AT Z=0

SUBSTRATE DIMENSIONS AND PHYSICAL CONSTANTS
A= 1.80000   B= 1.00000   T1= .04000   T2= .00100   T3= .06250   T4= .06250
H1= .6200E-02 H2= .6200E-01
K1= .8000E+00 K2= .7000E-02 K3= .4000E+01 K4= .4000E+01

NUMBER OF SOURCES= 4   NUMBER OF RES.= 5   NR1= 5   NR2= 0

LMAX= 40   MMAX= 40

TA= 0.0

SOURCE DATA
SOURCE NO. I	XS(I)	DXS(I)	YS(I)	DYS(I)	Q(I)
1	.3000	.2000	.6000	.2000	1.000
2	.3000	.2000	.2000	.2000	1.000
3	1.0000	.6000	.5500	.2500	2.000
4	1.2500	.1000	.2500	.1000	2.000

Fig. 7-25. TAMS standard output for heat sunk hybrid.

RES. DATA

RES. NO. I	XR(I)	DXR(I)	YR(I)	DYR(I)	R(I)	TØ(I)
1	0.0000	.1000	0.0000	1.0000	.101E+03	0.0
2	1.7000	.1000	0.0000	1.0000	.101E+03	0.0
3	.1500	.1000	0.0000	.1000	.115E+04	0.0
4	.4000	.1000	0.0000	.1000	.115E+04	0.0
5	1.4500	.1000	0.0000	.1000	.115E+04	0.0

THERMAL FLUX IN RES. AND TEMPERATURES CALCULATED AT RES. PAD CENTERS

RES. NO. I	F(I)	TR(I)
1	.397E+00	40.1
2	.409E+00	41.2
3	.350E-01	40.3
4	.354E-01	40.8
5	.362E-01	41.6

TEMPERATURES CALCULATED AT SOURCE CENTERS

SOURCE NO. I	TS(I) WITH SOURCES ONLY	TS(I) WITH SOURCES AND RES.
1	51.5	44.1
2	51.6	44.1
3	51.7	44.3
4	64.0	56.6

Fig. 7-25. (Continued)

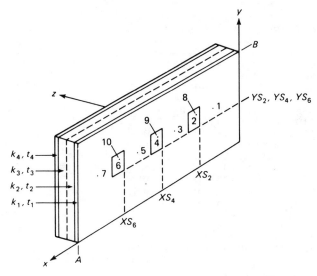

Fig. 7-26. View of source plane side of TAMS model of double-sided Cu-clad epoxy-glass circuit card.

The total heat transferred is:

$$Q = 2ABh_1(\bar{T}_s - 25)$$

$$= 2(6.0 \text{ in.})(3.0 \text{ in.})h_1(\bar{T}_s - 25)$$

Thus $h_1 = Q/[36.0(\bar{T}_s - 25)]$.

Both h_1 and Q are plotted in Fig. 7-27 vs. the average rise above ambient for two surface conditions: $\epsilon = 0.9$ for a flat black painted

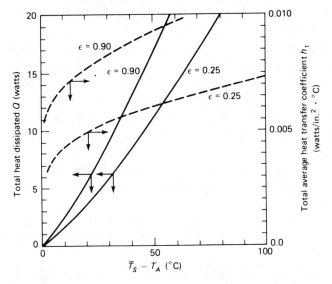

Fig. 7-27. Thermal characteristics for double-sided Cu-clad, epoxy-glass card.

```
TWO SIDED EPOXY-GLASS CARD ───────────────────── 1
1    0 ───────────────────────────────────────── 2
30    30 ──────────────────────────────────────── 3
6.0    3.0    0.0014    0.031    0.031    0.0014──── 4
0.0084    0.0084    0.0 ──────────────────────── 5
10.0    0.00737    0.00737    10.0 ───────────── 6
10    0    3 ─────────────────────────────────── 7
0.655    0.0    1.28    0.0    0.0─┐
1.310    0.4    0.98    0.6    5.0 │
2.265    0.0    1.28    0.0    0.0 │
2.820    0.4    0.98    0.6    2.5 │
3.675    0.0    1.28    0.0    0.0 │
4.310    0.4    0.98    0.6    2.5 ├─────────────── 8
5.355    0.0    1.28    0.0    0.0 │
1.560    0.0    1.48    0.0    0.0 │
2.870    0.0    1.48    0.0    0.0 │
4.360    0.0    1.48    0.0    0.0─┘
1    2    3    4    5    6    7    8    9    10 ──── 9
1.31    0.4    0.98    0.6    1.0E10    0.0─┐
2.82    0.4    0.98    0.6    1.0E10    0.0 ├──── 10
4.31    0.4    0.98    0.6    1.0E10    0.0─┘
```

Fig. 7-28. Standard TAMS input for Cu-clad epoxy-glass card.

surface and $\epsilon = 0.25$ for a rather severely oxidized copper surface. The specific total $Q = 10$ watts indicates:

$$\epsilon = 0.9: \quad \bar{T}_s - 25 \simeq 33°C, \quad h_1 = h_2 = 0.0084 \text{ watt/in.}^2 \cdot °C$$

$$\epsilon = 0.25: \quad \bar{T}_s - 25 \simeq 47°C, \quad h_1 = h_2 = 0.0060 \text{ watt/in.}^2 \cdot °C$$

The standard TAMS input, Fig. 7-28, also has seven lumped parameter resistors at the nonsource plane, which are used to obtain temperatures in that plane. The very large resistance, 1.0×10^{10}, inhibits heat flow through these elements. Heat sources 2, 4, and 6, dissipating 5.0, 2.5, and 2.5 watts, respectively, have coordinates in the input listing consistent with Fig. 7-26.

Line printer output is reprinted in Fig. 7-29. The nonsource plane temperature rises are found under the column TR(I) for the region immediately opposite the actual transistor sources 2, 4, 6. Note that the heat transferred through each resistance is ~10^{-8} watt—very small, as required. Each computed TR is for the point $XR + (DXR/2)$, $YR + (DYR/2)$. Resistances 1, 2, 3 correspond to source x, y locations 2, 4, 6, respectively. The temperature gradient through the board thickness is quite large, particularly at the 5-watt source, where a 39°C temperature difference is found between the front and rear surfaces.

Source 2 is used to make the convergence calculations shown in Fig. 7-30, from which it is clear that LMAX = MMAX = 30 is sufficient.

Temperatures are plotted in Figs. 7-31 and 7-32 for $\epsilon = 0.9$ and $\epsilon = 0.25$, respectively. The Y location for sources ($Q = 0$) 8, 9, 10

TWO SIDED EPOXY-GLASS CARD

TAMSIV-THERMAL ANALYZER FOR MULTILAYER SYSTEMS
NOV. 11, 1978 VERSION

THERMAL ANALYSIS FOR NEWTON'S LAW COOLING AT Z=0 AND C4
SOURCES AND LEADS AT Z=0

SUBSTRATE DIMENSIONS AND PHYSICAL CONSTANTS
A= 6.00000   B= 3.00000   T1= .00140   T2= .03100   T3= .03100   T4= .00140
H1= .8400E-02 H2= .8400E-02
K1= .1000E+02 K2= .7370E-02 K3= .7370E-02 K4= .1000E+02

NUMBER OF SOURCES= 10   NUMBER OF RES.= 3   NR1= 0   NR2= 3

LMAX= 30   MMAX= 30

TA= 0.0

SOURCE DATA

SOURCE NO. I	XS(I)	DXS(I)	YS(I)	DYS(I)	Q(I)
1	.6550	0.0000	1.2800	0.0000	0.000
2	1.3100	.4000	.9800	.6000	5.000
3	2.2650	0.0000	1.2800	0.0000	0.000
4	2.8200	.4000	.9800	.6000	2.500
5	3.6750	0.0000	1.2800	0.0000	0.000
6	4.3100	.4000	.9800	.6000	2.500

Fig. 7-29. Standard TAMS output for circuit card.

SOURCE DATA

SOURCE NO. I	XS(I)	DXS(I)	YS(I)	DYS(I)	Q(I)	TO(I)
7	5.3550	0.0000	1.2800	0.0000	0.000	
8	1.5600	0.0000	1.4800	0.0000	0.000	
9	2.8700	0.0000	1.4800	0.0000	0.000	
10	4.3600	0.0000	1.4800	0.0000	0.000	

RES. DATA

RES. NO. I	XR(I)	DXR(I)	YR(I)	DYR(I)	R(I)	TO(I)
1	1.3100	.4000	.9800	.6000	.100E+11	0.0
2	2.8200	.4000	.9800	.6000	.100E+11	0.0
3	4.3100	.4000	.9800	.6000	.100E+11	0.0

THERMAL FLUX IN RES. AND TEMPERATURES CALCULATED AT RES. PAD CENTERS

RES. NO. I	F(I)	TR(I)
1	.597E-08	59.7
2	.452E-08	45.2
3	.372E-08	37.2

TEMPERATURES CALCULATED AT SOURCE CENTERS

Fig. 7-29. (Continued)

SOURCE NO. I	TS(I) WITH SOURCES ONLY	TS(I) WITH SOURCES AND RES.
1	38.2	38.2
2	98.5	98.5
3	48.4	48.4
4	64.7	64.7
5	38.5	38.5
6	56.6	56.6
7	24.0	24.0
8	88.9	88.9
9	56.9	56.9
10	49.0	49.0

Fig. 7-29. (*Continued*)

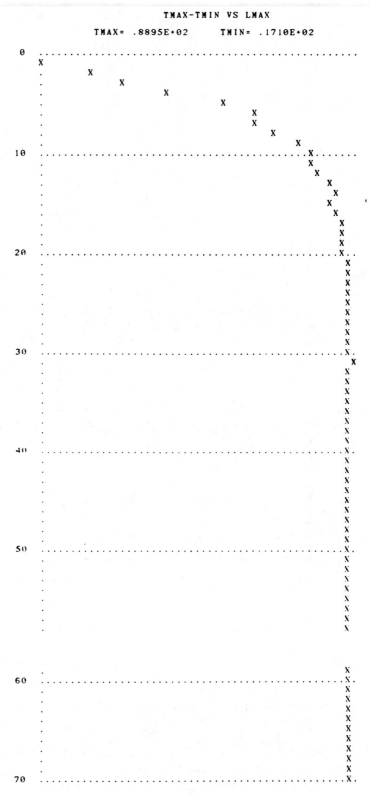

Fig. 7-30. TAMS convergence plot for circuit card, source no. 2 used.

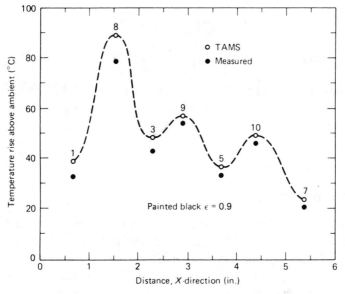

Fig. 7-31. Theoretical and experimental temperatures for double-sided, Cu-clad, epoxy-glass card. Number adjacent to TAMS result identifies TAMS source.

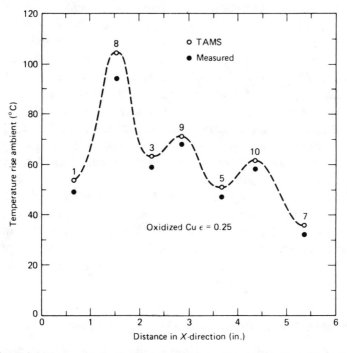

Fig. 7-32. Theoretical and experimental temperatures for double-sided, Cu-clad, epoxy-glass card. Number adjacent to TAMS result identifies TAMS source.

is slightly offset from 1, 3, 5, 7 to correspond with thermocouple mounting locations on the transistor tabs. The appropriate measurements are indicated in the figures. The experimental–theoretical correlation is rather good.

Chapter 8
Application of Digital Computer Methods II—Thermal Network Method

8.1 INTRODUCTION

Geometries not within the restrictions of the three-dimensional slab requirements of TAMS may be represented by a thermal network model. Such models may be used to describe adequately nearly all thermal packaging problems and have the advantage of being able to accommodate nonlinearities: temperature-dependent quantities such as thermal conductivity, natural convection heat transfer coefficients, radiation heat transfer coefficients, and the rather complex multiple surface radiation exchange phenomena. The program TNETFA (Chapter 10) includes these phenomena as well as others. Both steady state and time-dependent solutions are possible.

Nonthermal steady state network solutions are soluble with TNETFA if the proper variable identification is made. Two-dimensional laminar flow streamlines and velocity potentials may be calculated to provide semiquantitative estimates of airflow characteristics when the problem geometry is appropriate. Complex air pressure and volumetric flow rate solutions are possible for airflow/pressure networks that are sufficiently complex to make calculator solution difficult if not impossible.

This chapter is devoted to the illustration of thermal and airflow network modeling using TNETFA. A variety of problems are used to illustrate most TNETFA features. Those situations not addressed here may be studied by computing TNETFA solutions to problems also soluble by calculator and comparing answers to obtain user confidence.

8.2 PROBLEM—PIN FIN, STEADY STATE SOLUTION

This problem is sufficiently simple that it may be analyzed by the methods of Chapter 4, and therefore is interesting not because of complexity but because of the introductory aspect for basic input/output features of TNETFA. The pin fin geometry is illustrated in Fig. 8-1. A simple network model is shown in Fig. 8-2.

The pin is subdivided into six "nodes," all of equal length with

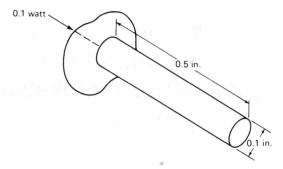

Fig. 8-1. Pin fin.

the exception of 1 and 6, which are half-length. The resultant conductive path lengths L are identical, however. The conduction conductances are for non-temperature-dependent copper and are computed to be:

$$C = kA_k/L$$

$$= k\pi r^2/L$$

$$= (10.0 \text{ watts/in.} \cdot {}^\circ\text{C})\pi(0.05 \text{ in.})^2/(0.1 \text{ in.})$$

$$= 0.785 \text{ watt/}^\circ\text{C}$$

This value is input as a string of conductances in line 9, Fig. 8-3.
The pin is cooled by natural convection and radiation, both
temperature-dependent elements, connected from the pin surface
to ambient node 7. Note that the temperature gradient through the
pin diameter is neglected; this is reflected in Fig. 8-2, which indicates
that no surface nodes are used. The program library element for
natural convection from a small device is used (the reader is cau-
tioned that the author has not actually empirically verified this element for devices this small—see Chapter 2). The surface areas are

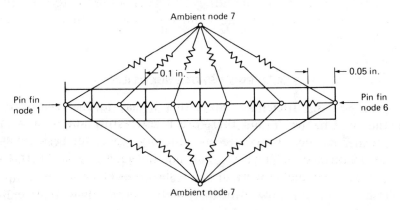

Fig. 8-2. Network model of pin fin.

DATA SET

```
1-D PIN FIN  ⎞──────────────────── 1
STEADY STATE ⎠
1    2    0  ─────────────────────── 2
7    1    1    0    3    4    0    1    0 ──── 3
20.0     0.0    ⎞
7    20.0       ⎟──────────────────── 4
1    20.0   0.1 ⎠
0    0  ──────────────────────────── 6
5    1    1    2    1    0.7850    0    ⎞
4    2    1    7    0    0.0314    101  ⎟──── 7
4    2    1    7    0    0.0251   -1    ⎠
1    7    0.0157    101  ⎞
6    7    0.0157    101  ⎟
1    7    0.0126   -1    ⎟──────────── 8
6    7    0.0126   -1    ⎠
6    0.1 ────────────────────────────── 11
200    1.9    0.001    200  ⎞
1.0    1.0                  ⎟────────── 14
200    1                    ⎠
```

Fig. 8-3.   TNETFA input for pin fin.

input in a string (line 10, Fig. 8-3) for nodes 2–5:

$$A_s = \pi d \Delta x$$

$$= \pi (0.1 \text{ in.})(0.1 \text{ in.})$$

$$= 0.0314 \text{ in.}^2$$

and individually input (lines 12, 13, Fig. 8-3) for nodes 1, 6:

$$A_s = 0.0314/2 = 0.0157 \text{ in.}^2$$

Note that the type of free convection and significant dimension are indicated in line 16, Fig. 8-3.

Simple two-surface radiation exchange to ambient is input using:

$$\mathcal{F}A_s = \epsilon A_s$$

$$= (0.8)(0.0314)$$

$$= 0.0251 \text{ in.}^2, \text{ nodes 2–5}$$

and:

$$\mathcal{F}A_s = 0.0126 \text{ in.}^2, \text{ nodes 1, 6}$$

The principal advantages of using a digital computation method are the speed and ease of obtaining a solution for a nonlinear problem. The reader may recall that the scientific calculator methods emphasized in previous chapters required computing the watts dissipated for several values of prescribed surface temperatures and then interpolating at proper power dissipation to obtain the corresponding temperature.

The iteration limits and output requests are specified in the last three lines of Fig. 8-3: a maximum of 200 iterations, temperature and energy balance output at 200 iterations (or at problem termination if earlier), problem termination at a maximum temperature change per iteration of 0.001°C. An overrelaxation constant of 1.9 was used. All nodes were started at 20.0°C, and node 7 is a constant-temperature node. Line 4 of the input specifies the number of lines, etc., per data set.

The pin fin problem output is reproduced in Fig. 8-4. The problem went the full 200 iterations with final energy balance of (0.002144/0.1)(100) = 2.1%. This is the percent energy (watts) unaccounted for in the iteration scheme. A perfect solution (mathematically) would have a zero energy balance. A few percent is usually adequate. One peculiarity of this particular solution is the very large number of iterations required of so few nodes; but sometimes this happens, and it is not always expected. Perhaps a better choice of the overrelaxation constant would have helped. Certainly a starting temperature closer to the final answer would have been better.

8.3 PROBLEM—HEAT SINKING BRACKET, STEADY STATE SOLUTION

A bracket used to heat five, 2-watt transistors is attached to a side panel 10 in. in length and 7 in. high in the manner shown in Fig. 8-5. The bracket side of the panel faces air at 40°C for a 20°C ambient on the exterior side of the panel. The approximate bracket temperatures in the vicinity of the transistor bases are to be estimated for natural convection cooling.

The thermal network model is shown in Fig. 8-6 with the nodal points indicated, but with conductance elements omitted for clarity. The lower half of the bracket is not included in the calculation because the symmetry permits the center of the middle transistor to be considered an adiabatic barrier. The two upper transistors each contain two nodes, the center (half a transistor) a single node. Nodes 1, 4, 5, 8, 9 therefore dissipate one watt each.

Conduction conductances connecting nodes 1–13 are:

$$C = kA_k/L$$

$$= (4.0 \text{ watts/in.} \cdot {}^\circ\text{C})(1.0 \text{ in.})(0.05 \text{ in.})/(0.25 \text{ in.})$$

$$= 0.80 \text{ watt/}{}^\circ\text{C}$$

for aluminum.

Node 13 is connected to the exterior ambient node 14 (T(14) = T_E = 20°C) via a conductance that represents a series addition of a spreading resistance at the bracket–panel contact and a panel to exterior convection resistance. Figure 4-17(a) is used to find kR_{sp}:

$$kR_{sp} \simeq 14 \text{ for } l = 0.25 \text{ in., } t = 0.04 \text{ in.}$$

$$R_{sp} = 14/(4.0) = 3.5°C/\text{watt}$$

NETWORK THERMAL ANALYSIS

1-D PIN FIN
STEADY STATE

UNITS=2

NUMBER OF NODES=   7   NUMBER OF CONDUCTORS=   34

NLOOP= 200   TPRINT= 200   NPRINT=   1   LOOPEN= 200   ALDT= .1000E-02   BETA= 1.90

NATURAL CONVECTION PARAMETER

1   SMALL SURFACE, VERTICAL   P= .1000E+00

                    LOOPCT=   0   MAXDT=   -1

                          TEMPERATURES

T(1)= .2000E+02   T(2)= .2000E+02   T(3)= .2000E+02   T(4)= .2000E+02   T(5)= .2000E+02   T(6)= .2000E+02
T(7)= .2000E+02

                    LOOPCT= 200   MAXDT= .1210E-01

                          TEMPERATURES

T(1)= .5152E+02   T(2)= .5142E+02   T(3)= .5133E+02   T(4)= .5128E+02   T(5)= .5124E+02   T(6)= .5124E+02
T(7)= .2000E+02

ENERGY BALANCE =-.2144E-02

                    LOOPCT= 200   MAXDT= .1210E-01

Fig. 84.   TNETFA output for pin fin.

```
DETAIL OF NODE   1        TEMPERATURE= .5152E+02    POWER= .1000E+00    STABILITY CONSTANT =    -I    CAP= .1000E-19

NODE  CTYPE  CMODE       C       CONDUCTANCE    FLUX     REYN NO.   BNDRY LAYER   HYDR DIA   HT TRANS COEF
 7     -1            .1260E-01   .5440E-04    .1715E-02                                        .4317E-02
 7    101     6      .1570E-01   .2587E-03    .8157E-02                                        .1648E-01
 2      0            .7850E+00   .7850E-02    .8524E-02
                                 NET TOTAL =  .9511E-01


DETAIL OF NODE   2        TEMPERATURE= .5142E+02    POWER=0.            STABILITY CONSTANT =    -I    CAP= .1000E-19

NODE  CTYPE  CMODE       C       CONDUCTANCE    FLUX     REYN NO.   BNDRY LAYER   HYDR DIA   HT TRANS COEF
 7     -1            .2510E-01   .1083E-03    .3402E-02                                        .4315E-02
 7    101     6      .3140E-01   .5168E-03    .1624E-01                                        .1646E-01
 3      0            .7850E+00   .7850E+00    .6522E-01
 1      0            .7850E+00   .7850E+00   -.8524E-01
                                 NET TOTAL =-.3843E-03


DETAIL OF NODE   3        TEMPERATURE= .5133E+02    POWER=0.            STABILITY CONSTANT =    -I    CAP= .1000E-19

NODE  CTYPE  CMODE       C       CONDUCTANCE    FLUX     REYN NO.   BNDRY LAYER   HYDR DIA   HT TRANS COEF
 7     -1            .2510E-01   .1083E-03    .3392E-02                                        .4313E-02
 7    101     6      .3140E-01   .5164E-03    .1618E-01                                        .1644E-01
 4      0            .7850E+00   .7850E+00    .4526E-01
 2      0            .7850E+00   .7850E+00   -.6522E-01
                                 NET TOTAL =-.3800E-03


DETAIL OF NODE   4        TEMPERATURE= .5128E+02    POWER=0.            STABILITY CONSTANT =    -I    CAP= .1000E-19

NODE  CTYPE  CMODE       C       CONDUCTANCE    FLUX     REYN NO.   BNDRY LAYER   HYDR DIA   HT TRANS COEF
 7     -1            .2510E-01   .1082E-03    .3385E-02                                        .4312E-02
 7    101     6      .3140E-01   .5160E-03    .1614E-01                                        .1643E-01
 5      0            .7850E+00   .7850E+00    .2536E-01
 3      0            .7850E+00   .7850E+00   -.4526E-01
                                 NET TOTAL =-.3758E-03


DETAIL OF NODE   5        TEMPERATURE= .5124E+02    POWER=0.            STABILITY CONSTANT =    -I    CAP= .1000E-19

NODE  CTYPE  CMODE       C       CONDUCTANCE    FLUX     REYN NO.   BNDRY LAYER   HYDR DIA   HT TRANS COEF
 7     -1            .2510E-01   .1082E-03    .3381E-02                                        .4311E-02
 7    101     6      .3140E-01   .5158E-03    .1612E-01                                        .1643E-01
```

Fig. 8-4. (Continued)

```
6    0         .7850E+00   .7850E+00    .5495E-02
4    0         .7850E+00   .7850E+00   -.2536E-01
                            NET TOTAL =-.3716E-03
```

DETAIL OF NODE 6   TEMPERATURE= .5124E+02   POWER=0.   STABILITY CONSTANT = -1   CAP= .1000E-19

NODE	CTYPE	CMODE	C	CONDUCTANCE	FLUX	REYN NO.	BNDRY LAYER	HYDR DIA	HT TRANS COEF
7	-1		.1260E-01	.5432E-04	.1697E-02				.4311E-02
7	101	6	.1570E-01	.2579E-03	.8056E-02				.1643E-01
5		6	.7850E+00	.7850E+00	-.5495E-02				
				NET TOTAL =	.4257E-02				

DETAIL OF NODE -7   TEMPERATURE= .2000E+02   POWER=0.   STABILITY CONSTANT = -1   CAP= .1000E-19

THIS IS A CONSTANT TEMPERATURE NODE

NODE	CTYPE	CMODE	C	CONDUCTANCE	FLUX	REYN NO.	BNDRY LAYER	HYDR DIA	HT TRANS. COEF
6	-1		.1260E-01	.5432E-04	-.1697E-02				.4311E-02
1	-1		.1260E-01	.5440E-04	-.1715E-02				.4317E-02
6	101	6	.1570E-01	.2579E-03	-.8056E-02				.1643E-01
1	101	6	.1570E-01	.2587E-03	-.8157E-02				.1648E-01
5	-1		.2510E-01	.1082E-03	-.3381E-02				.4311E-02
4	-1		.2510E-01	.1082E-03	-.3385E-02				.4312E-02
3	-1		.2510E-01	.1083E-03	-.3392E-02				.4313E-02
2	-1		.2510E-01	.1083E-03	-.3402E-02				.4315E-02
5	101	6	.3140E-01	.5158E-03	-.1612E-01				.1643E-01
4	101	6	.3140E-01	.5160E-03	-.1614E-01				.1643E-01
3	101	6	.3140E-01	.5164E-03	-.1618E-01				.1644E-01
2	101	6	.3140E-01	.5168E-03	-.1624E-01				.1646E-01
				NET TOTAL =	-.9786E-01				

ENERGY BALANCE =-.2144E-02

Fig. 8-4. (Continued)

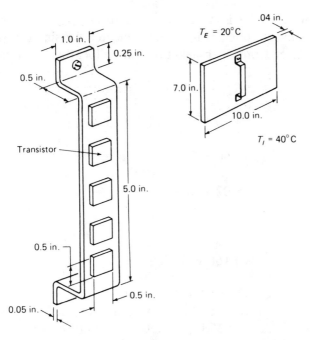

Fig. 8-5. Heat sinking bracket with five, 2-watt transistors.

Assuming a panel rise of about 10°C above T_E for a first guess:

$$R_c = 1/h_c A_s$$

$$\simeq 1/0.0024\left(\tfrac{10}{7}\right)^{0.25}(10\text{ in.})(7\text{ in.})$$

$$= 5.44°C/\text{watt}$$

$$R_r = 1/\epsilon h_r A_s$$

$$\simeq 1/(0.8)(0.0038)(70)$$

$$= 4.70°C/\text{watt}$$

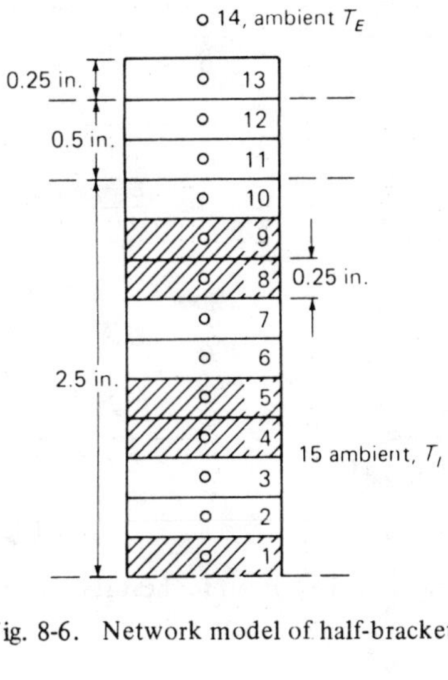

Fig. 8-6. Network model of half-bracket.

An emissivity of 0.8 is used for a thin vinyl-covered surface. Radiation and convection losses to T_I are neglected for simplification. The total resistance from the bracket mounting location is estimated:

$$R = R_{sp} + R_c R_r/(R_c + R_r)$$

$$= 3.50 + (5.44)(4.70)/(5.44 + 4.70)$$

$$= 3.50 + 2.52$$

$$= 6.02°C/watt$$

$$C = 1/6.02$$

$$= 0.166 \text{ watt}/°C$$

which is used between nodes 13 and 14.

Each side of the bracket contributes to the convection nodal area, $A_s = 2(1.0 \text{ in.})(0.25 \text{ in.}) = 0.50 \text{ in.}^2$. A surface emissivity $\epsilon = 0.1$ is sufficiently small to be neglected.

The problem input listing in Fig. 8-7 indicates a maximum of 50 iterations, an energy balance output every 10 iterations, problem termination when the maximum temperature change per iteration is $0.001°C$, and an overrelaxation constant $\beta = 1.8$.

The output is reproduced in Fig. 8-8. Note an energy balance at termination (43 iterations), $(0.1731 \times 10^{-4}/5)(100) = 3.5 \times 10^{-4}\%$. This is far fewer iterations than in the previous problem. The total flux of 4.24 watts to node 14, the external ambient, indicates that most of the transistor heat is carried to the outside, not convected to the cabinet interior, node 15.

DATA SET

```
1-D BRACKET WITH DISTRIBUTED TRANSISTORS          1
STEADY STATE
 1    2    0                                       2
15    2    5    0    2    1    0    1    0          3
50.0    0.0
14    20.0
15    40.0
 1    50.0    1.0
 4    50.0    1.0                                  4
 5    50.0    1.0
 8    50.0    1.0
 9    50.0    1.0
 0    0                                            6
12    1    1     2    1    0.80      0             7
12    1    1    15    0    0.50    101
13   14    0.166     0                             8
 1    5.0                                         11
50    1.8    0.001    10                          14
1.0    1.0
50    1
```

Fig. 8-7. TNETFA input for heat sinking bracket.

NETWORK THERMAL ANALYSIS

1-D BRACKET WITH DISTRIBUTED TRANSISTORS
STEADY STATE

UNITS=2

NUMBER OF NODES= 15   NUMBER OF CONDUCTORS= 50

NLOOP= 50   TPRINT= 50   NPRINT= 1   LOOPEN= 10   ALDT= .1000E-02   BETA= 1.80

NATURAL CONVECTION PARAMETER

1   VERTICAL FLAT PLATE OR CYLINDER   P = .5000E+01

LOOPCT= 0   MAXDT= -1

TEMPERATURES

T(1) = .5000E+02   T(2) = .5000E+02   T(3) = .5000E+02   T(4) = .5000E+02   T(5) = .5000E+02   T(6)F .5000E+02
T(7) = .5000E+02   T(8) = .5000E+02   T(9) = .5000E+02   T(10) = .5000E+02   T(11) = .5000E+02   T(12) = .5000E+02
T(13) = .5000E+02   T(14) = .2000E+02   T(15) = .4000E+02

ENERGY BALANCE = -.1399E+00

ENERGY BALANCE = -.1484E-01

ENERGY BALANCE = -.1163E-02

ENERGY BALANCE = -.1182E-03

LOOPCT= 43   MAXDT= .6409E-03

Fig. 8-8. TNETFA output for heat sinking bracket.

TEMPERATURES

```
T( 1)= .8559E+02   T( 2)= .8446E+02   T( 3)= .8344E+02   T( 4)= .8253E+02   T( 5)= .8047E+02   T( 6)= .7722E+02
T( 7)= .7415E+02   T( 8)= .7112E+02   T( 9)= .6691E+02   T(10)= .6151E+02   T(11)= .5616E+02   T(12)= .5084E+02
T(13)= .4554E+02   T(14)= .2000E+02   T(15)= .4000E+02
```

LOOPCT=   43   MAXDT= .6409E-03

DETAIL OF NODE   1   TEMPERATURE= .8559E+02   POWER= .1000E+01   STABILITY CONSTANT =   -1

```
NODE  CTYPE  CMODE      C        CONDUCTANCE      FLUX       REYN NO.    BNDRY LAYER   HYDR DIA    CAP= .1000E-19
 15    101     1    .5000E+00    .2041E-02     .9304E-01                                          HT TRANS COEF
  2     0           .8000E+00    .8000E+00     .9068E+00                                            .4081E-02
        0                                      NET TOTAL =   .9998E+00
```

DETAIL OF NODE   2   TEMPERATURE= .8446E+02   POWER=0.   STABILITY CONSTANT =   -1

```
NODE  CTYPE  CMODE      C        CONDUCTANCE      FLUX       REYN NO.    BNDRY LAYER   HYDR DIA    CAP= .1000E-19
 15    101     1    .5000E+00    .2029E-02     .9021E-01                                          HT TRANS COEF
  3     0           .8000E+00    .8000E+00     .8167E+00                                            .4058E-02
  1     0           .8000E+00    .8000E+00    -.9068E+00
                                              NET TOTAL =   .1100E-03
```

DETAIL OF NODE   3   TEMPERATURE= .8344E+02   POWER=0.   STABILITY CONSTANT =   -1

```
NODE  CTYPE  CMODE      C        CONDUCTANCE      FLUX       REYN NO.    BNDRY LAYER   HYDR DIA    CAP= .1000E-19
 15    101     1    .5000E+00    .2019E-02     .8768E-01                                          HT TRANS COEF
  4     0           .8000E+00    .8000E+00     .7292E+00                                            .4037E-02
  2     0           .8000E+00    .8000E+00    -.8167E+00
                                              NET TOTAL =   .1777E-03
```

DETAIL OF NODE   4   TEMPERATURE= .8253E+02   POWER= .1000E+01   STABILITY CONSTANT =   -1

```
NODE  CTYPE  CMODE      C        CONDUCTANCE      FLUX       REYN NO.    BNDRY LAYER   HYDR DIA    CAP= .1000E-19
 15    101     1    .5000E+00    .2009E-02     .8543E-01                                          HT TRANS COEF
  5     0           .8000E+00    .8000E+00     .1644E+01                                            .4018E-02
  3     0           .8000E+00    .8000E+00    -.7292E+00
                                              NET TOTAL =   .1000E+01
```

Fig. 8-8. (Continued)

DETAIL OF NODE 5   TEMPERATURE= .8047E+02   POWER= .1000E+01   STABILITY CONSTANT =   -1   CAP= .1000E-19

NODE	CTYPE	CMODE	C	CONDUCTANCE	FLUX	REYN NO.	BNDRY LAYER	HYDR DIA	HT TRANS COEF
15	101	1	.5000E+00	.1986E-02	.8039E-01				.3973E-02
6	0		.8000E+00	.8000E+00	.2563E+01				
4	0		.8000E+00	.8000E+00	-.1644E+01				
				NET TOTAL =	.1000E+01				

DETAIL OF NODE 6   TEMPERATURE= .7727E+02   POWER=0.   STABILITY CONSTANT =   -1   CAP= .1000E-19

NODE	CTYPE	CMODE	C	CONDUCTANCE	FLUX	REYN NO.	BNDRY LAYER	HYDR DIA	HT TRANS COEF
15	101	1	.5000E+00	.1949E-02	.7264E-01				.3898E-02
7	0		.8000E+00	.8000E+00	.2491E+01				
5	0		.8000E+00	.8000E+00	-.2563E+01				
				NET TOTAL =	.3871E-03				

DETAIL OF NODE 7   TEMPERATURE= .7415E+02   POWER=0.   STABILITY CONSTANT =   -1   CAP= .1000E-19

NODE	CTYPE	CMODE	C	CONDUCTANCE	FLUX	REYN NO.	BNDRY LAYER	HYDR DIA	HT TRANS COEF
15	101	1	.5000E+00	.1910E-02	.6524E-01				.3821E-02
8	0		.8000E+00	.8000E+00	.2425E+01				
6	0		.8000E+00	.8000E+00	-.2491E+01				
				NET TOTAL =	-.8124E-03				

DETAIL OF NODE 8   TEMPERATURE= .7112E+02   POWER= .1000E+01   STABILITY CONSTANT =   -1   CAP= .1000E-19

NODE	CTYPE	CMODE	C	CONDUCTANCE	FLUX	REYN NO.	BNDRY LAYER	HYDR DIA	HT TRANS COEF
15	101	1	.5000E+00	.1870E-02	.5819E-01				.3739E-02
9	0		.8000E+00	.8000E+00	.3367E+01				
7	0		.8000E+00	.8000E+00	-.2425E+01				
				NET TOTAL =	.9999E+00				

DETAIL OF NODE 9   TEMPERATURE= .6691E+02   POWER= .1000E+01   STABILITY CONSTANT =   -1   CAP= .1000E-19

NODE	CTYPE	CMODE	C	CONDUCTANCE	FLUX	REYN NO.	BNDRY LAYER	HYDR DIA	HT TRANS COEF
15	101	1	.5000E+00	.1807E-02	.4864E-01				.3614E-02
10	0		.8000E+00	.8000E+00	.4318E+01				
8	0		.8000E+00	.8000E+00	-.3367E+01				
				NET TOTAL =	.1000E+01				

Fig. 8-8. (Continued)

DETAIL OF NODE 10   TEMPERATURE= .6151E+02   POWER=0.   STABILITY CONSTANT =   -I   CAP= .1000E-19

NODE	CTYPE	CMODE	C	CONDUCTANCE	FLUX	REYN NO.	BNDRY LAYER	HYDR DIA	HT TRANS COEF
15	101	1	.5000E+00	.1714E-02	.3688E-01				.3428E-02
11	0	0	.8000E+00	.8000E+00	.4282E+01				
9	0	0	.8000E+00	.8000E+00	-.4318E+01				
					NET TOTAL = .1376E-03				

DETAIL OF NODE 11   TEMPERATURE= .5616E+02   POWER=0.   STABILITY CONSTANT =   -I   CAP= .1000E-19

NODE	CTYPE	CMODE	C	CONDUCTANCE	FLUX	REYN NO.	BNDRY LAYER	HYDR DIA	HT TRANS COEF
15	101	1	.5000E+00	.1601E-02	.2587E-01				.3201E-02
12	0	0	.8000E+00	.8000E+00	.4256E+01				
10	0	0	.8000E+00	.8000E+00	-.4282E+01				
					NET TOTAL = .1124E-03				

DETAIL OF NODE 12   TEMPERATURE= .5084E+02   POWER=0.   STABILITY CONSTANT =   -I   CAP= .1000E-19

NODE	CTYPE	CMODE	C	CONDUCTANCE	FLUX	REYN NO.	BNDRY LAYER	HYDR DIA	HT TRANS COEF
15	101	1	.5000E+00	.1453E-02	.1575E-01				.2906E-02
13	0	0	.8000E+00	.8000E+00	.4240E+01				
11	0	0	.8000E+00	.8000E+00	-.4256E+01				
					NET TOTAL = .4098E-04				

DETAIL OF NODE 13   TEMPERATURE= .4554E+02   POWER=0.   STABILITY CONSTANT =   -I   CAP= .1000E-19

NODE	CTYPE	CMODE	C	CONDUCTANCE	FLUX	REYN NO.	BNDRY LAYER	HYDR DIA	HT TRANS COEF
14	0	0	.1660E+00	.1660E+00	.4240E+01				
12	0	0	.8000E+00	.8000E+00	-.4240E+01				
					NET TOTAL =-.3733E-06				

DETAIL OF NODE -14   TEMPERATURE= .2000E+02   POWER=0.

THIS IS A CONSTANT TEMPERATURE NODE

NODE	CTYPE	CMODE	C	CONDUCTANCE	FLUX	REYN NO.	BNDRY LAYER	HYDR DIA	HT TRANS COEF
13	0		.1660E+00	.1660E+00	-.4240E+01				
					NET TOTAL =-.4240E+01				

Fig. 8-8. (*Continued*)

DETAIL OF NODE -15   TEMPERATURE= .4000E+02   POWER=0.   STABILITY CONSTANT =   -1   CAP= .1000E-19

THIS IS A CONSTANT TEMPERATURE NODE

NODE	CTYPE	CMODE	C	CONDUCTANCE	FLUX	REYN NO.	BNDRY LAYER	HYDR DIA	HT TRANS COEF
12	101	1	.5000E+00	.1453E-02	-.1575E-01				.2906E-02
11	101	1	.5000E+00	.1601E-02	-.2587E-01				.3201E-02
10	101	1	.5000E+00	.1714E-02	-.3688E-01				.3428E-02
9	101	1	.5000E+00	.1807E-02	-.4864E-01				.3614E-02
8	101	1	.5000E+00	.1870E-02	-.5819E-01				.3739E-02
7	101	1	.5000E+00	.1910E-02	-.6524E-01				.3821E-02
6	101	1	.5000E+00	.1949E-02	-.7264E-01				.3898E-02
5	101	1	.5000E+00	.1986E-02	-.8039E-01				.3973E-02
4	101	1	.5000E+00	.2009E-02	-.8543E-01				.4018E-02
3	101	1	.5000E+00	.2019E-02	-.8768E-01				.4037E-02
2	101	1	.5000E+00	.2029E-02	-.9021E-01				.4058E-02
1	101	1	.5000E+00	.2041E-02	-.9304E-01				.4081E-02

NET TOTAL =-.7599E+00

ENERGY BALANCE = .1731E-04

Fig. 8-8. (Continued)

8.4 PROBLEM—NATURAL CONVECTION COOLED PLASTIC CABINET

A very simple thermal network model for a portable electronic enclosure is illustrated in Fig. 8-9. The exterior surfaces transfer heat to a 20°C surrounding ambient via natural flat plate convection and simple radiation from a surface with an emissivity $\epsilon = 0.8$.

Circuit cards, components, and other hardware are distributed in a manner that is not expected seriously to impede the interior flow of convective air currents.* Natural flat plate convection is used to simulate heat transfer between the internal air and the cabinet surfaces. Radiation shielding by adjacent circuit cards usually permits the total neglect of this mechanism of heat transfer within a cabinet.

Wall conductors are used to account for a small temperature difference between the inner and outer walls. Interpanel heat conduction is neglected because of the very small conductances involved. A word of caution is in order here for thermal systems with relatively very large conductance elements such as one would have between interior and exterior surface nodes on metal walled cabinets. These elements should usually be neglected with programs such as TNETFA that use a Gauss-Seidel iterative method. This is best explained by considering the iteration equation E1.15 applied to a wall node:

$$T_i = \sum_{j \neq i} C_{ij} T_j \Big/ \sum_{j \neq i} C_{ij}$$

A thin metal wall conductance is usually very large compared with radiation and natural convection conductances. Suppose then that i is an interior surface node and two elements are involved: an interior convection element C_c and a wall conduction element C_k. Then:

$$T_i = (C_k T_w + C_c T_0)/(C_k + C_c)$$

where T_w and T_0 are the outer wall and interior air nodal temperatures. Since $C_k \gg C_c$ and $C_k T_w \gg C_c T_0$, each iteration of T_i will be determined very nearly as:

$$T_i \simeq \frac{C_k T_w}{C_k} = T_w$$

The problem with this and similar problems with some extremely dominant conductances is that it results in *very* slow convergence to a solution: *TNETFA does not like a mix of very large and very small conductances.*

---

*Systems with several vertical circuit cards that do not fully extend to top or bottom panels have been found to be described adequately by this rather simplistic model.

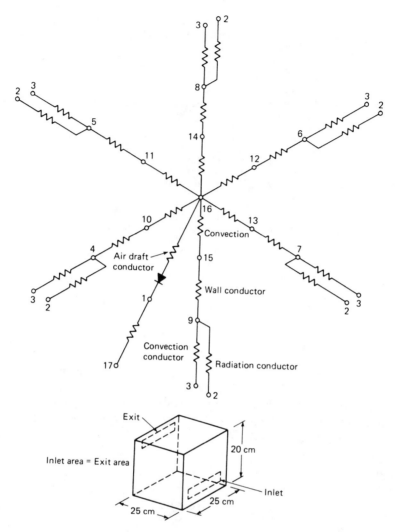

Fig. 8-9. Thermal network model for plastic walled cabinet. Reprinted from TNETFA program manual. Courtesy Tektronix, Inc., P.O. Box 500, Beaverton, Oregon 97077.

One of the few exceptions to this is demonstrated by the fixed-temperature node 17 connected to node 1 as in Fig. 8-9. In this case, node 17 is a fixed ambient air node, and the intervening element is intentionally given an arbitrary but very large value ($C = 100.0$).

The internal air node (16) is connected to node 1, but node 1 is not connected to node 16. This is emphasized by the symbol for a diode in Fig. 8-9, which permits heat transfer from node 16 to node 1. The rule for achieving the proper hook-up is followed in Fig. 8-10, line 21, the nodal, airflow conductance description. The downstream (in the airflow direction) node is listed first; the upstream, second.

Since component and circuit card details are neglected, the total power dissipation of 20 watts is considered to be generated at node 16, the interior air.

DATA SET

```
PLASTIC CABINET,  20 WATTS )____ 1
NON-VENTED
1   1   0--------------------------------- 2
17  3   1   0   10  2   0   3   0 ---- 3
30.0    0.0
2    20.0
3    20.0                )
17   20.0                )_____ 4
16   35.0   20.0         )
0    0  ------------------------------------- 6
4   2   0   4   1   400.0   -1
2   2   0   8   1   500.0   -1
4   3   0   4   1   500.0   101
1   3   0   8   0   625.0   102
1   3   0   9   0   625.0   103    )____ 7
4   4   1   10  1   6.670   0
2   8   1   14  1   8.330   0
4  16   0   10  1   500.0   101
1  16   0   14  0   625.0   102
1  16   0   15  0   625.0   103
16  1   0.001   30 1 )_____ 8
17  1   100.0   0
1    20.0
2    6.25  )_____ 11
3    6.25
100  1.0   0.0001   100 )
1    1                   )_____ 14
100  1
```

Fig. 8-10. TNETFA input for nonvented plastic cabinet.

The following description of conductance input requirements is consistent with the TNETFA rules of Chapter 10 and the input listings in Figs. 8-10 and 8-13:

Side wall radiation:

$$\epsilon A_s = (0.8)(500)$$

$$= 400 \text{ cm}^2$$

Top, bottom panel radiation:

$$\epsilon A_s = (0.8)(625) = 500 \text{ cm}^2$$

Side panel convection, interior, exterior:

$$A_s = 500 \text{ cm}^2$$

Top, bottom panel convection, interior, exterior:

$$A_s = 625 \text{ cm}^2$$

Note: side panel convection is via vertical flat plates; top panel exterior is facing upward and $T_s > T_A$, whereas interior is facing down-ward and $T_s < T_I$ (interior air); bottom panel exterior is facing down-ward and $T_s > T_A$, whereas interior is facing upward and $T_s < T_I$. The vertical plate dimensional parameter is 20 cm; the horizontal panel dimensional parameter is $(25)(25)/[2(25 + 25)] = 6.25$ cm.

The wall conduction elements for the vertical panels are obtained from:

$$C_k = kA_k/t$$

$$= (0.004 \text{ watt/cm} \cdot {}^\circ\text{C})(25 \text{ cm})(20 \text{ cm})/(0.3 \text{ cm})$$

$$= 6.67 \text{ watts/}^\circ\text{C}$$

The horizontal panel conduction elements are found in a similar manner to be $C_k = 8.33$ watts/°C. An airflow rate 0.001 (i.e., ≃ zero) cm³/sec between nodes 16 and 1 is used for the sealed enclosure.

The output for the sealed case is reproduced in Fig. 8-11, which shows a computed internal air temperature of 42.40°C in a 20°C am-bient. The detailed node output indicates that 11.25 watts were radiated and 8.76 watts convected from the cabinet exterior.

The ventilated version of this cabinet is analyzed in the following manner. The inlet/exit dominated airflow resistance is:

$$R = \frac{2.4 \times 10^{-3}}{A_I^2} + \frac{2.4 \times 10^{-3}}{A_E^2}$$

$$= 2.4 \times 10^{-3}[(1/A_I^2) + (1/A_E^2)]$$

If the vents are 4 cm × 20 cm and are 50% open:

$$A_I = [(4 \text{ cm})(20 \text{ cm})/(2.54 \text{ cm/in.})^2](0.5)$$

$$= 6.2 \text{ in.}^2$$

$$A_E = A_I = 6.2 \text{ in.}^2$$

$$R = 4.8 \times 10^{-3}/(6.2 \text{ in.}^2)^2$$

$$= 1.25 \times 10^{-4} \text{ in. } H_2O/(ft^3/min.)^2$$

which is used in the air draft iteration equation E6.15:

$$G' = G - \frac{(293RG + 0.909RQ)G^2 - 1.09 \times 10^{-3}dQ}{(909RG + 1.82RQ)G}$$

for a source dissipation heat distance $d \simeq 18$ cm $= 7.1$ in. This is used to compute an air draft G ft³/min. vs. Q watts heat transferred

NETWORK THERMAL ANALYSIS

PLASTIC CABINET, 20 WATTS
NON-VENTED

UNITS=1

NUMBER OF NODES= 17   NUMBER OF CONDUCTORS= 52

NLOOP= 100   TPRINT= 100   NPRINT= 100   LOOPEN= 1   ALDT= .1000E-03   BETA= 1.00

NATURAL CONVECTION PARAMETER

```
1  VERTICAL FLAT PLATE OR CYLINDER                                                         P= .2000E+02
2  HORIZONTAL FLAT PLATE OR CYLINDER, HEATED SIDE FACING UP OR COOLED SIDE FACING DOWN     P= .6250E+01
3  HORIZONTAL FLAT PLATE OR CYLINDER, HEATED SIDE FACING DOWN OR COOLED SIDE FACING UP     P= .6250E+01
```

```
                       LOOPCT=    0        MAXDT=      -1

                              TEMPERATURES

T(  1)= .3000E+02   T(  2)= .2000E+02   T(  3)= .2000E+02   T(  4)= .2000E+02   T(  5)= .3000E+02   T(  6)= .3000E+02
T(  7)= .3000E+02   T(  8)= .3000E+02   T(  9)= .3000E+02   T( 10)= .3000E+02   T( 11)= .3000E+02   T( 12)= .3000E+02
T( 13)= .3000E+02   T( 14)= .3000E+02   T( 15)= .3000E+02   T( 16)= .3500E+02   T( 17)= .2000E+02
```

```
                       LOOPCT=    100      MAXDT= .3069E-03

                              TEMPERATURES

T(  1)= .2000E+02   T(  2)= .2000E+02   T(  3)= .2000E+02   T(  4)= .2745E+02   T(  5)= .2745E+02   T(  6)= .2745E+02
T(  7)= .2745E+02   T(  8)= .2788E+02   T(  9)= .2627E+02   T( 10)= .2793E+02   T( 11)= .2793E+02   T( 12)= .2793E+02
T( 13)= .2793E+02   T( 14)= .2843E+02   T( 15)= .2659E+02   T( 16)= .4240E+02   T( 17)= .2000E+02
```

Fig. 8-11. TNETFA output for nonvented plastic cabinet.

ENERGY BALANCE = .8543E-02

LOOPCT= 100   MAXDT= .3069E-03

DETAIL OF NODE 1

TEMPERATURE= .2000E+02   POWER=0.   STABILITY CONSTANT = -1   CAP= .1000E-19

NODE	CTYPE	CMODE	C	CONDUCTANCE	FLUX	REYN NO.	BNDRY LAYER	HYDR DIA	HT TRANS COEF
17	0		.1000E+03	.1000E+03	0.				
				NET TOTAL =0.					

DETAIL OF NODE -2

TEMPERATURE= .2000E+02   POWER=0.   STABILITY CONSTANT = -1   CAP= .1000E-19

THIS IS A CONSTANT TEMPERATURE NODE

NODE	CTYPE	CMODE	C	CONDUCTANCE	FLUX	REYN NO.	BNDRY LAYER	HYDR DIA	HT TRANS COEF
9	-1		.5000E+03	.2945E+00	-.1846E+01				.5890E-03
8	-1		.5000E+03	.2969E+00	-.2339E+01				.5938E-03
7	-1		.4000E+03	.2370E+00	-.1767E+01				.5925E-03
6	-1		.4000E+03	.2370E+00	-.1767E+01				.5925E-03
5	-1		.4000E+03	.2370E+00	-.1767E+01				.5925E-03
4	-1		.4000E+03	.2370E+00	-.1767E+01				.5925E-03
				NET TOTAL =-.1125E+02					

DETAIL OF NODE -3

TEMPERATURE= .2000E+02   POWER=0.   STABILITY CONSTANT = -1   CAP= .1000E-19

THIS IS A CONSTANT TEMPERATURE NODE

NODE	CTYPE	CMODE	C	CONDUCTANCE	FLUX	REYN NO.	BNDRY LAYER	HYDR DIA	HT TRANS COEF
9	103	3	.6250E+03	.1377E+00	-.8632E+00				.2203E-03
8	102	2	.6250E+03	.2913E+00	-.2295E+01				.4661E-03
7	101	1	.5000E+03	.1878E+00	-.1400E+01				.3757E-03
6	101	1	.5000E+03	.1878E+00	-.1400E+01				.3757E-03
5	101	1	.5000E+03	.1878E+00	-.1400E+01				.3757E-03
4	101	1	.5000E+03	.1878E+00	-.1400E+01				.3757E-03
				NET TOTAL =-.8758E+01					

Fig. 8-11. (Continued)

DETAIL OF NODE 4

TEMPERATURE= .2745E+02   POWER=0.   STABILITY CONSTANT =   -I   CAP= .1000E-19

NODE	CTYPE	CMODE	C	CONDUCTANCE	FLUX	REYN NO.	BNDRY LAYER	HYDR DIA	HT TRANS COEF
10	0		.6670E+01	.6670E+01	-.3165E+01				
3	101	1	.5000E+03	.1878E+00	.1400E+01				.3757E-03
2		-1	.4000E+03	.2370E+00	.1767E+01				.5925E-03
				NET TOTAL =	.1143E-02				

DETAIL OF NODE 5

TEMPERATURE= .2745E+02   POWER=0.   STABILITY CONSTANT =   -I   CAP= .1000E-19

NODE	CTYPE	CMODE	C	CONDUCTANCE	FLUX	REYN NO.	BNDRY LAYER	HYDR DIA	HT TRANS COEF
11	0		.6670E+01	.6670E+01	-.3165E+01				
3	101	1	.5000E+03	.1878E+00	.1400E+01				.3757E-03
2		-1	.4000E+03	.2370E+00	.1767E+01				.5925E-03
				NET TOTAL =	.1143E-02				

DETAIL OF NODE 6

TEMPERATURE= .2745E+02   POWER=0.   STABILITY CONSTANT =   -I   CAP= .1000E-19

NODE	CTYPE	CMODE	C	CONDUCTANCE	FLUX	REYN NO.	BNDRY LAYER	HYDR DIA	HT TRANS COEF
12	0		.6670E+01	.6670E+01	-.3165E+01				
3	101	1	.5000E+03	.1878E+00	.1400E+01				.3757E-03
2		-1	.4000E+03	.2370E+00	.1767E+01				.5925E-03
				NET TOTAL =	.1143E-02				

DETAIL OF NODE 7

TEMPERATURE= .2745E+02   POWER=0.   STABILITY CONSTANT =   -I   CAP= .1000E-19

NODE	CTYPE	CMODE	C	CONDUCTANCE	FLUX	REYN NO.	BNDRY LAYER	HYDR DIA	HT TRANS COEF
13	0		.6670E+01	.6670E+01	-.3165E+01				
3	101	1	.5000E+03	.1878E+00	.1400E+01				.3757E-03
2		-1	.4000E+03	.2370E+00	.1767E+01				.5925E-03
				NET TOTAL =	.1143E-02				

DETAIL OF NODE 8

TEMPERATURE= .2788E+02   POWER=0.   STABILITY CONSTANT =   -I   CAP= .1000E-19

NODE	CTYPE	CMODE	C	CONDUCTANCE	FLUX	REYN NO.	BNDRY LAYER	HYDR DIA	HT TRANS COEF
14	0		.8330E+01	.8330E+01	-.4632E+01				
3	102	2	.6250E+03	.2913E+00	.2295E+01				.4661E-03
2		-1	.5000E+03	.2969E+00	.2339E+01				.5938E-03
				NET TOTAL =	.1265E-02				

Fig. 8-11. (*Continued*)

DETAIL OF NODE 9   STABILITY CONSTANT =   -I   CAP= .1000E-19

TEMPERATURE= .2627E+02   POWER=0.

NODE	CTYPE	CMODE	C	CONDUCTANCE	FLUX	REYN NO.	BNDRY LAYER	HYDR DIA	HT TRANS COEF
15	0		.8330E+01	.8330E+01	-.2706E+01				.2203E-03
3	103	3	.6250E+03	.1377E+00	.8632E+00				.5890E-03
2	-1		.5000E+03	.2945E+00	.1846E+01				
				NET TOTAL =	.2519E-02				

DETAIL OF NODE 10   STABILITY CONSTANT =   -I   CAP= .1000E-19

TEMPERATURE= .2793E+02   POWER=0.

NODE	CTYPE	CMODE	C	CONDUCTANCE	FLUX	REYN NO.	BNDRY LAYER	HYDR DIA	HT TRANS COEF
16	101	1	.5000E+03	.2187E+00	-.3165E+01				.4374E-03
4			.6670E+01	.6670E+01	.3165E+01				
				NET TOTAL =	.4033E-04				

DETAIL OF NODE 11   STABILITY CONSTANT =   -I   CAP= .1000E-19

TEMPERATURE= .2793E+02   POWER=0.

NODE	CTYPE	CMODE	C	CONDUCTANCE	FLUX	REYN NO.	BNDRY LAYER	HYDR DIA	HT TRANS COEF
16	101	1	.5000E+03	.2187E+00	-.3165E+01-				.4374E-03
5			.6670E+01	.6670E+01	.3165E+01				
				NET TOTAL =	.4033E-04				

DETAIL OF NODE 12   STABILITY CONSTANT =   -I   CAP= .1000E-19

TEMPERATURE= .2793E+02   POWER=0.

NODE	CTYPE	CMODE	C	CONDUCTANCE	FLUX	REYN NO.	BNDRY LAYER	HYDR DIA	HT TRANS COEF
16	101	1	.5000E+03	.2187E+00	-.3165E+01				.4374E-03
6			.6670E+01	.6670E+01	.3165E+01				
				NET TOTAL =	.4033E-04				

DETAIL OF NODE 13   STABILITY CONSTANT =   -I   CAP= .1000E-19

TEMPERATURE= .2793E+02   POWER=0.

NODE	CTYPE	CMODE	C	CONDUCTANCE	FLUX	REYN NO.	BNDRY LAYER	HYDR DIA	HT TRANS COEF
16	101	1	.5000E+03	.2187E+00	-.3165E+01				.4374E-03
7	0		.6670E+01	.6670E+01	.3165E+01				
				NET TOTAL =	.4033E-04				

Fig. 8-11. (Continued)

DETAIL OF NODE 14   TEMPERATURE= .2843E+02   POWER=0.   STABILITY CONSTANT =   -1   CAP= .1000E-19

NODE	CTYPE	CMODE	C	CONDUCTANCE	FLUX	REYN NO.	BNDRY LAYER	HYDR DIA	HT TRANS COEF
16	102	2	.6250E+03	.3316E+00	-.4632E+01				.5305E-03
8	0		.8330E+01	.8330E+01	.4632E+01				
				NET TOTAL =	.6284E-04				

DETAIL OF NODE 15   TEMPERATURE= .2659E+02   POWER=0.   STABILITY CONSTANT =   -1   CAP= .1000E-19

NODE	CTYPE	CMODE	C	CONDUCTANCE	FLUX	REYN NO.	BNDRY LAYER	HYDR DIA	HT TRANS COEF
16	103	3	.6250E+03	.1712E+00	-.2706E+01				.2739E-03
9	0		.8330E+01	.8330E+01	.2706E+01				
				NET TOTAL =	.2569E-04				

DETAIL OF NODE 16   TEMPERATURE= .4240E+02   POWER= .2000E+02   STABILITY CONSTANT =   -1   CAP= .1000E-19

NODE	CTYPE	CMODE	C	CONDUCTANCE	FLUX	REYN NO.	BNDRY LAYER	HYDR DIA	HT TRANS COEF
1	301		.1000E-02	.1176E-05	.2635E-04				
15	103	3	.6250E+03	.1712E+00	.2706E+01				.2739E-03
14	102	2	.6250E+03	.3316E+00	.4632E+01				.5305E-03
13	101	1	.5000E+03	.2187E+00	.3165E+01				.4374E-03
12	101	1	.5000E+03	.2187E+00	.3165E+01				.4374E-03
11	101	1	.5000E+03	.2187E+00	.3165E+01				.4374E-03
10	101	1	.5000E+03	.2187E+00	.3165E+01				.4374E-03
				NET TOTAL =	.2000E+02				

DETAIL OF NODE -17   TEMPERATURE= .2000E+02   POWER=0.   STABILITY CONSTANT =   -1   CAP= .1000E-19
THIS IS A CONSTANT TEMPERATURE NODE

NODE	CTYPE	CMODE	C	CONDUCTANCE	FLUX	REYN NO.	BNDRY LAYER	HYDR DIA	HT TRANS COEF
1	0		.1000E+03	.1000E+03	0.				
				NET TOTAL	=0.				

ENERGY BALANCE = .8543E-02

Fig. 8-11. (Continued)

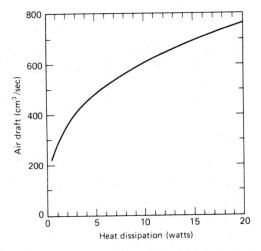

Fig. 8-12. Natural convection air draft vs. heat dissipated into air draft for plastic walled cabinet.

to the airstream. Using a conversion of:

$$G \text{ (cm}^3/\text{sec)} = \frac{G \text{ (ft}^3/\text{min.)}}{2.1 \times 10^{-3}}$$

Fig. 8-12 is plotted.

TNETFA and Fig. 8-12 are used in the following iterative fashion. A first guess of the air draft heat dissipation is made, say 5 watts. This implies a 480 cm³/sec air draft, and this value is used in the 21st line of the input (Fig. 8-13). TNETFA is then executed. The de-

DATA SET

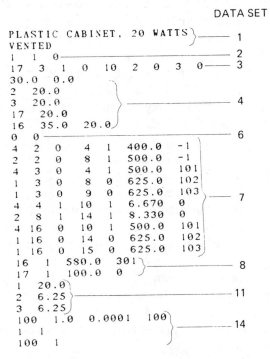

Fig. 8-13. TNETFA input for vented plastic cabinet.

tailed output for node 16 is searched for the flux (watts) from node 16 to node 1. This leads to a revised air draft. This procedure is repeated only a couple of times to arrive at a consistent air draft, heat input combination. The reproduced output in Fig. 8-14 indicates a final value of 580 cm³/sec air draft and an internal air at 33.33°C.

8.5 PROBLEM—SIMPLE HEAT SINK

This example is used to illustrate additional TNETFA features and modeling methods not used in the preceding examples. A simple natural convection, radiation cooled two-fin heat sink is shown in Fig. 8-15. All radiation to ambient is collected at node 2 (20°C), and all natural convection to ambient is collected at node 1 (20°C). Nodes 1 and 2 are defined as constant temperatures in the TNETFA input.

Nodes 3, 4, and 5 are the heat sink nodes, which are interconnected by conduction elements:

$$R_k = [0.25/(4.0)(0.2)(2.0)] + [0.5/(4.0)(0.1)(2.0)]$$

$$C_k = 1/R_k$$

$$= 1.28 \text{ watts}/°C$$

Convection from the exterior sides of nodes 3, 4, and 5 is accomplished using vertical flat plate library elements from TNETFA. The surface area is required for input. Convection from the interior surfaces is somewhat more complex because of the fin spacing effects. The ratio h_c/h_H from Figs. 5-14 through 5-17 for $\overline{\Delta T_s} = 10, 25, 50, 100°C$ is estimated, and from this, h_c computed. This interior convection is entered in TNETFA as a temperature-dependent variable array. TNETFA uses an average temperature of two nodes (in this problem, the relevant sink node and node 1) to linearly interpolate the array. Thus the array data must be entered in a consistent fashion. Table 8-1 summarizes the relevant data. Columns 3 and 6 contain the data used in the four data pairs in line 33, Fig. 8-16. h_H was computed from E2.13 using a fin height $H = 2.0$ in.

Several radiation shape factors must be obtained before the multisurface radiation exchange input is described. Referring to Fig. 8-15, it is clear that the geometry consists of plates parallel and perpendicular to each other, the shape factors for which may be obtained with sufficient accuracy from Figs. 3-6 and 3-7. Surfaces a, b, c represent the ambient (node 2) to which the interior of the sink radiates. The nonzero F_{ij} are listed in Table 8-2. $F_{ii} = 0$ for $i = 3, 4, 5$. The reader is advised to work out the details of the radiation input as an exercise.

NETWORK THERMAL ANALYSIS

PLASTIC CABINET, 20 WATTS
VENTED

UNITS=1

NUMBER OF NODES= 17   NUMBER OF CONDUCTORS= 52

NLOOP= 100   TPRINT= 100   NPRINT= 1   LOOPEN= 100   ALDT= .1000E-03   BETA= 1.00

NATURAL CONVECTION PARAMETER

```
1   VERTICAL FLAT PLATE OR CYLINDER                                                          P= .2000E+02
2   HORIZONTAL FLAT PLATE OR CYLINDER, HEATED SIDE FACING UP OR COOLED SIDE FACING DOWN      P= .6250E+01
3   HORIZONTAL FLAT PLATE OR CYLINDER, HEATED SIDE FACING DOWN OR COOLED SIDE FACING UP      P= .6250E+01
```

LOOPCT=   0   MAXDT=   -1

TEMPERATURES

```
T(  1)= .3000E+02   T(  2)= .2000E+02   T(  3)= .2000E+02   T(  4)= .3000E+02   T(  5)= .3000E+02   T(  6)= .3000E+02
T(  7)= .3000E+02   T(  8)= .3000E+02   T(  9)= .3000E+02   T( 10)= .3000E+02   T( 11)= .3000E+02   T( 12)= .3000E+02
T( 13)= .3000E+02   T( 14)= .3000E+02   T( 15)= .3000E+02   T( 16)= .3500E+02   T( 17)= .2000E+02
```

LOOPCT=   100   MAXDT= .5933E-03

TEMPERATURES

```
T(  1)= .2000E+02   T(  2)= .2000E+02   T(  3)= .2000E+02   T(  4)= .2430E+02   T(  5)= .2430E+02   T(  6)= .2430E+02
T(  7)= .2430E+02   T(  8)= .2457E+02   T(  9)= .2357E+02   T( 10)= .2456E+02   T( 11)= .2456E+02   T( 12)= .2456E+02
T( 13)= .2456E+02   T( 14)= .2487E+02   T( 15)= .2374E+02   T( 16)= .3333E+02   T( 17)= .2000E+02
```

Fig. 8-14. TNETFA output for vented plastic cabinet.

ENERGY BALANCE = .1328E-01   MAXDT= .5933E-03

LOOPCT= 100

DETAIL OF NODE   1   TEMPERATURE= .2000E+02   POWER=0.   STABILITY CONSTANT =   -1   CAP= .1000E-19

NODE	CTYPE	CMODE	C	CONDUCTANCE	FLUX	REYN NO.	BNDRY LAYER	HYDR DIA	HT TRANS COEF
17	0		.1000E+03	.1000E+03	0.				
				NET TOTAL =0.					

DETAIL OF NODE -2   TEMPERATURE= .2000E+02   POWER=0.   STABILITY CONSTANT =   -1   CAP= .1000E-19

THIS IS A CONSTANT TEMPERATURE NODE

NODE	CTYPE	CMODE	C	CONDUCTANCE	FLUX	REYN NO.	BNDRY LAYER	HYDR DIA	HT TRANS COEF
9	-1		.5000E+03	.2904E+00	-.1036E+01				.5809E-03
8	-1		.5000E+03	.2919E+00	-.1333E+01				.5839E-03
7	-1		.4000E+03	.2332E+00	-.1003E+01				.5831E-03
6	-1		.4000E+03	.2332E+00	-.1003E+01				.5831E-03
5	-1		.4000E+03	.2332E+00	-.1003E+01				.5831E-03
4	-1		.4000E+03	.2332E+00	-.1003E+01				.5831E-03
				NET TOTAL =-.6380E+01					

DETAIL OF NODE -3   TEMPERATURE= .2000E+02   POWER=0.   STABILITY CONSTANT =   -1   CAP= .1000E-19

THIS IS A CONSTANT TEMPERATURE NODE

NODE	CTYPE	CMODE	C	CONDUCTANCE	FLUX	REYN NO.	BNDRY LAYER	HYDR DIA	HT TRANS COEF
9	103	3	.6250E+03	.1198E+00	-.4275E+00				.1917E-03
8	102	2	.6250E+03	.2547E+00	-.1163E+01				.4076E-03
7	101	1	.5000E+03	.1640E+00	-.7051E+00				.3280E-03
6	101	1	.5000E+03	.1640E+00	-.7051E+00				.3280E-03
5	101	1	.5000E+03	.1640E+00	-.7051E+00				.3280E-03
4	101	1	.5000E+03	.1640E+00	-.7051E+00				.3280E-03
				NET TOTAL =-.4411E+01					

DETAIL OF NODE   4   TEMPERATURE= .2430E+02   POWER=0.   STABILITY CONSTANT =   -1   CAP= .1000E-19

NODE	CTYPE	CMODE	C	CONDUCTANCE	FLUX	REYN NO.	BNDRY LAYER	HYDR DIA	HT TRANS COEF
10	0		.6670E+01	.6670E+01	-.1706E+01				

Fig. 8-14. (Continued)

```
  3  101   1      .5000E+03    .1640E+00    .7051E+00                                          .3280E-03
  2   -1           .4000E+03    .2332E+00    .1003E+01                                          .5831E-03
                                NET TOTAL =  .1655E-02


DETAIL OF NODE  5      TEMPERATURE= .2430E+02     POWER=0.     STABILITY CONSTANT =  -1     CAP = .1000E-19

NODE CTYPE CMODE        C        CONDUCTANCE      FLUX      REYN NO.   BNDRY LAYER   HYDR DIA   HT TRANS COEF
 11    0             .6670E+01    .6670E+01    -.1706E+01
  3  101   1         .5000E+03    .1640E+00     .7051E+00                                          .3280E-03
  2   -1             .4000E+03    .2332E+00     .1003E+01                                          .5831E-03
                                  NET TOTAL =   .1655E-02


DETAIL OF NODE  6      TEMPERATURE= .2430E+02     POWER=0.     STABILITY CONSTANT =  -1     CAP = .1000E-19

NODE CTYPE CMODE        C        CONDUCTANCE      FLUX      REYN NO.   BNDRY LAYER   HYDR DIA   HT TRANS COEF
 12    0             .6670E+01    .6670E+01    -.1706E+01
  3  101   1         .5000E+03    .1640E+00     .7051E+00                                          .3280E-03
  2   -1             .4000E+03    .2332E+00     .1003E+01                                          .5831E-03
                                  NET TOTAL =   .1655E-02


DETAIL OF NODE  7      TEMPERATURE= .2430E+02     POWER=0.     STABILITY CONSTANT =  -1     CAP = .1000E-19

NODE CTYPE CMODE        C        CONDUCTANCE      FLUX      REYN NO.   BNDRY LAYER   HYDR DIA   HT TRANS COEF
 13    0             .6670E+01    .6670E+01    -.1706E+01
  3  101   1         .5000E+03    .1640E+00     .7051E+00                                          .3280E-03
  2   -1             .4000E+03    .2332E+00     .1003E+01                                          .5831E-03
                                  NET TOTAL =   .1655E-02


DETAIL OF NODE  8      TEMPERATURE= .2457E+02     POWER=0.     STABILITY CONSTANT =  -1     CAP = .1000E-19

NODE CTYPE CMODE        C        CONDUCTANCE      FLUX      REYN NO.   BNDRY LAYER   HYDR DIA   HT TRANS COEF
 14    0             .8330E+01    .8330E+01    -.2495E+01
  3  102   2         .6250E+03    .2547E+00     .1163E+01                                          .4076E-03
  2   -1             .5000E+03    .2919E+00     .1333E+01                                          .5839E-03
                                  NET TOTAL =   .1658E-02
```

Fig. 8-14. (*Continued*)

DETAIL OF NODE 9

STABILITY CONSTANT = -1 CAP= .1000E-19

REYN NO.	BNDRY LAYER	HYDR DIA	HT TRANS COEF
			.1917E-03
			.5809E-03

TEMPERATURE= .2357E+02 POWER=0.

NODE	CTYPE	CMODE	C	CONDUCTANCE	FLUX
15	0	3	.8330E+01	.8330E+01	-.1459E+01
3	103		.6250E+03	.1198E+00	.4275E+00
2	-1		.5000E+03	.2904E+00	.1036E+01
				NET TOTAL =	.4847E-02

DETAIL OF NODE 10

STABILITY CONSTANT = -1 CAP= .1000E-19

REYN NO.	BNDRY LAYER	HYDR DIA	HT TRANS COEF
			.3888E-03

TEMPERATURE= .2456E+02 POWER=0.

NODE	CTYPE	CMODE	C	CONDUCTANCE	FLUX
16	101	1	.5000E+03	.1944E+00	-.1706E+01
4	0		.6670E+01	.6670E+01	.1706E+01
				NET TOTAL =	.3439E-04

DETAIL OF NODE 11

STABILITY CONSTANT = -1 CAP= .1000E-19

REYN NO.	BNDRY LAYER	HYDR DIA	HT TRANS COEF
			.3888E-03

TEMPERATURE= .2456E+02 POWER=0.

NODE	CTYPE	CMODE	C	CONDUCTANCE	FLUX
16	101	1	.5000E+03	.1944E+00	-.1706E+01
5	0		.6670E+01	.6670E+01	.1706E+01
				NET TOTAL =	.3439E-04

DETAIL OF NODE 12

STABILITY CONSTANT = -1 CAP= .1000E-19

REYN NO.	BNDRY LAYER	HYDR DIA	HT TRANS COEF
			.3888E-03

TEMPERATURE= .2456E+02 POWER=0.

NODE	CTYPE	CMODE	C	CONDUCTANCE	FLUX
16	101	1	.5000E+03	.1944E+00	-.1706E+01
6	0		.6670E+01	.6670E+01	.1706E+01
				NET TOTAL =	.3439E-04

DETAIL OF NODE 13

STABILITY CONSTANT = -1 CAP= .1000E-19

REYN NO.	BNDRY LAYER	HYDR DIA	HT TRANS COEF
			.3888E-03

TEMPERATURE= .2456E+02 POWER=0.

NODE	CTYPE	CMODE	C	CONDUCTANCE	FLUX
16	101	1	.5000E+03	.1944E+00	-.1706E+01
7	0		.6670E+01	.6670E+01	.1706E+01
				NET TOTAL =	.3439E-04

Fig. 8-14. (*Continued*)

DETAIL OF NODE 14   TEMPERATURE= .2487E+02   POWER=0.   STABILITY CONSTANT =   -1   CAP= .1000E-19

NODE	CTYPE	CMODE	C	CONDUCTANCE	FLUX	REYN NO.	BNDRY LAYER	HYDR DIA	HT TRANS COEF
16	102	2	.6250E+03	.2948E+00	-.2495E+01				.4716E-03
8	0		.8330E+01	.8330E+01	.2495E+01				
				NET TOTAL =	.5586E-04				

DETAIL OF NODE 15   TEMPERATURE= .2374E+02   POWER=0.   STABILITY CONSTANT =   -1   CAP= .1000E-19

NODE	CTYPE	CMODE	C	CONDUCTANCE	FLUX	REYN NO.	BNDRY LAYER	HYDR DIA	HT TRANS COEF
16	103	3	.6250E+03	.1521E+00	-.1459E+01				.2434E-03
9	0		.8330E+01	.8330E+01	.1459E+01				
				NET TOTAL =	.1386E-04				

DETAIL OF NODE 16   TEMPERATURE= .3333E+02   POWER= .2000E+02   STABILITY CONSTANT =   -1   CAP= .1000E-19

NODE	CTYPE	CMODE	C	CONDUCTANCE	FLUX	REYN NO.	BNDRY LAYER	HYDR DIA	HT TRANS COEF
1	301		.5800E+03	.6918E+00	.9222E+01				
15	103	3	.6250E+03	.1521E+00	.1459E+01				.2434E-03
14	102	2	.6250E+03	.2948E+00	.2495E+01				.4716E-03
13	101	1	.5000E+03	.1944E+00	.1706E+01				.3888E-03
12	101	1	.5000E+03	.1944E+00	.1706E+01				.3888E-03
11	101	1	.5000E+03	.1944E+00	.1706E+01				.3888E-03
10	101	1	.5000E+03	.1944E+00	.1706E+01				.3888E-03
				NET TOTAL =	.2000E+02				

DETAIL OF NODE -17   TEMPERATURE= .2000E+02   POWER=0.   STABILITY CONSTANT =   -1   CAP= .1000E-19
THIS IS A CONSTANT TEMPERATURE NODE

NODE	CTYPE	CMODE	C	CONDUCTANCE	FLUX	REYN NO.	BNDRY LAYER	HYDR DIA	HT TRANS COEF
1	0		.1000E+03	.1000E+03	0.			.1328E-01	
				NET TOTAL =	=0.				

ENERGY BALANCE = .1328E-01

Fig. 8-14. (Continued)

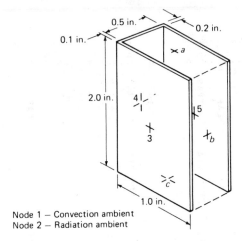

Node 1 — Convection ambient
Node 2 — Radiation ambient

Fig. 8-15. Simple heat sink.

Actual TNETFA input requires the angle factor, surface area product, which is trivially obtained from Table 8-2. The surface areas and emissivities are listed in lines 28–31, Fig. 8-16. The surface area of node 2 is the sum of surfaces a, b, c, or 2.0 in.^{2}. Note that multi-surface radiation exchange is coded into the node connection scheme by a CTYPE $= -2$.

Three different solutions were obtained from TNETFA:

1. Steady state; input Fig. 8-16, output Fig. 8-17.
2. Transient, power on at $t = 0$ sec; input Fig. 8-18, output Fig. 8-19.
3. Transient, time variable power; input Fig. 8-21, partial output Fig. 8-22.

The total heat dissipation is 5 watts at node 4 in all three cases.

The transient solutions require a capacitance for each non-constant-temperature node. The capacitance is also included in the steady state input although it is not used. From Chapter 10, the

Table 8-1. Convection input for simple heat sink.

$\Delta \bar{T}_s$ (°C)	$\bar{T}_s$ (°C)	$\dfrac{\bar{T}_s + T_A}{2}$ (°C)	$\dfrac{h_c}{h_H}$	h_H (watts/in.$^2 \cdot$°C)	h_c (watts/in.$^2 \cdot$°C)
10.0	30.0	25.0	0.92	0.0036	0.0033
25.0	45.0	32.5	0.96	0.0045	0.0043
50.0	70.0	45.0	0.97	0.0054	0.0052
100.0	120.0	70.0	0.98	0.0064	0.0063

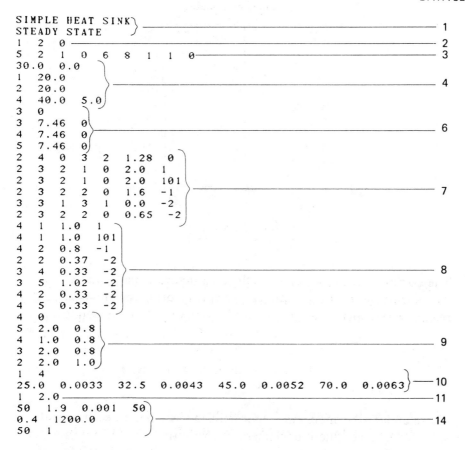

```
SIMPLE HEAT SINK⎫
STEADY STATE    ⎭ ─────────────────────────────────────────── 1
1   2   0 ──────────────────────────────────────────────────── 2
5   2   1   0   6   8   1   1   0─────────────────────────────── 3
30.0    0.0
1   20.0                ⎫
2   20.0                ⎬─────────────────────────────────────── 4
4   40.0    5.0         ⎭
3   0
3   7.46    0 ⎫
4   7.46    0 ⎬───────────────────────────────────────────────── 6
5   7.46    0 ⎭
2   4   0   3   2   1.28    0  ⎫
2   3   2   1   0   2.0     1  ⎪
2   3   2   1   0   2.0     101⎪
2   3   2   2   0   1.6    -1  ⎬──────────────────────────────── 7
3   3   1   3   1   0.0    -2  ⎪
2   3   2   2   0   0.65   -2  ⎭
4   1   1.0    1   ⎫
4   1   1.0    101 ⎪
4   2   0.8   -1   ⎪
2   2   0.37  -2   ⎪
3   4   0.33  -2   ⎬─────────────────────────────────────────── 8
3   5   1.02  -2   ⎪
4   2   0.33  -2   ⎪
4   5   0.33  -2   ⎭
4   0             ⎫
5   2.0    0.8    ⎪
4   1.0    0.8    ⎬─────────────────────────────────────────── 9
3   2.0    0.8    ⎪
2   2.0    1.0    ⎭
1   4
25.0    0.0033    32.5    0.0043    45.0    0.0052    70.0    0.0063⎫─ 10
1   2.0 ──────────────────────────────────────────────────────────── 11
50   1.9    0.001    50⎫
0.4  1200.0           ⎬───────────────────────────────────────────── 14
50   1                ⎭
```

Fig. 8-16. TNETFA input for simple heat sink–steady state.

TNETFA manual, the capacitance is specified as:

$$CAP \, (\text{joules}/°C) = \text{density} \times \text{specific heat} \times \text{volume}$$

$$= (2.71 \text{ gm/cm}^3)(0.84 \text{ joules}/°C \cdot \text{gm}) \Delta V(\text{in.}^3)$$

$$\cdot (2.54 \text{ cm/in.})^3$$

$$= 37.3 \Delta V(\text{in.}^3)$$

$$= 7.46 \text{ joules}/°C$$

for nodes 3, 4, 5. This is input in lines 10–12, Fig. 8-16.

A maximum time step Δt is estimated from the stability criteria, E1.21:

Table 8-2. Nonzero shape factors
for simple heat sink.

Surface i	Surface j	F_{ij}
3	4	0.17
3	5	0.51
3	a	0.079
3	b	0.17
3	c	0.079
4	5	0.33
4	a	0.084
4	b	0.17
4	c	0.084
4	2, i.e., $a + b + c$	0.33

$$\Delta t \leqslant CAP \Big/ \sum_{j \neq i} C_{ij}$$

The complex nature of some of the conductances, particularly radiation exchange, precludes an easy estimate of a maximum Δt. A
recommended procedure is to start with a very rough estimate, e.g.:

$$\Delta t \sim 7.46/6(1.28)$$

$$= 0.97 \text{ sec}$$

where about six conductances per node are assumed to each have the
value of the conduction element 1.28 watts/°C. The program is
then run for a couple of time steps and a detailed node printout requested. The stability constant $STAB$ is checked to ensure that it is
less than unity for each non-constant-temperature node. If $STAB$ is
greater than 1.0 for any node, Δt may be proportionately reduced
to ensure $STAB \leqslant 1.0$. The complete solution may then be computed. An actual time step of 0.4 sec was used. The largest $STAB$
occurs at node 4 as 0.14.

The first transient problem starts all nodes at the 20°C ambient.
The 5.0-watt source is turned on at the beginning of the problem.
Problem termination is set at 1200.0 sec. Temperature printouts are
requested every 400 time steps or (400)(0.4) = 160 sec. The final
temperature of node 4, 82.26°C, is a slight overshoot of the previously computed 81.52°C steady state solution.

The appropriate input for a time variable source (node 4) is listed
in Fig. 8-21 via a source vs. time array using lines 9 and 10. Line 9
specifies node 4 and three time, power pairs. The source is 5.0 watts
from $t = 0$ through $t = 60$ sec. At 61 sec, the source is set at 0.0
watts. Since all time and temperature arrays in TNETFA use linear
interpolation between specified time intervals or temperature values,
a ramp shutoff between $t = 60$ and 61 sec is defined by this array.

NETWORK THERMAL ANALYSIS

SIMPLE HEAT SINK
STEADY STATE

UNITS=2

NUMBER OF NODES= 5   NUMBER OF CONDUCTORS= 38

NLOOP= 50   TPRINT= 50   NPRINT= 1   LOOPEN= 50   ALDT= .1000E-02   BETA= 1.90

ARRAY DATA

ARRAY 1   4 X-Y PAIRS

.2500E+02, .3300E-02   .3250E+02, .4300E-02   .4500E+02, .5200E-02   .7000E+02, .6300E-02

NATURAL CONVECTION PARAMETER

1   VERTICAL FLAT PLATE OR CYLINDER   P = .2000E+01

LOOPCT= 0   MAXDT=   -I

TEMPERATURES

T(1)= .2000E+02   T(2)= .2000E+02   T(3)= .3000E+02   T(4)= .4000E+02   T(5)= .3000E+02

LOOPCT= 50   MAXDT= .1977E+00

TEMPERATURES

T(1)= .2000E+02   T(2)= .2000E+02   T(3)= .7986E+02   T(4)= .8152E+02   T(5)= .7992E+02

ENERGY BALANCE =-.3533E-01

LOOPCT= 50   MAXDT= .1977E+00

Fig. 8-17. TNETFA output for simple heat sink—steady state.

DETAIL OF NODE -1

TEMPERATURE= .2000E+02   POWER=0.   STABILITY CONSTANT =   -1   CAP= .1000E-19

THIS IS A CONSTANT TEMPERATURE NODE

NODE	CTYPE	CMODE	C	CONDUCTANCE	FLUX	REYN NO.	BNDRY LAYER	HYDR DIA	HT TRANS COEF
4	101	1	.1000E+01	.5605E-02	-.3449E+00				.5605E-02
4	1		.1000E+01	.5453E-02	-.3355E+00				
5	101	1	.2000E+01	.1115E-01	-.6679E+00				.5573E-02
3	101	1	.2000E+01	.1114E-01	-.6672E+00				.5572E-02
5	1		.2000E+01	.1084E-01	-.6493E+00				
3	1		.2000E+01	.1083E-01	-.6486E+00				
				NET TOTAL =-.3313E+01					

DETAIL OF NODE -2

TEMPERATURE= .2000E+02   POWER=0.   STABILITY CONSTANT =   -1   CAP= .1000E-19

THIS IS A CONSTANT TEMPERATURE NODE

MULTI-SURFACE RADIATION EXCHANGE AREA= .2000E+01   EMISSIVITY=1.000

NODE	CTYPE	CMODE	C	CONDUCTANCE	FLUX	REYN NO.	BNDRY LAYER	HYDR DIA	HT TRANS COEF
4	-2		.3300E+00	.1515E-02	-.9318E-01				.4590E-02
2	-2		.3700E+00	0.	0.				0.
4	-1		.8000E+00	.4007E-02	-.2465E+00				.5009E-02
5	-2		.6500E+00	.2990E-02	-.1792E+00				.4600E-02
3	-2		.6500E+00	.2990E-02	-.1790E+00				.4600E-02
5	-1		.1600E+01	.7952E-02	-.4764E+00				.4970E-02
3	-1		.1600E+01	.7950E-02	-.4759E+00				.4969E-02
				NET TOTAL =-.1650E+01					

DETAIL OF NODE 3

TEMPERATURE= .7986E+02   POWER=0.   STABILITY CONSTANT =   -1   CAP= .7460E+01

MULTI-SURFACE RADIATION EXCHANGE AREA= .2000E+01   EMISSIVITY= .800

NODE	CTYPE	CMODE	C	CONDUCTANCE	FLUX	REYN NO.	BNDRY LAYER	HYDR DIA	HT TRANS COEF
5	-2		.1020E+01	.4760E-02	-.2478E-03				.4667E-02
4	-2		.3300E+00	.1390E-02	-.2305E-02				.4211E-02
2	-2		.6500E+00	.2990E-02	.1790E+00				.4600E-02
2	-1		.1600E+01	.7950E-02	.4759E+00				.4969E-02
1	101	1	.2000E+01	.1114E-01	.6672E+00				.5572E-02
1	0		.2000E+01	.1083E-01	.6486E+00				
4	0		.1280E+01	.1280E+01	-.2123E+01				
				NET TOTAL =-.1552E+01					

Fig. 8-17. (Continued)

DETAIL OF NODE 4   TEMPERATURE= .8152E+02   POWER= .5000E+01   STABILITY CONSTANT =   -1   CAP= .7460E+01

MULTI-SURFACE RADIATION EXCHANGE AREA= .1000E+01   EMISSIVITY= .800

NODE	CTYPE	CMODE	C	CONDUCTANCE	FLUX	REYN NO.	BNDRY LAYER	HYDR DIA	HT TRANS COEF
5	-2		.3300E+00	.1385E-02	.2225E-02				.4196E-02
2	-2		.3300E+00	.1515E-02	.9318E-01				.4590E-02
3	-2		.3300E+00	.1390E-02	.2305E-02				.4211E-02
2	-1		.8000E+00	.4007E-02	.2465E+00				.5009E-02
1	101	1	.1000E+01	.5605E-02	.3449E+00				.5605E-02
1	-1		.1000E+01	.5453E-02	.3355E+00				
5	0		.1280E+01	.1280E+01	.2057E+01				
3	0		.1280E+01	.1280E+01	.2123E+01				
				NET TOTAL =	.5205E+01				

DETAIL OF NODE 5   TEMPERATURE= .7992E+02   POWER=0.   STABILITY CONSTANT =   -1   CAP= .7460E+01

MULTI-SURFACE RADIATION EXCHANGE AREA= .2000E+01   EMISSIVITY= .800

NODE	CTYPE	CMODE	C	CONDUCTANCE	FLUX	REYN NO.	BNDRY LAYER	HYDR DIA	HT TRANS COEF
4	-2		.3300E+00	.1385E-02	-.2225E-02				.4196E-02
3	-2		.1020E+01	.4760E-02	.2478E-03				.4667E-02
2	-2		.6500E+00	.2990E-02	.1792E+00				.4600E-02
2	-1		.1600E+01	.7952E-02	.4764E+00				.4970E-02
1	101	1	.2000E+01	.1115E-01	.6679E+00				.5573E-02
1	-1		.2000E+01	.1084E-01	.6493E+00				
4	0		.1280E+01	.1280E+01	-.2057E+01				
				NET TOTAL =-.8589E-01					

ENERGY BALANCE =-.3658E-01

Fig. 8-17. (Continued)

DATA SET

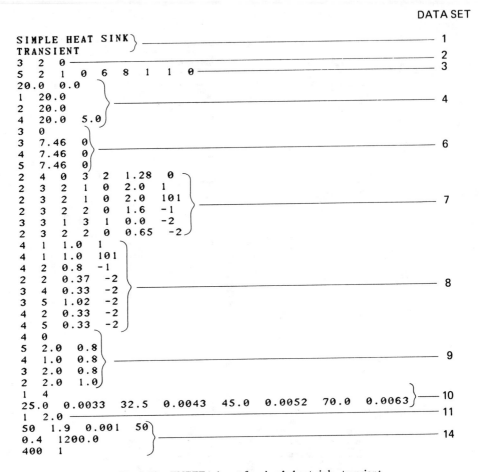

```
SIMPLE HEAT SINK ⟩ ────────────────────────────────────────── 1
TRANSIENT                                                        2
3   2   0   ──────────────────────────────────────────────────  3
5   2   1   0   6   8   1   1   0 ─────────────────────────────
20.0    0.0     ⟍
1   20.0        │
2   20.0        ⟩ ──────────────────────────────────────────── 4
4   20.0    5.0 ⟋
3   0           ⟍
3   7.46    0   │
4   7.46    0   ⟩ ──────────────────────────────────────────── 6
5   7.46    0   ⟋
2   4   0   3   2   1.28    0   ⟍
2   3   2   1   0   2.0     1   │
2   3   2   1   0   2.0     101 ⟩ ───────────────────────────── 7
2   3   2   2   0   1.6     -1  │
3   3   1   3   1   0.0     -2  │
2   3   2   2   0   0.65    -2  ⟋
4   1   1.0     1   ⟍
4   1   1.0     101 │
4   2   0.8     -1  │
2   2   0.37    -2  │
3   4   0.33    -2  ⟩ ───────────────────────────────────────── 8
3   5   1.02    -2  │
4   2   0.33    -2  │
4   5   0.33    -2  ⟋
4   0           ⟍
5   2.0     0.8 │
4   1.0     0.8 ⟩ ────────────────────────────────────────────── 9
3   2.0     0.8 │
2   2.0     1.0 ⟋
1   4                                                           ⟍
25.0    0.0033   32.5   0.0043   45.0   0.0052   70.0   0.0063  ⟩ ── 10
1   2.0 ─────────────────────────────────────────────────────── 11
50   1.9    0.001    50  ⟍
0.4    1200.0            ⟩ ──────────────────────────────────── 14
400    1                ⟋
```

Fig. 8-18. TNETFA input for simple heat sink–transient.

Another TNETFA characteristic of t, T-dependent arrays is that if t is outside the bounds specified by the lower and upper array limits, T is set at the value indicated by the nearest t, T array pair.* In this problem, then, at time t greater than 61 sec node 4 has a dissipation at the last specified time, which is 0.0 watts at 61 sec.

The time-dependent source problem was also set at a 1200 sec (20 min.) termination time with printout intervals every 25 time steps or $(25)(0.4) = 10$ sec. Results from the transient cases are plotted (Figs. 8-20 and 8-23).

8.6 PROBLEM—HEAT SUNK TRANSISTORS IN SEALED PLASTIC BOX

The geometry for this problem is shown in Fig. 8-24. Two TO-3 transistors are attached to a natural convection cooled heat sink. The upper and lower transistors dissipate 1 and 3 watts, respectively.

_____
*The temperature-dependent h_c for interior fin surface convection will similarly be $h_c = 0.0033$ for $\overline{T} < 25.0°$ C and $h_c = 0.0063$ for $\overline{T} > 70.0°$ C.

```
NETWORK  THERMAL  ANALYSIS


SIMPLE HEAT SINK
TRANSIENT

UNITS=2

NUMBER OF NODES=   5    NUMBER OF CONDUCTORS=  38

NLOOP=  50    TPRINT= 400    NPRINT=   1    LOOPEN=   50    ALDT= .1000E-02    BETA= 1.90

ARRAY DATA

ARRAY  1      4 X-Y PAIRS

 .2500E+02,  .3300E-02     .3250E+02,  .4300E-02     .4500E+02,  .5200E-02     .7000E+02,  .6300E-02

NATURAL CONVECTION PARAMETER

  1    VERTICAL FLAT PLATE OR CYLINDER                                              P =  .2000E+01


                          LOOPCT=    0        TIME=  0.

                                        TEMPERATURES

T(  1) = .2000E+02    T(  2) = .2000E+02    T(  3) = .2000E+02    T(  4) = .2000E+02    T(  5) = .2000E+02


                          LOOPCT=  400        TIME=  .1600E+03

                                        TEMPERATURES

T(  1) = .2000E+02    T(  2) = .2000E+02    T(  3) = .4841E+02    T(  4) = .4980E+02    T(  5) = .4841E+02


                          LOOPCT=  800        TIME=  .3200E+03
```

Fig. 8-19. TNETFA output for simple heat sink—transient.

```
                              TEMPERATURES
T(  1)= .2000E+02   T(  2)= .2000E+02   T(  3)= .6467E+02   T(  4)= .6614E+02   T(  5)= .6467E+02
                    LOOPCT= 1200        TIME= .4800E+03
                              TEMPERATURES
T(  1)= .2000E+02   T(  2)= .2000E+02   T(  3)= .7302E+02   T(  4)= .7453E+02   T(  5)= .7302E+02
                    LOOPCT= 1600        TIME= .6400E+03
                              TEMPERATURES
T(  1)= .2000E+02   T(  2)= .2000E+02   T(  3)= .7718E+02   T(  4)= .7872E+02   T(  5)= .7718E+02
                    LOOPCT= 2000        TIME= .8000E+03
                              TEMPERATURES
T(  1)= .2000E+02   T(  2)= .2000E+02   T(  3)= .7936E+02   T(  4)= .8077E+02   T(  5)= .7923E+02
                    LOOPCT= 2400        TIME= .9600E+03
                              TEMPERATURES
T(  1)= .2000E+02   T(  2)= .2000E+02   T(  3)= .8023E+02   T(  4)= .8178E+02   T(  5)= .8023E+02
                    LOOPCT= 2800        TIME= .1120E+04
                              TEMPERATURES
T(  1)= .2000E+02   T(  2)= .2000E+02   T(  3)= .8071E+02   T(  4)= .8226E+02   T(  5)= .8071E+02
                    LOOPCT= 3000        MAXDT= .1200E+04

DETAIL OF NODE  -1   POWER=0.   TEMPERATURE= .2000E+02   STABILITY CONSTANT =   -1   CAP= .1000E-19
```

Fig. 8-19. (Continued)

THIS IS A CONSTANT TEMPERATURE NODE

NODE	CTYPE	CMODE	C	CONDUCTANCE	FLUX	REYN NO.	BNDRY LAYER	HYDR DIA	HT TRANS COEF
4	101	1	.1000E+01	.5622E-02	-.3509E+00				.5622E-02
4	1		.1000E+01	.5473E-02	-.3415E+00				
5	101	1	.2000E+01	.1118E-01	-.6806E+00				.5592E-02
3	101	1	.2000E+01	.1118E-01	-.6806E+00				.5592E-02
5	1		.2000E+01	.1088E-01	-.6619E+00				
3	1		.2000E+01	.1088E-01	-.6619E+00				
				NET TOTAL =	-.3377E+01				

DETAIL OF NODE -2   TEMPERATURE= .2000E+02   POWER=0.   STABILITY CONSTANT =   -1   CAP= .1000E-19

THIS IS A CONSTANT TEMPERATURE NODE

MULTI-SURFACE RADIATION EXCHANGE AREA= .2000E+01   EMISSIVITY=1.000

NODE	CTYPE	CMODE	C	CONDUCTANCE	FLUX	REYN NO.	BNDRY LAYER	HYDR DIA	HT TRANS COEF
4	-2		.3300E+00	.1522E-02	-.9495E-01				.4611E-02
2	-2		.3700E+00	0.	0.				0.
4	-1		.8000E+00	.4025E-02	-.2511E+00				.5031E-02
5	-2		.6500E+00	.3004E-02	-.1828E+00				.4622E-02
3	-2		.6500E+00	.3004E-02	-.1828E+00				.4622E-02
5	-1		.1600E+01	.7988E-02	-.4861E+00				.4993E-02
3	-1		.1600E+01	.7988E-02	-.4861E+00				.4993E-02
				NET TOTAL =	-.1684E+01				

DETAIL OF NODE   3   TEMPERATURE= .8085E+02   POWER=0.   STABILITY CONSTANT = .71E-01   CAP= .7460E+01

MULTI-SURFACE RADIATION EXCHANGE AREA= .2000E+01   EMISSIVITY= .800

NODE	CTYPE	CMODE	C	CONDUCTANCE	FLUX	REYN NO.	BNDRY LAYER	HYDR DIA	HT TRANS COEF
5	-2		.1020E+01	.1594E-01	-.7248E-14				.1563E-01
4	-2		.3300E+00	.1382E-02	-.2142E-02				.4189E-02
2	-2		.6500E+00	.3004E-02	-.1828E+00				.4622E-02
2	-1		.1600E+01	.7988E-02	.4861E+00				.4622E-02
1	101	1	.2000E+01	.1118E-01	.6806E+00				.4993E-02
1	1		.2000E+01	.1088E-01	.6619E+00				.5592E-02
4	0		.1280E+01	.1280E+01	-.1984E+01				
				NET TOTAL =	-.2549E-01				

Fig. 8-19. (Continued)

DETAIL OF NODE 4

TEMPERATURE= .8240E+02   POWER= .5000E+01   STABILITY CONSTANT = .14E+00   CAP = .7460E+01

MULTI-SURFACE RADIATION EXCHANGE AREA= .1000E+01   EMISSIVITY= .800

NODE	CTYPE	CMODE	C	CONDUCTANCE	FLUX	REYN NO.	BNDRY LAYER	HYDR DIA	HT TRANS COEF
5	-2		.3300E+00	.1382E-02	.2142E-02				.4189E-02
2	-2		.3300E+00	.1522E-02	.9495E-01				.4611E-02
3	-2		.3300E+00	.1382E-02	.2142E-02				.4189E-02
2	-1		.8000E+00	.4025E-02	.2511E+00				.5031E-02
1	101	1	.1000E+01	.5622E-02	.3509E+00				.5622E-02
1	-1		.1000E+01	.5473E-02	.3415E+00				
5	0		.1280E+01	.1280E+01	.1984E+01				
3	0		.1280E+01	.1280E+01	.1984E+01				

NET TOTAL = .5010E+01

DETAIL OF NODE 5

TEMPERATURE= .8085E+02   POWER=0.   STABILITY CONSTANT = .71E-01   CAP = .7460E+01

MULTI-SURFACE RADIATION EXCHANGE AREA= .2000E+01   EMISSIVITY= .800

NODE	CTYPE	CMODE	C	CONDUCTANCE	FLUX	REYN NO.	BNDRY LAYER	HYDR DIA	HT TRANS COEF
4	-2		.3300E+00	.1382E-02	-.2142E-02				.4189E-02
3	-2		.1020E+01	.1594E-01	.7248E-14				.1563E-01
2	-2		.6500E+00	.3004E-02	.1828E+00				.4622E-02
2	-1		.1600E+01	.7988E-01	.4861E+00				.4993E-02
1	101	1	.2000E+01	.1118E-01	.6806E+00				.5592E-02
1	-1		.2000E+01	.1088E-01	.6619E+00				
4	0		.1280E+01	.1280E+01	-.1984E+01				

NET TOTAL = .2549E-01

ENERGY BALANCE = .6125E-01

Fig. 8-19. (*Continued*)

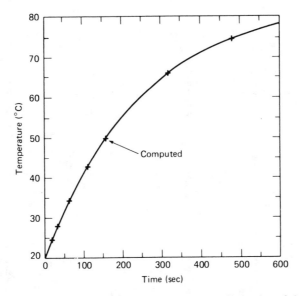

Fig. 8-20. Temperature of node 4 for constant 5-watt heat dissipation of simple heat sink.

DATA SET

```
SIMPLE HEAT SINK                                          1
TRANSIENT - TIME VARIABLE INPUT                          2
3   2   0                                                 2
5   2   1   1   6   8   1   1   0                         3
20.0    0.0
1    20.0                                                 4
2    20.0
4    20.0    5.0
4    3
0.0    5.0    60.0    5.0    61.0    0.0                  5
3    0
3    7.46    0
4    7.46    0                                            6
5    7.46    0
2    4    0    3    2    1.28    0
2    3    2    1    0    2.0    1
2    3    2    1    0    2.0    101
2    3    2    2    0    1.6    -1                         7
3    3    1    3    1    0.0    -2
2    3    2    2    0    0.65    -2
4    1    1.0    1
4    1    1.0    101
4    2    0.8    -1
2    2    0.37    -2
3    4    0.33    -2                                      8
3    5    1.02    -2
4    2    0.33    -2
4    5    0.33    -2
4    0
5    2.0    0.8
4    1.0    0.8                                           9
3    2.0    0.8
2    2.0    1.0
1    4
25.0    0.0033    32.5    0.0043    45.0    0.0052    70.0    0.0063    10
1    2.0                                                  11
50    1.9    0.001    50
0.4    1200.0                                             14
25    0
```

Fig. 8-21. TNETFA input for simple heat sink—transient with time variable heat source.

NETWORK THERMAL ANALYSIS

SIMPLE HEAT SINK
TRANSIENT - TIME VARIABLE INPUT

UNITS=2

NUMBER OF NODES=   5   NUMBER OF CONDUCTORS=   38

NLOOP= 50   TPRINT= 25   NPRINT=   0   LOOPEN=   50   ALDT= .1000E-02   BETA= 1.90

ARRAY DATA

ARRAY 1   4 X-Y PAIRS

.2500E+02, .3300E-02, .3250E+02, .4300E-02, .4500E+02, .5200E-02, .7000E+02, .6300E-02

TIME, POWER ARRAY

NODE   4   3 T-Q PAIRS

0.   .5000E+01   .6000E+02, .5000E+01   .6100E+02, 0.

NATURAL CONVECTION PARAMETER

1   VERTICAL FLAT PLATE OR CYLINDER   P = .2000E+01

                              LOOPCT=   0   TIME= 0.

                                      TEMPERATURES

T(1)= .2000E+02   T(2)= .2000E+02   T(3)= .2000E+02   T(4)= .2000E+02   T(5)= .2000E+02

                              LOOPCT=   25   TIME= .1000E+02

Fig. 8-22. TNETFA output for simple heat sink with time variable heat source.

```
                             TEMPERATURES
T(  1) = .2000E+02   T(  2) = .2000E+02   T(  3) = .2178E+02   T(  4) = .2308E+02   T(  5) = .2178E+02
                   LOOPCT =   50       TIME = .2000E+02
                             TEMPERATURES
T(  1) = .2000E+02   T(  2) = .2000E+02   T(  3) = .2394E+02   T(  4) = .2525E+02   T(  5) = .2394E+02
                   LOOPCT =   75       TIME = .3000E+02
                             TEMPERATURES
T(  1) = .2000E+02   T(  2) = .2000E+02   T(  3) = .2606E+02   T(  4) = .2737E+02   T(  5) = .2606E+02
                   LOOPCT =  100       TIME = .4000E+02
                             TEMPERATURES
T(  1) = .2000E+02   T(  2) = .2000E+02   T(  3) = .2811E+02   T(  4) = .2943E+02   T(  5) = .2811E+02
                   LOOPCT =  125       TIME = .5000E+02
```

Fig. 8-22. (Continued)

TEMPERATURES

T(1) = .2000E+02   T(2) = .2000E+02   T(3) = .3012E+02   T(4) = .3144E+02   T(5) = .3012E+02

LOOPCT= 150   TIME= .6000E+02

TEMPERATURES

T(1) = .2000E+02   T(2) = .2000E+02   T(3) = .3207E+02   T(4) = .3340E+02   T(5) = .3207E+02

LOOPCT= 175   TIME= .7000E+02

TEMPERATURES

T(1) = .2000E+02   T(2) = .2000E+02   T(3) = .3225E+02   T(4) = .3229E+02   T(5) = .3225E+02

LOOPCT= 200   TIME= .8000E+02

TEMPERATURES

T(1) = .2000E+02   T(2) = .2000E+02   T(3) = .3195E+02   T(4) = .3198E+02   T(5) = .3195E+02

Fig. 8-22. (*Continued*)

```
LOOPCT=  225      TIME=  .9000E+02
                TEMPERATURES
T(  1) = .2000E+02   T(  2) = .2000E+02   T(  3) = .3165E+02   T(  4) = .3168E+02   T(  5) = .3165E+02

LOOPCT=  250      TIME=  .1000E+03
                TEMPERATURES
T(  1) = .2000E+02   T(  2) = .2000E+02   T(  3) = .3136E+02   T(  4) = .3139E+02   T(  5) = .3136E+02

LOOPCT=  275      TIME=  .1100E+03
                TEMPERATURES
T(  1) = .2000E+02   T(  2) = .2000E+02   T(  3) = .3108E+02   T(  4) = .3111E+02   T(  5) = .3108E+02
```

Fig. 8-22. (Continued)

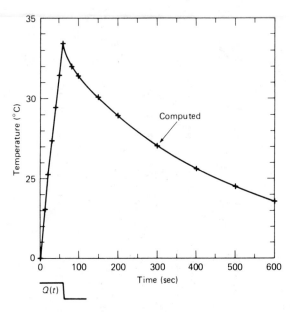

Fig. 8-23. Temperature of node 4 for time-dependent heat dissipation of simple heat sink.

The protective plastic box walls are 0.05 in. thick with a thermal conductivity $k = 0.01$ watt/in. $\cdot$ °C. The aluminum heat sink uses $k = 4.0$ watts/in. $\cdot$ °C. The box exterior and finned sink have an emissivity $\epsilon = 0.8$.

The network model in Fig. 8-25 shows the system based on a two-node heat sink, one node per transistor. Fixed ambient temperature nodes 13 and 14 collect convection and radiation heat transfer, respectively, from the fins. Fixed ambient temperature nodes 1 and 2 collect radiation and convection, respectively, from the various flat plate box surfaces. Node 15 is the interior air node to or from which all interior box surfaces convect. The various FA products for the interior multiple-surface reflection and radiation are not indi-

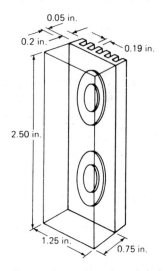

Fig. 8-24. Heat sunk transistors in sealed plastic box.

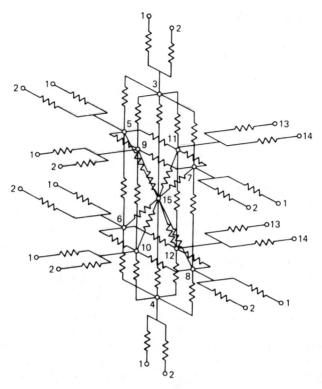

Fig. 8-25. Thermal network model of heat sunk transistors in sealed plastic box.

cated in Fig. 8-25 because it would make the illustration even more difficult to understand.

Calculation of the conduction input is straightforward if perhaps a little tedious. For example, between nodes 3 and 11:

$$R = R_{\text{plastic}} + R_{\text{Al}}$$
$$= (L/kwt)_{\text{plastic}} + (L/kwt)_{\text{Al}}$$
$$= [0.375/(0.01)(1.25)(0.05)] + [0.625/(4.0)(1.25)(0.05)]$$
$$= 602.50°C/\text{watt}$$
$$C = 1/602.50$$
$$= 0.0016 \text{ watt}/°C$$

The convection conductances for the interior of the finned heat sink are accounted for in a manner identical to that of the preceding problem. $(T_s + T_A)/2$, h_c pairs are input as a temperature-dependent array. The tabulated intermediate results (using the methods of Chapter 4) as well as the required h_c are found in Table 8-3. The area A_f is found from:

$$A_f = WH \left[1 + \frac{2(N-1)}{W} L\right]$$
$$= (1.25)^2 \left[1 + \frac{2(5)(0.2)}{1.25}\right]$$
$$= 4.065 \text{ in.}^2$$

Table 8-3. Convection input for heat sink on plastic box.

$\overline{\Delta T}_s$ (°C)	$\overline{T}_s$ (°C)	$\dfrac{\overline{T}_s + T_A}{2}$ (°C)	$\dfrac{h_c}{h_H}$	h_H (watts/in.$^2 \cdot$°C)	h_c (watts/in.$^2 \cdot$°C)
10.0	30.0	25.0	0.62	0.0034	0.0021
25.0	45.0	32.5	0.76	0.0043	0.0033
50.0	70.0	45.0	0.82	0.0051	0.0042
100.0	120.0	70.0	0.84	0.0060	0.0050

The sink to ambient $\mathcal{F}A$ product is $(0.32)(4.065) = 1.30$ in.^{2} for each of nodes 11 and 12. The TNETFA library element for a simple two-surface radiation exchange (CTYPE = -1) heat transfer coefficient is used. The two outer unshielded fin surfaces use $\mathcal{F} = \epsilon$.

The interior radiation shape factor, surface area products are determined from the angle factor methods and graphs for parallel and perpendicular surfaces of Chapter 3. For example, for the opposing surfaces of nodes 9, 10 and 11, 12, F_{9-11} and F_{10-12} are determined from Fig. 3-6. However, the cross terms $F_{9-12} = F_{10-11}$ require a bit of algebra:

$$(A_9 + A_{10})F_{(9+10)-(11+12)} = A_9 F_{9-(11+12)} + A_{10} F_{10-(11+12)}$$

$$= 2A_9 F_{9-(11+12)}$$

But:

$$A_9 F_{9-(11+12)} = A_9 F_{9-11} + A_9 F_{9-12}$$

Then:

$$(A_9 + A_{10})F_{(9+10)-(11+12)} = 2A_9 (F_{9-11} + F_{9-12})$$

$$F_{9-12} = \frac{(A_9 + A_{10})}{2A_9} F_{(9+10)-(11+12)} - F_{9-11}$$

$$= F_{(9+10)-(11+12)} - F_{9-11}$$

$$= 0.45 - 0.35 = 0.10$$

The many other cross terms between the parallel and perpendicular surfaces are similarly found. The FA products are used in the node connection input scheme for CTYPE = -2.

The input listing in Fig. 8-26 is constructed to permit two parameter runs, i.e., one extra case; hence, ICSE = 1, the third quantity in line 3. The thermal network model is identical for the two cases, but the plastic box interior is simulated for emissivity $\epsilon = 0.05$ and 0.80 for the first and second cases, respectively. The second case is obtained by repeating the last 11 input lines, but with the necessary modification of the emissivity input. This constitutes repeat of data sets 9–14 as indicated in the program instructions (Chapter 10).

The full output for this problem is reproduced in Fig. 8-27. An

```
TRANSISTORS IN PLASTIC BOX WITH HEAT SINK ⎫──── 1
STEADY STATE - EM=0.05, EM=0.80
 1   2   1 ──────────────────────────────────── 2
15   4   2   0   47   0   1   3   0 ──────────── 3
30.0   0.0
 1   20.0
 2   20.0
13   20.0
14   20.0                                         ⎫──────── 4
11   40.0   1.0
12   40.0   3.0 ⎭
 0   0 ──────────────────────────────────────── 6
 2   11   1   13   0   4.065      1    ⎫
 2   11   1   13   0   0.625    101
 2   11   1   14   0   1.300     -1
 2   11   1   14   0   0.500     -1
 2    3   1    1   0   0.800     -1
 4    5   1    1   0   0.800     -1
 2    9   1    1   0   1.25      -1
 2    3   0    2  13   1.000    102
 2    4   0    2  13   1.000    103
 4    5   1    2   0   1.000    101
 2    9   1    2   0   1.56     101
 4   15   0    9   1   1.56     101
 4   15   0    5   1   1.00     101
 2    3   0    5   2   3.00E-4    0
 2    3   1    9   1   6.25E-4    0
 2    3   1   11   1   1.66E-3    0
 2    4   0    6   2   3.00E-4    0
 2    5   2    6   2   3.00E-4    0
 1    9   0   10   0   5.00E-4    0
 1   11   0   12   0   0.200      0
 2    9   0    5   2   6.25E-4    0
 2   10   0    6   2   6.25E-4    0
 2   11   0    5   2   1.66E-3    0
 2   12   0    6   2   1.66E-3    0   ⎬──────── 7
10    3   1    3   1   0.00      -2
 4    5   2    6   2   0.00      -2
 1    3   0    4   0   0.044     -2
 2    5   1    7   1   0.128     -2
 2    9   1   11   1   0.554     -2
 2    5   1   11   1   0.2522    -2
 2    7   1   11   1   0.2522    -2
 2    5   1    8  -1   0.0530    -2
 2    5   1    9   1   0.2522    -2
 2    7   1    9   1   0.2522    -2
 2    9   1   12  -1   0.1500    -2
 2    5   2   10   0   0.0394    -2
 2    5   2   12   0   0.0394    -2
 2    6   2    9   0   0.0394    -2
 2    6   2   11   0   0.0394    -2
 2    3   1    9   1   0.2520    -2
 2    3   1   11   1   0.2520    -2
 2    3   1    5   1   0.153     -2
 2    3   1    7   1   0.153     -2
 2    3   0    6   2   0.0197    -2
 2    4   0    5   2   0.0197    -2
 2    3   0   10   2   0.0234    -2
 2    4   0    9   2   0.0234    -2 ⎭
 0   2
 3    8   0.938   0.05  ⎫──────── 9
 9   12   1.563   0.05  ⎭
 1    4                  ⎫──────── 10
25.0   0.0021   32.5   0.0033
           45.0   0.0042   70.0   0.005 ⎭
           6    2.5  ⎫
           7    0.23 ⎬──────── 11
           8    0.23 ⎭
           50   1.0   0.001   50 ⎫
           1.0   1.0            ⎬──────── 14
           50   1 ⎭
           0    2
           3    8   0.938   0.8 ⎫──────── 9
           9   12   1.563   0.8 ⎭
           1    4 ⎫
           25.0   0.0021   32.5   0.0033 ⎬── 10
           45.0   0.0042   70.0   0.005 ⎭
           6    2.5  ⎫
           7    0.23 ⎬──────── 11
           8    0.23 ⎭
           100   1.0   0.001   100 ⎫
           1.0   1.0             ⎬──────── 14
           100   1 ⎭
```

Fig. 8-26.   TNETFA input for transistors in plastic box.

NETWORK THERMAL ANALYSIS

TRANSISTORS IN PLASTIC BOX WITH HEAT SINK
STEADY STATE - EM=0.05, EM=0.80

UNITS=2

NUMBER OF NODES= 15   NUMBER OF CONDUCTORS= 208

NLOOP= 50   TPRINT= 50   NPRINT= 1   LOOPEN= 50   ALDT= .1000E-02   BETA= 1.00

ARRAY DATA

ARRAY 1   4   X-Y PAIRS

.2500E+02, .2100E-02   .3250E+02, .3300E-02   .4500E+02, .4200E-02   .7000E+02, .5000E-02

NATURAL CONVECTION PARAMETER

1   SMALL SURFACE, VERTICAL   P= .2500E+01
2   SMALL HORIZONTAL SURFACE, HEATED SIDE FACING UP OR COOLED SIDE FACING DOWN   P= .2300E+00
3   SMALL HORIZONTAL SURFACE, HEATED SIDE FACING DOWN OR COOLED SIDE FACING UP   P= .2300E+00

LOOPCT= 0   MAXDT= -1

TEMPERATURES

T(1)= .2000E+02   T(2)= .2000E+02   T(3)= .3000E+02   T(4)= .3000E+02   T(5)= .3000E+02   T(6)= .3000E+02
T(7)= .3000E+02   T(8)= .3000E+02   T(9)= .3000E+02   T(10)= .3000E+02   T(11)= .4000E+02   T(12)= .4000E+02
T(13)= .2000E+02   T(14)= .2000E+02   T(15)= .3000E+02

LOOPCT= 23   MAXDT= .9618E-03

TEMPERATURES

T(1)= .2000E+02   T(2)= .2000E+02   T(3)= .3343E+02   T(4)= .3455E+02   T(5)= .3397E+02   T(6)= .3442E+02
T(7)= .3397E+02   T(8)= .3442E+02   T(9)= .3010E+02   T(10)= .3016E+02   T(11)= .7044E+02   T(12)= .7486E+02
T(13)= .2000E+02   T(14)= .2000E+02   T(15)= .4476E+02

LOOPCT= 23   MAXDT= .9618E-03

Fig. 8-27. TNETFA output for transistors in plastic box.

DETAIL OF NODE -1   TEMPERATURE= .2000E+02   POWER=0.   STABILITY CONSTANT =   -1   CAP= .1000E-19

THIS IS A CONSTANT TEMPERATURE NODE

NODE	CTYPE	CMODE	C	CONDUCTANCE	FLUX	REYN NO.	BNDRY LAYER	HYDR DIA	HT TRANS COEF
10	-1		.1250E+01	.4844E-02	-.4921E-01				.3875E-02
9	-1		.1250E+01	.4843E-02	-.4893E-01				.3874E-02
8	-1		.8000E+00	.3168E-02	-.4569E-01				.3960E-02
7	-1		.8000E+00	.3161E-02	-.4415E-01				.3951E-02
6	-1		.8000E+00	.3168E-02	-.4569E-01				.3960E-02
5	-1		.8000E+00	.3161E-02	-.4415E-01				.3951E-02
4	-1		.8000E+00	.3170E-02	-.4612E-01				.3963E-02
3	-1		.8000E+00	.3152E-02	-.4235E-01				.3940E-02
				NET TOTAL =	-.3663E+00				

DETAIL OF NODE -2   TEMPERATURE= .2000E+02   POWER=0.   STABILITY CONSTANT =   -1   CAP= .1000E-19

THIS IS A CONSTANT TEMPERATURE NODE

NODE	CTYPE	CMODE	C	CONDUCTANCE	FLUX	REYN NO.	BNDRY LAYER	HYDR DIA	HT TRANS COEF
10	101	6	.1560E+01	.5606E-01	-.5695E-01				.3594E-02
9	101	6	.1560E+01	.5595E-02	-.5653E-01				.3587E-02
8	101	6	.1000E+01	.4062E-02	-.5858E-01				.4062E-02
7	101	6	.1000E+01	.4017E-02	-.5612E-01				.4017E-02
6	101	6	.1000E+01	.4062E-02	-.5858E-01				.4062E-02
5	101	6	.1000E+01	.4017E-02	-.5612E-01				.4017E-02
4	103	8	.1000E+01	.3537E-02	-.5145E-01				.3537E-02
3	102	7	.1000E+01	.6890E-02	-.9256E-01				.6890E-02
				NET TOTAL =	-.4869E+00				

DETAIL OF NODE 3   TEMPERATURE= .3343E+02   POWER=0.   STABILITY CONSTANT =   -1   EMISSIVITY= .050

MULTI-SURFACE RADIATION EXCHANGE AREA= .9380E+00

NODE	CTYPE	CMODE	C	CONDUCTANCE	FLUX	REYN NO.	BNDRY LAYER	HYDR DIA	HT TRANS COEF
12	-2		.2340E-01	.4770E-05	-.1976E-03				.2039E-03
10	-2		.2340E-01	-.3163E-05	-.1036E-04				-.1352E-03
8	-2		.1970E-01	.1370E-05	-.1351E-05				.6952E-04
6	-2		.1970E-01	.1370E-05	-.1351E-05				.6952E-04

Fig. 8-27. *(Continued)*

NODE	CTYPE	CMODE	C	CONDUCTANCE	FLUX	REYN NO.	BNDRY LAYER	HYDR DIA	HT TRANS COEF
7	-2		.1530E+00	-.2434E-03	.1301E-03				-.1591E-02
5	-2		.1530E+00	-.2434E-03	.1301E-03				-.1591E-02
11	-2		.2520E+00	-.4430E-04	-.1639E-02				.1758E-03
9	-2		.2520E+00	.3278E-04	.1092E-03				.1301E-03
4	-2		.4400E-01	.4298E-04	-.4783E-04				.9769E-03
11	0		.1660E-02	.1660E-02	-.6143E-01				
9	0		.6250E-03	.6250E-03	.2082E-02				
7	0		.3000E-03	.3000E-03	-.1604E-03				
5	0		.3000E-03	.3000E-03	-.1604E-03				
15	102	7	.1000E+01	.6512E-02	-.7372E-01				.6512E-02
2	102	7	.1000E+01	.6890E-02	-.9256E-01				.6890E-02
1	-1		.8000E+00	.3152E-02	-.4235E-01				.3940E-02
				NET TOTAL	=-.3938E-05				

DETAIL OF NODE 4 TEMPERATURE= .3455E+02 POWER=0. STABILITY CONSTANT = -1 CAP= .1000E-19

MULTI-SURFACE RADIATION EXCHANGE AREA= .9380E+00 EMISSIVITY= .050

NODE	CTYPE	CMODE	C	CONDUCTANCE	FLUX	REYN NO.	BNDRY LAYER	HYDR DIA	HT TRANS COEF
11	-2		.2340E-01	.3533E-05	-.1268E-03				.1510E-03
9	-2		.2340E-01	.8007E-05	.3558E-04				.3422E-03
7	-2		.1970E-01	.6604E-04	.3817E-04				.3352E-02
5	-2		.1970E-01	.6604E-04	.3817E-04				.3352E-02
8	-2		.1530E+00	.1235E-02	-.1558E-03				.8074E-02
6	-2		.1530E+00	.1235E-02	-.1558E-03				.8074E-02
12	-2		.2520E+00	.4600E-04	-.1854E-02				.1825E-03
10	-2		.2520E+00	.3699E-04	.1623E-03				.1468E-03
3	-2		.4400E-01	.4298E-04	-.4783E-04				.9769E-03
8	0		.3000E-03	.3000E-03	-.3785E-04				
6	0		.3000E-03	.3000E-03	-.3785E-04				
12	0		.1660E-02	.1660E-02	-.6692E-01				
10	0		.6250E-03	.6250E-03	-.2743E-02				
15	103	8	.1000E+01	.3147E-02	-.3212E-01				.3147E-02
2	103	8	.1000E+01	.3537E-02	-.5145E-01				.3537E-02
1	-1		.8000E+00	.3170E-02	-.4612E-01				.3963E-02
				NET TOTAL	=-.3474E-05				

DETAIL OF NODE 5 TEMPERATURE= .3397E+02 POWER=0. STABILITY CONSTANT = -1 CAP= .1000E-19

MULTI-SURFACE RADIATION EXCHANGE AREA= .9380E+00 EMISSIVITY= .050

Fig. 8-27. (Continued)

NODE	CTYPE	CNODE	C	CONDUCTANCE	FLUX	REYN NO.	BNDRY LAYER	HYDR DIA	HT TRANS COEF
4	-2		.1970E-01	.660 E-04	-.3817E-04				.3352E-02
3	-2		.1530E+00	-.2434E-03	-.1301E-03				-.1591E-02
12	-2		.3940E-01	.8952E-05	-.3662E-03				.2273E-03
10	-2		.3940E-01	-.1337E-04	-.5097E-04				-.3394E-03
9	-2		.2522E+00	-.2723E-04	-.1053E-03				-.1080E-03
8	-2		.5300E-01	.1078E-03	-.4872E-04				.2034E-02
11	-2		.2522E+00	.5087E-04	-.1855E-02				.2017E-03
7	-2		.1280E+00	.3537E-04	-.1523E-10				.2763E-03
11	0		.1660E-02	.1660E-02	-.6054E-01				
9	0		.6250E-03	.6250E-03	.2416E-02				
6	0		.3000E-03	.3000E-03	-.1356E-03				
3	0		.3000E-03	.3000E-03	.1604E-03				
15	101	6	.1000E+01	.3670E-02	-.3959E-01				.3670E-02
2	101	6	.1000E+01	.4017E-02	.5612E-01				.4017E-02
1	-1		.8000E+00	.3161E-02	.4415E-01				.3951E-02
				NET TOTAL =-.2906E-05					

DETAIL OF NODE 6   TEMPERATURE= .3442E+02   POWER=0.   STABILITY CONSTANT =   -1   CAP= .1000E-19

MULTI-SURFACE RADIATION EXCHANGE AREA= .9380E+00   EMISSIVITY= .050

NODE	CTYPE	CNODE	C	CONDUCTANCE	FLUX	REYN NO.	BNDRY LAYER	HYDR DIA	HT TRANS COEF
3	-2		.1970E-01	.1370E-05	.1351E-05				.6952E-04
4	-2		.1530E+00	.1235E-02	-.1558E-03				.8074E-02
11	-2		.3940E-01	.7041E-05	-.2536E-03				.1787E-03
9	-2		.3940E-01	.4580E-05	.1977E-04				.1162E-03
10	-2		.2522E+00	.2215E-04	-.9440E-04				-.8781E-04
7	-2		.5300E-01	-.1078E-03	.4872E-04				.2034E-02
12	-2		.2522E+00	.5224E-04	-.2112E-02				.2071E-03
8	-2		.1280E+00	.3531E-04	-.1489E-10				.2759E-03
12	0		.1660E-02	.1660E-02	-.6713E-01				
10	0		.6250E-03	.6250E-03	-.2664E-02				
5	0		.3000E-03	.3000E-03	-.1356E-03				
4	0		.3000E-03	.3000E-03	-.3785E-04				
15	101	6	.1000E+01	.3615E-02	-.3736E-01				.3615E-02
2	101	6	.1000E+01	.4062E-02	.5858E-01				.4062E-02
1	-1		.8000E+00	.3168E-02	.4569E-01				.3960E-02
				NET TOTAL =-.2978E-05					

Fig. 8-27. (Continued)

DETAIL OF NODE 7   TEMPERATURE= .3397E+02   POWER=0.   STABILITY CONSTANT = -1   CAP= .1000E-19

MULTI-SURFACE RADIATION EXCHANGE AREA= .9380E+00   EMISSIVITY= .050

NODE	CTYPE	CMODE	C	CONDUCTANCE	FLUX	REYN NO.	BNDRY LAYER	HYDR DIA	HT TRANS COEF
4	-2		.1970E-01	.6604E-04	-.3817E-04				.3352E-02
3	-2		.1530E+00	-.2434E-03	-.1301E-03				-.1591E-02
12	-2		.3940E-01	.8957E-05	-.3662E-03				-.2273E-03
10	-2		.3940E-01	-.1337E-04	.5097E-04				-.3394E-03
9	-2		.2522E+00	.2723E-04	-.1053E-03				-.1080E-03
6	-2		.5300E-01	.1078E-03	-.4872E-04				.2034E-02
11	-2		.2522E+00	.5087E-04	-.1855E-02				.2017E-03
5	-2		.1280E+00	.3537E-04	-.1523E-10				.2763E-03
11	0		.1660E-02	.1660E-02	-.6054E-01				
9	0		.6250E-03	.6250E-03	-.2416E-02				
8	0		.3000E-03	.3000E-03	-.1356E-03				
3	0		.3000E-03	.3000E-03	-.1604E-03				
15	101		.1000E+01	.3670E-02	.3959E-01				.3670E-02
2	101		.1000E+01	.4017E-01	.5612E-01				.4017E-02
1	-1		.8000E+00	.3161E-02	-.4415E-01				.3951E-02
				NET TOTAL =-.2899E-05					

DETAIL OF NODE 8   TEMPERATURE= .3442E+02   POWER=0.   STABILITY CONSTANT = -1   CAP= .1000E-19

MULTI-SURFACE RADIATION EXCHANGE AREA= .9380E+00   EMISSIVITY= .050

NODE	CTYPE	CMODE	C	CONDUCTANCE	FLUX	REYN NO.	BNDRY LAYER	HYDR DIA	HT TRANS COEF
3	-2		.1970E-01	.1370E-05	.1351E-05				.6952E-04
4	-2		.1530E+00	.1235E-02	-.1558E-03				.8074E-02
11	-2		.3940E-01	.7041E-05	-.2536E-03				.1787E-03
9	-2		.3940E-01	.4580E-05	.1977E-04				.1162E-03
10	-2		.2522E+00	-.2215E-04	.9440E-04				-.8781E-04
5	-2		.5300E-01	.1078E-03	-.4872E-04				.2034E-02
12	-2		.2522E+00	.5224E-04	-.2112E-02				.2071E-03
6	-2		.1280E+00	.3531E-04	-.1489E-10				.2759E-03
12	0		.1660E-02	.1660E-02	-.6713E-01				
10	0		.6250E-03	.6250E-03	-.2664E-02				
7	0		.3000E-03	.3000E-03	-.1356E-03				
4	0		.3000E-03	.3000E-03	-.3785E-04				
15	101		.1000E+01	.3615E-02	-.3736E-01				.3615E-02
2	101		.1000E+01	.4062E-02	.5858E-01				.4062E-02
1	-1		.8000E+00	.3168E-02	-.4569E-01				.3960E-02
				NET TOTAL =-.2971E-05					

Fig. 8-27. (Continued)

DETAIL OF NODE   9

TEMPERATURE= .3010E+02   POWER=0.   STABILITY CONSTANT = -1   CAP= .1000E-19

MULTI-SURFACE RADIATION EXCHANGE AREA= .1563E+01   EMISSIVITY= .050

NODE	CTYPE	CMODE	C	CONDUCTANCE	FLUX	REYN NO.	BNDRY LAYER	HYDR DIA	HT TRANS COEF
4	-2		.2340E-01	.8007E-05	-.3558E-04				.3422E-03
3	-2		.2520E+00	.3278E-03	-.1092E-03				.1301E-03
8	-2		.3940E-01	.4580E-05	-.1977E-04				.1162E-03
6	-2		.3940E-01	.4580E-05	-.1977E-04				.1162E-03
12	-2		.1500E+00	.2976E-04	-.1332E-03				.1984E-03
7	-2		.2522E+00	-.2723E-04	.1053E-03				-.1080E-03
5	-2		.2522E+00	-.2723E-04	.1053E-03				-.1080E-03
11	-2		.5540E+00	.9530E-04	-.3844E-02				.1720E-03
7	0		.6250E-03	.6250E-03	-.2416E-02				
5	0		.6250E-03	.6250E-03	-.2416E-02				
10	0		.5000E-03	.5000E-03	-.2724E-04				
3	0		.6250E-03	.6250E-03	-.2082E-02				
15	101	6	.1560E+01	.6373E-02	-.9338E-01				.4085E-02
2	101	6	.1560E+01	.5595E-02	.5653E-01				.3587E-02
1	-1		.1250E+01	.4843E-02	.4893E-01				.3874E-02

NET TOTAL =-.2613E-05

DETAIL OF NODE   10

TEMPERATURE= .3016E+02   POWER=0.   STABILITY CONSTANT = -1   CAP= .1000E-19

MULTI-SURFACE RADIATION EXCHANGE AREA= .1563E+01   EMISSIVITY= .050

NODE	CTYPE	CMODE	C	CONDUCTANCE	FLUX	REYN NO.	BNDRY LAYER	HYDR DIA	HT TRANS COEF
3	-2		.2340E-01	-.3163E-05	.1036E-04				-.1352E-03
4	-2		.2520E+00	.3699E-04	-.1623E-03				.1468E-03
7	-2		.3940E-01	-.1337E-04	.5097E-04				-.3394E-03
5	-2		.3940E-01	-.1337E-04	.5097E-04				-.3394E-03
11	-2		.1500E+00	.2258E-04	-.9094E-03				.1505E-03
8	-2		.2522E+00	-.2215E-04	.9440E-04				-.8781E-04
12	-2		.2522E+00	-.2215E-04	.9440E-04				-.8781E-04
8	0		.5540E+00	.9917E-04	-.4433E-02				.1790E-03
6	0		.6250E-03	.6250E-03	-.2664E-02				
9	0		.6250E-03	.6250E-03	-.2664E-02				
4	0		.5000E-03	.5000E-03	-.2724E-02				
7	0		.6250E-03	.6250E-03	-.2743E-02				
15	101	6	.1560E+01	.6365E-02	-.9291E-01				.4080E-02

Fig. 8-27.   (Continued)

NODE	CTYPE	CMODE	C	CONDUCTANCE	FLUX	REYN NO.	BNDRY LAYER	HYDR DIA	HT TRANS COEF
2	101	6	.1560E+01	.5600E-02	.5695E-01				.3594E-02
1	-1		.1250E+01	.4841E-02	.4921E-01				.3875E-02
				NET TOTAL	= -.2490E-05				

DETAIL OF NODE 11

TEMPERATURE= .70?E+02   POWER= .1000E+01   STABILITY CONSTANT = -1   CAP= .1000E-19

MULTI-SURFACE RADIATION EXCHANGE AREA= .1563E+01   EMISSIVITY= .050

NODE	CTYPE	CMODE	C	CONDUCTANCE	FLUX	REYN NO.	BNDRY LAYER	HYDR DIA	HT TRANS COEF
4	-2		.2340E-01	.3533E-05	.1268E-03				.1510E-03
3	-2		.2520E+00	.4430E-04	.1639E-02				.1758E-03
8	-2		.3940E-01	.7041E-05	.2536E-03				.1787E-03
6	-2		.3940E-01	.7041E-05	.2536E-03				.1787E-03
10	-2		.1500E+00	.2258E-04	.9094E-03				.1505E-03
7	-2		.2522E+00	.5087E-04	.1855E-02				.2017E-03
5	-2		.2522E+00	.5087E-04	.1855E-02				.2017E-03
9	-2		.5540E+00	.9530E-04	.3844E-02				.1720E-03
7	0		.1660E+02	.1660E-02	.6054E-01				
5	0		.1660E-02	.1660E-02	.6054E-01				
12	0		.1660E-02	.1660E-02	-.8843E+00				
3	0		.2000E+00	.2000E+00	.6143E-01				
15	101	6	.1660E-02	.1756E-02	.1992E+00				.4972E-02
14	-1		.1560E+01	.7756E-02	.1196E+00				.4743E-02
14	-1		.5000E+00	.2372E-02	.3110E+00				.4743E-02
13	101	6	.1300E+01	.6166E-02	.1985E+00				.6297E-02
13	1		.6250E+00	.3935E-02	.8626E+00				
			.4065E+01	.1710E-01	.9998E+00				
				NET TOTAL	=				

DETAIL OF NODE 12

TEMPERATURE= .7486E+02   POWER= .3000E+02   STABILITY CONSTANT = -1   CAP= .1000E-19

MULTI-SURFACE RADIATION EXCHANGE AREA= .7486E+02   EMISSIVITY= .050

NODE	CTYPE	CMODE	C	CONDUCTANCE	FLUX	REYN NO.	BNDRY LAYER	HYDR DIA	HT TRANS COEF
3	-2		.2340E-01	.4770E-05	.1976E-03				.2039E-03
4	-2		.2520E+00	.4600E-04	.1854E-02				.1825E-03
7	-2		.3940E-01	.8957E-05	.3662E-03				.2273E-03
5	-2		.3940E-01	.8957E-05	.3662E-03				.2273E-03
9	-2		.1500E+00	.2976E-04	.1332E-02				.1984E-03
8	-2		.2522E+00	.5224E-04	.2112E-02				.2071E-03
6	-2		.2522E+00	.5224E-04	.2112E-02				.2071E-03
10	-2		.5540E+00	.9917E-04	.4433E-02				.1790E-03

Fig. 8-27. (Continued)

(continuation of preceding node detail)

```
NODE  CTYPE  CMODE       C        CONDUCTANCE      FLUX       REYN NO.  BNDRY LAYER  HYDR DIA  HT TRANS COEF
 8      0                         .1660E-02      .6713E-01
 6      0                         .1660E-02      .6713E-01
11      0                         .2000E+00      .8843E+00
 4      0                         .1660E-02      .6692E-01
15    101      6     .1560E+01    .8199E-02      .2468E+00                                     .5256E-02
14     -1           .5000E+00     .2424E-02      .1330E+00                                     .4848E-02
14     -1           .1300E+01     .6302E-02      .3457E+00                                     .4848E-02
13    101      6    .6250E+00     .4053E-02      .2223E+00                                     .6485E-02
13      1          .4065E+01      .1739E-01      .9539E+00
                         NET TOTAL = .3000E+01
```

STABILITY CONSTANT =   -1   CAP= .1000E-19

DETAIL OF NODE -13

TEMPERATURE= .2000E+02   POWER=0.

THIS IS A CONSTANT TEMPERATURE NODE

```
NODE  CTYPE  CMODE       C        CONDUCTANCE      FLUX       REYN NO.  BNDRY LAYER  HYDR DIA  HT TRANS COEF
12    101      6    .6250E+00    .4053E-02     -.2223E+00                                      .6485E-02
11    101      6    .6250E+00    .3935E-02     -.1985E+00                                      .6297E-02
12      1          .4065E+01     .1739E-01     -.9539E+00
11      1          .4065E+01     .1710E-01     -.8626E+00
                         NET TOTAL =-.2237E+01
```

STABILITY CONSTANT =   -1   CAP= .1000E-19

DETAIL OF NODE -14

TEMPERATURE= .2000E+02   POWER=0.

THIS IS A CONSTANT TEMPERATURE NODE

```
NODE  CTYPE  CMODE       C        CONDUCTANCE      FLUX       REYN NO.  BNDRY LAYER  HYDR DIA  HT TRANS COEF
12     -1          .5000E+00     .2424E-02     -.1330E+00                                      .4848E-02
11     -1          .5000E+00     .2372E-02     -.1196E+00                                      .4743E-02
12     -1          .1300E+01     .6302E-02     -.3457E+00                                      .4848E-02
11     -1          .1300E+01     .6166E-02     -.3110E+00                                      .4743E-02
                         NET TOTAL =-.9093E+01
```

STABILITY CONSTANT =   -1   CAP= .1000E-19

DETAIL OF NODE   15

TEMPERATURE= .4476E+02   POWER=0.

```
NODE  CTYPE  CMODE       C        CONDUCTANCE      FLUX       REYN NO.  BNDRY LAYER  HYDR DIA  HT TRANS COEF
 8    101          .1000E+01     .3615E-02      .3736E-01                                      .3615E-02
 7    101          .1000E+01     .3670E-02      .3959E-01                                      .3670E-02
 6    101          .1000E+01     .3615E-02      .3736E-01                                      .3615E-02
 5    101          .1000E+01     .3670E-02      .3959E-01                                      .3670E-02
```

Fig. 8-27. *(Continued)*

NODE	CTYPE	CMODE	C	CONDUCTANCE	FLUX	REYN NO.	BNDRY LAYER	HYDR DIA	HT TRANS COEF
12	101	6	.1560E+01	.8199E-02	-.2468E+00				.5256E-02
11	101	6	.1560E+01	.7756E-02	-.1992E+00				.4972E-02
10	101	6	.1560E+01	.6365E-02	.9291E-01				.4080E-02
9	101	6	.1560E+01	.6373E-02	.9338E-01				.4085E-02
4	103	8	.1000E+01	.3147E-02	.3212E-01				.3147E-02
3	102	7	.1000E+01	.6512E-02	.7372E-01				.6512E-02

NET TOTAL = .7599E-05

ENERGY BALANCE = -.1641E-03

NETWORK THERMAL ANALYSIS

TRANSISTORS IN PLASTIC BOX WITH HEAT SINK
STEADY STATE - EM=0.05, EM=0.80

UNITS=2

NUMBER OF NODES= 15   NUMBER OF CONDUCTORS= 208

NLOOP= 100   TPRINT= 100   NPRINT= 1   LOOPEN= 100   ALDT= .1000E-02   BETA= 1.00

ARRAY DATA

ARRAY 1   4 X-Y PAIRS

.2500E+02, .2100E-02, .3250E+02, .3300E-02, .4500E+02, .4200E-02, .7000E+02, .5000E-02

NATURAL CONVECTION PARAMETER

1   SMALL SURFACE, VERTICAL   P= .2500E+01
2   SMALL HORIZONTAL SURFACE, HEATED SIDE FACING UP OR COOLED SIDE FACING DOWN   P= .2300E+00
3   SMALL HORIZONTAL SURFACE, HEATED SIDE FACING DOWN OR COOLED SIDE FACING UP   P= .2300E+00

LOOPCT= 0   MAXDT= .9618E-03

TEMPERATURES

T(1)= .2000E+02   T(2)= .2000E+02   T(3)= .3343E+02   T(4)= .3455E+02   T(5)= .3397E+02   T(6)= .3442E+02
T(7)= .3397E+02   T(8)= .3442E+02   T(9)= .3010E+02   T(10)= .3016E+02   T(11)= .7044E+02   T(12)= .7486E+02
T(13)= .2000E+02   T(14)= .2000E+02   T(15)= .4476E+02

LOOPCT= 17   MAXDT= .8897E-03

Fig. 8-27. (Continued)

TEMPERATURES

```
T(  1) = .2000E+02   T(  2) = .2000E+02   T(  3) = .3518E+02   T(  4) = .3696E+02   T(  5) = .3610E+02   T(  6) = .3674E+02
T(  7) = .3610E+02   T(  8) = .3674E+02   T(  9) = .3674E+02   T( 10) = .3427E+02   T( 11) = .3464E+02   T( 12) = .7172E+02
T( 13) = .2000E+02   T( 14) = .2000E+02   T( 15) = .4582E+02
```

LOOPCT=   17   MAXDT= .8897E-03

DETAIL OF NODE -1

TEMPERATURE= .2000E+02   POWER=0.   STABILITY CONSTANT = -1   CAP = .1000E-19

THIS IS A CONSTANT TEMPERATURE NODE

NODE	CTYPE	CMODE	C	CONDUCTANCE	FLUX	REYN NO.	BNDRY LAYER	HYDR DIA	HT TRANS COEF
10	-1		.1250E+01	.4956E-02	-.7258E-01				.3965E-02
9	-1		.1250E+01	.4946E-02	-.7057E-01				.3957E-02
8	-1		.8000E+00	.3206E-02	-.5368E-01				.4007E-02
7	-1		.8000E+00	.3195E-02	-.5145E-01				.3994E-02
6	-1		.8000E+00	.3206E-02	-.5368E-01				.4007E-02
5	-1		.8000E+00	.3195E-02	-.5145E-01				.3994E-02
4	-1		.8000E+00	.3209E-02	-.5444E-01				.4012E-02
3	-1		.8000E+00	.3180E-02	-.4827E-01				.3975E-02

NET TOTAL =-.4561E+00

DETAIL OF NODE -2

TEMPERATURE= .2000E+02   POWER=0.   STABILITY CONSTANT = -1   CAP = .1000E-19

THIS IS A CONSTANT TEMPERATURE NODE

NODE	CTYPE	CMODE	C	CONDUCTANCE	FLUX	REYN NO.	BNDRY LAYER	HYDR DIA	HT TRANS COEF
10	101	6	.1560E+01	.6372E-02	-.9331E-01				.4084E-02
9	101	6	.1560E+01	.6314E-02	-.9008E-01				.4047E-02
8	101	6	.1000E+01	.4280E-02	-.7167E-01				.4280E-02
7	101	6	.1000E+01	.4222E-02	-.6799E-01				.4222E-02
6	101	6	.1000E+01	.4280E-02	-.7167E-01				.4280E-02
5	101	6	.1000E+01	.4222E-02	-.6799E-01				.4222E-02
4	103	8	.1000E+01	.3721E-02	-.6311E-01				.3721E-02
3	102	7	.1000E+01	.7173E-02	-.1089E+00				.7173E-02

NET TOTAL =-.6347E+00

Fig. 8-27. (Continued)

DETAIL OF NODE 3   TEMPERATURE= .3518E+02   POWER=0.   STABILITY CONSTANT =   -1   CAP= .1000E-19

MULTI-SURFACE RADIATION EXCHANGE AREA= .9380E+00   EMISSIVITY= .800

NODE	CTYPE	CMODE	C	CONDUCTANCE	FLUX	REYN NO.	BNDRY LAYER	HYDR DIA	HT TRANS COEF
12	-2		.2340E-01	.9119E-04	-.3332E-02				.3897E-02
10	-2		.2340E-01	-.1547E-03	.8230E-04				-.6611E-02
8	-2		.1970E-01	.8648E-04	-.1355E-03				.4390E-02
6	-2		.1970E-01	.8648E-04	-.1355E-03				.4390E-02
7	-2		.1530E+00	.5463E-03	-.5054E-03				.3570E-02
5	-2		.1530E+00	.5463E-03	-.5054E-03				.3570E-02
11	-2		.2520E+00	.9502E-03	-.3058E-01				.3771E-02
9	-2		.2520E+00	-.2025E-03	-.1842E-03				-.8037E-03
4	-2		.4400E-01	.1839E-03	-.3286E-03				.4180E-02
11	0		.1660E-02	.1660E-02	-.5343E-01				
9	0		.6250E-03	.6250E-03	.5686E-03				
7	0		.3000E-03	.3000E-03	-.2775E-03				
5	0		.3000E-03	.3000E-03	-.2775E-03				
15	102	7	.1000E+01	.6380E-02	.6791E-01				.6380E-02
2	102	7	.1000E+01	.7173E-02	.1089E+00				.7173E-02
1	-1		.8000E+00	.3180E-01	-.4827E-01				.3975E-02
				NET TOTAL =	.4791E-05				

DETAIL OF NODE 4   TEMPERATURE= .3696E+02   POWER=0.   STABILITY CONSTANT =   -1   CAP= .1000E-19

MULTI-SURFACE RADIATION EXCHANGE AREA= .9380E+00   EMISSIVITY= .800

NODE	CTYPE	CMODE	C	CONDUCTANCE	FLUX	REYN NO.	BNDRY LAYER	HYDR DIA	HT TRANS COEF
11	-2		.2340E-01	.8767E-04	-.2665E-02				.3747E-02
9	-2		.2340E-01	.5847E-04	.1577E-03				.2499E-02
7	-2		.1970E-01	.9523E-04	.8207E-04				.4834E-02
5	-2		.1970E-01	.9523E-04	.8207E-04				.4834E-02
8	-2		.1530E+00	.4101E-03	.9007E-04				.2681E-02
6	-2		.1530E+00	.4101E-03	.9006E-04				.2681E-02
12	-2		.2520E+00	.9784E-03	-.3401E-01				.3883E-02
10	-2		.2520E+00	.4294E-03	.9958E-03				.1704E-02
3	-2		.4400E-01	.1839E-03	-.3286E-03				.4180E-02
8	0		.3000E-03	.3000E-03	.6588E-04				
6	0		.3000E-03	.3000E-03	.6588E-04				
12	0		.1660E-02	.1660E-02	-.5769E-01				
10	0		.6250E-03	.6250E-03	-.1449E-02				
15	103	8	.1000E+01	.3002E-02	-.2659E-01				.3002E-02

Fig. 8-27. (Continued)

2	103	8	.1000E+01	.3721E-02	.6311E-01	.3721E-02
1	-1		.8000E+00	.3209E-02	.5444E-01	.4012E-02
				NET TOTAL =	.3577E-05	

DETAIL OF NODE 5

TEMPERATURE= .3610E+02   POWER=0.   STABILITY CONSTANT =   -1   CAP= .1000E-19

MULTI-SURFACE RADIATION EXCHANGE AREA= .9380E+00   EMISSIVITY= .800

NODE	CTYPE	CMODE	C	CONDUCTANCE	FLUX	REYN NO.	BNDRY LAYER	HYDR DIA	HT TRANS COEF
4	-2		.1970E-01	.9523E-04	.8207E-04				.4834E-02
3	-2		.1530E+00	.5463E-03	.5054E-03				.3570E-02
12	-2		.3940E-01	.1539E-01	.5481E-02				.3906E-02
10	-2		.3940E-01	-.5789E-05	-.8436E-05				-.1469E-03
9	-2		.2522E+00	.3535E-03	.6486E-03				.1402E-02
8	-2		.5300E-01	.2952E-03	-.1896E-03				.5570E-02
11	-2		.2522E+00	.9525E-03	-.2977E-01				.3777E-02
7	-2		.1280E+00	.6725E-03	.5728E-08				.5254E-02
11	0		.1660E-02	.6600E-02	.5189E-01				
9	0		.6250E-03	.6250E-03	-.1147E-02				
6	0		.3000E-03	.3000E-03	-.1927E-03				
3	0		.3000E-03	.2775E-03	-.2775E-03				
15	101		.1000E+01	.3538E-02	-.3439E-01				.3538E-02
2	101	6	.1000E+01	.4222E-02	.6799E-01				.4222E-02
1	-1	6	.8000E+00	.3195E-02	.5145E-01				.3994E-02
				NET TOTAL =	.3934E-05				

DETAIL OF NODE 6

TEMPERATURE= .3674E+02   POWER=0.   STABILITY CONSTANT =   -1   CAP= .1000E-19

MULTI-SURFACE RADIATION EXCHANGE AREA= .9380E+00   EMISSIVITY= .800

NODE	CTYPE	CMODE	C	CONDUCTANCE	FLUX	REYN NO.	BNDRY LAYER	HYDR DIA	HT TRANS COEF
3	-2		.1970E-01	.8648E-04	.1355E-03				.4390E-02
4	-2		.1530E+00	.4101E-03	-.9006E-04				.2681E-02
11	-2		.3940E-01	.1473E-03	-.4510E-02				.3739E-02
9	-2		.3940E-01	.9781E-04	-.2423E-03				.2483E-02
10	-2		.2522E+00	.4040E-03	.8482E-03				.1602E-02
7	-2		.5300E-01	.2952E-03	-.1896E-03				.5570E-02
12	-2		.2522E+00	.9773E-03	-.3418E-01				.3875E-02
8	-2		.1280E+00	.6748E-03	.5251E-08				.5272E-02

Fig. 8-27. (Continued)

NODE	CTYPE	CMODE	C	CONDUCTANCE	FLUX	REYN NO.	BNDRY LAYER	HYDR DIA	HT TRANS COEF
12	0		.1660E-02	.1660E-02	-.5806E-01				
10	0		.6250E-03	.6250E-03	.1312E-02				
5	0		.3000E-03	.3000E-03	.1927E-03				
4	0		.3000E-03	.3000E-03	-.6588E-04				
15	101	6	.1000E+01	.3455E-02	-.3136E-01				.3455E-02
2	101	6	.1000E+01	.4280E-02	.7167E-01				.4280E-02
1	-1		.8000E+00	.3206E-02	.5368E-01				.4007E-02
				NET TOTAL =	.3484E-05				

DETAIL OF NODE 7   TEMPERATURE= .3610E+02   POWER=0.   STABILITY CONSTANT =   -1   CAP= .1000E-19

MULTI-SURFACE RADIATION EXCHANGE AREA= .9380E+00   EMISSIVITY= .800

NODE	CTYPE	CMODE	C	CONDUCTANCE	FLUX	REYN NO.	BNDRY LAYER	HYDR DIA	HT TRANS COEF
4	-2		.1970E-01	.9523E-04	-.8207E-04				.4834E-02
3	-2		.1530E+00	.5463E-03	.5054E-03				.3570E-02
12	-2		.3940E-01	.1539E-03	-.5481E-02				.3906E-02
10	-2		.3940E-01	-.5791E-05	-.8437E-05				-.1470E-03
9	-2		.2522E+00	.3535E-03	.6486E-03				.1402E-02
6	-2		.5300E-01	.2952E-03	-.1896E-03				.5570E-02
11	-2		.2522E+00	.9525E-03	-.2977E-01				.3777E-02
5	-2		.1280E+00	.6725E-03	-.5728E-08				.5254E-02
11	0		.1660E-02	.1660E-02	-.5189E-01				
9	0		.6250E-03	.6250E-03	-.1147E-02				
8	0		.3000E-03	.3000E-03	-.1927E-03				
3	0		.3000E-03	.3000E-03	-.2775E-03				
15	101	6	.1000E+01	.3538E-02	-.3439E-01				.3538E-02
2	101	6	.1000E+01	.4222E-02	.6799E-01				.4222E-02
1	-1		.8000E+00	.3195E-02	-.5145E-01				.3994E-02
				NET TOTAL =	.3743E-05				

DETAIL OF NODE 8   TEMPERATURE= .367 E+02   POWER=0.   STABILITY CONSTANT =   -1   CAP= .1000E-19

MULTI-SURFACE RADIATION EXCHANGE AREA= .9380E+00   EMISSIVITY= .800

NODE	CTYPE	CMODE	C	CONDUCTANCE	FLUX	REYN NO.	BNDRY LAYER	HYDR DIA	HT TRANS COEF
3	-2		.1970E-01	.8658E-04	-.1358E-03				.4390E-02
4	-2		.1530E+00	.4101E-03	-.5007E-04				.2681E-02
11	-2		.3940E-01	.1478E-03	-.4510E-02				.3739E-02
9	-2		.3940E-01	.9781E-04	-.2423E-03				.2483E-02
10	-2		.2522E+00	.4040E-03	.8482E-03				.1602E-02

Fig. 8-27. (Continued)

NODE	CTYPE	CMODE	C	CONDUCTANCE	FLUX	HT TRANS COEF
5	-2		.5300E-01	.2952E-03	.1896E-03	.5570E-02
12	-2		.2522E+00	.9773E-03	-.3418E-01	.3875E-02
6	-2		.1280E+00	.6748E-03	-.5251E-08	.5272E-02
12	0		.1660E-02	.1660E-02	.5806E-01	
10	0		.6250E-03	.6250E-03	.1312E-02	
7	0		.3000E-03	.3000E-03	.1927E-03	
4	0		.3000E-03	.3000E-03	-.6588E-04	
15	101	6	.1000E+01	.3455E-01	-.3136E-01	.3455E-02
2	101	6	.1000E+01	.4280E-02	.7167E-01	.4280E-02
1	-1		.8000E+00	.3206E-02	.5368E-01	.4007E-02

NET TOTAL = .3310E-05

TEMPERATURE= .3427E+02   POWER=0.   STABILITY CONSTANT = -1   CAP= .1000E-19

MULTI-SURFACE RADIATION EXCHANGE AREA= .1563E+01   EMISSIVITY= .800

DETAIL OF NODE   9

NODE	CTYPE	CMODE	C	CONDUCTANCE	FLUX	REYN NO.	BNDRY LAYER	HYDR DIA	HT TRANS COEF
4	-2		.2340E-01	.5847E-04	-.1577E-03				.2499E-02
3	-2		.2520E+00	-.2025E-03	-.1842E-03				-.8037E-03
8	-2		.3940E-01	.9781E-04	-.2423E-03				.2483E-02
6	-2		.3940E-01	.9781E-04	-.2423E-03				.2483E-02
12	-2		.1500E+00	.5674E-03	-.2125E-01				.3783E-02
7	-2		.2522E+00	.3535E-03	-.6486E-03				.1402E-02
5	-2		.2522E+00	.3535E-03	-.6486E-03				.1402E-02
11	-2		.5540E+00	.2019E-02	-.6683E-01				.3645E-02
7	0		.6250E-03	.6250E-03	-.1147E-02				
5	0		.6250E-03	.6250E-03	-.1147E-02				
10	0		.5000E-03	.5000E-03	-.1888E-03				
3	0		.6250E-03	.6250E-03	-.5686E-03				
15	101	6	.1560E+01	.5864E-02	-.6776E-01				.3759E-02
2	101	6	.1560E+01	.6314E-02	-.9008E-01				.4047E-02
1	-1		.1250E+01	.4946E-02	-.7057E-01				.3957E-02

NET TOTAL = .3284E-05

TEMPERATURE= .3464E+02   POWER=0.   STABILITY CONSTANT = -1   CAP= .1000E-19

MULTI-SURFACE RADIATION EXCHANGE AREA= .1563E+01   EMISSIVITY= .800

DETAIL OF NODE   10

Fig. 8-27. (Continued)

NODE	CTYPE	CMODE	C	CONDUCTANCE	FLUX	REYN NO.	BNDRY LAYER	HYDR DIA	HT TRANS COEF
3	-2		.2340E-01	-.1547E-03	.8230E-04				-.6611E-02
4	-2		.2520E+00	-.4294E-03	-.9958E-03				-.1704E-02
7	-2		.3940E-01	-.5791E-05	.8437E-05				-.1470E-03
5	-2		.3940E-01	.5789E-05	.8436E-05				-.1469E-03
11	-2		.1500E+00	.5403E-03	.1768E-01				.3602E-02
8	-2		.2522E+00	.4040E-03	.8482E-03				.1602E-02
6	-2		.2522E+00	.4040E-03	-.8482E-03				.1602E-02
12	-2		.5540E+00	.2075E-02	-.7695E-01				.3746E-02
8	0		.6250E-03	.6250E-03	-.1312E-02				
8	0		.6250E-03	.6250E-03	-.1312E-02				
9	0		.5000E-03	.5000E-03	.1888E-03				
4	0		.6250E-03	.6250E-03	-.1449E-02				.3716E-02
15	101	6	.1560E+01	.5797E-02	.6478E-01				.4084E-02
2	101	6	.1560E+01	.6372E-02	.9331E-01				.3965E-02
1	-1		.1250E+01	.4956E-02	-.7258E-01				

NET TOTAL = .3204E-05

DETAIL OF NODE 11

TEMPERATURE= .6736E+02   POWER= .1000E+01   STABILITY CONSTANT = -1   CAP= .1000E-19

MULTI-SURFACE RADIATION EXCHANGE AREA= .1563E+01   EMISSIVITY= .800

NODE	CTYPE	CMODE	C	CONDUCTANCE	FLUX	REYN NO.	BNDRY LAYER	HYDR DIA	HT TRANS COEF
4	-2		.2340E-01	.8767E-04	.2665E-02				.3747E-02
3	-2		.2520E+00	.9502E-03	.3058E-01				.3771E-02
8	-2		.3940E-01	.1473E-03	.4510E-02				.3739E-02
6	-2		.3940E-01	.1473E-03	.4510E-02				.3739E-02
10	-2		.1500E+00	.5403E-03	.1768E-01				.3602E-02
7	-2		.2522E+00	.9525E-03	.2977E-01				.3777E-02
5	-2		.2522E+00	.9525E-03	.2977E-01				.3777E-02
9	-2		.5540E+00	.2019E-02	.6683E-01				.3645E-02
7	0		.1660E-02	.1660E-02	.5189E-01				
5	0		.1660E-02	.1660E-02	.5189E-01				
12	0		.2000E+00	.2000E+00	-.8714E+00				
3	0		.1660E-02	.1660E-02	.5343E-01				
15	101	6	.1560E+01	.7293E-02	.1571E+00				.4675E-02
14	-1		.5000E+00	.2336E-02	.1106E+00				.4672E-02
14	-1		.1300E+01	.6073E-02	.2876E+00				

Fig. 8-27. (Continued)

NODE	CTYPE	CMODE	C	CONDUCTANCE	FLUX	REYN NO.	BNDRY LAYER	HYDR DIA	HT TRANS COEF
13	101	6	.6250E+00	.3850E-02	.1823E+00				.4672E-02
13	1		.4065E+01	.1669E-01	.7903E+00				.6159E-02
				NET TOTAL =	.1000E+01				

DETAIL OF NODE 12   TEMPERATURE= .7172E+02   POWER= .3000E+02   STABILITY CONSTANT =   -1   CAP= .1000E-19

MULTI-SURFACE RADIATION EXCHANGE AREA= .1563E+01   EMISSIVITY= .800

NODE	CTYPE	CMODE	C	CONDUCTANCE	FLUX	REYN NO.	BNDRY LAYER	HYDR DIA	HT TRANS COEF
3	-2		.2340E-01	.9119E-04	.3332E-02				.3897E-02
4	-2		.2520E+00	.9784E-03	.3401E-01				.3883E-02
7	-2		.3940E-01	.1539E-03	.5481E-02				.3906E-02
5	-2		.3940E-01	.1539E-03	.5481E-02				.3906E-02
9	-2		.1500E+00	.5674E-03	.2125E-01				.3783E-02
8	-2		.2522E+00	.9773E-03	.3418E-01				.3875E-02
6	-2		.2522E+00	.9773E-03	.3418E-01				.3875E-02
10	-2		.5540E+00	.2075E-02	.7695E-01				.3746E-02
8	0		.1660E-02	.1660E-02	.5806E-01				
6	0		.1660E-02	.1660E-02	.5806E-01				
11	0		.2000E+00	.2000E+00	.8714E+00				
4	0		.1660E-02	.1660E-02	.5769E-01				
15	101	6	.1560E+01	.7779E-02	.2015E+00				.4986E-02
14	-1		.5000E+00	.2387E-02	.1234E+00				.4773E-02
14	-1		.1300E-02	.6205E-02	.3209E+00				.4773E-02
13	101	6	.6250E+00	.3970E-02	.2053E+00				.6352E-02
13	1		.4065E+01	.1718E-01	.8888E+00				
				NET TOTAL =	.3000E+01				

DETAIL OF NODE -13   TEMPERATURE= .2000E+02   POWER=0.   STABILITY CONSTANT =   -1   CAP= .1000E-19

THIS IS A CONSTANT TEMPERATURE NODE

NODE	CTYPE	CMODE	C	CONDUCTANCE	FLUX	REYN NO.	BNDRY LAYER	HYDR DIA	HT TRANS COEF
12	101	6	.6250E+00	.3970E-02	-.2053E+00				.6352E-02
11	101	6	.6250E+00	.3850E-02	-.1823E+00				.6159E-02
12	1		.4065E+01	.1718E-01	-.8888E+00				
11	1		.4065E+01	.1669E-01	-.7903E+00				
				NET TOTAL =	-.2067E+01				

Fig. 8-27. (Continued)

DETAIL OF NODE -14 TEMPERATURE= .2000E+02 POWER=0. STABILITY CONSTANT = -1 CAP= .1000E-19

THIS IS A CONSTANT TEMPERATURE NODE

NODE	CTYPE	CMODE	C	CONDUCTANCE	FLUX	REYN NO.	BNDRY LAYER	HYDR DIA	HT TRANS COEF
12	-1		.5000E+00	.23?7E-02	-.1?34E+00				.4773E-02
11	-1		.5000E+00	.233?E-02	-.1106E+00				.4672E-02
12	-1		.1300E+01	.6205E-02	-.3?09E+00				.4773E-02
11	-1		.1300E+01	.607?E-02	-.?876E+00				.4672E-02
					NET TOTAL =-.8126E+00				

DETAIL OF NODE 15 TEMPERATURE= .4582E+02 POWER=0. STABILITY CONSTANT = -1 CAP= .1000E-19

NODE	CTYPE	CMODE	C	CONDUCTANCE	FLUX	REYN NO.	BNDRY LAYER	HYDR DIA	HT TRANS COEF
8	101	6	.1000E+01	.3455E-02	.3136E-01				.3455E-02
7	101	6	.1000E+01	.3538E-02	.3439E-01				.3538E-02
6	101	6	.1000E+01	.3455E-02	.3136E-01				.3455E-02
5	101	6	.1000E+01	.3538E-02	.3439E-01				.3538E-02
12	101	6	.1560E+01	.7779E-02	.2015E+00				.4986E-02
11	101	6	.1560E+01	.7293E-02	-.1571E+00				.4675E-02
10	101	6	.1560E+01	.5792E-02	.6478E-01				.3716E-02
9	101	6	.1560E+01	.5864E-02	.6776E-01				.3759E-02
4	103	8	.1000E+01	.3002E-02	.2659E-01				.3002E-02
3	102	7	.1000E+01	.6380E-02	.6791E-01				.6380E-02
					NET TOTAL =-.7868E-05				

ENERGY BALANCE = .1655E-03

Fig. 8-27. (Continued)

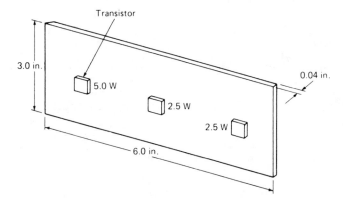

Fig. 8-28. Transistors attached to aluminum panel.

output listing is produced according to the print interval and node detail instructions specified in the final three lines of data sets 9–14 for each case. The results show that the 3-watt transistor case temperature is reduced from 74.86°C to 71.72°C, owing to the increased emissivity within the box.

8-7 PROBLEM—ALUMINUM PANEL WITH THREE TRANSISTORS

A solid bulkhead, rear or side panel may often be used to heat sink components. Such a system is illustrated in Fig. 8-28, where both front and rear surfaces are in a 25°C ambient. Simple radiation to a nonreflecting ambient is presumed. The panel is rather ordinary aluminum with an emissivity of 0.1 and a thermal conductivity $k = 3.5$ watts/in. · °C.

The uniform planar nodal subdivision in Fig. 8-29 is established by the approximate size of the transistor bases, nodes 49, 53, and 57. The panel is not subdivided through the thickness because the temperature gradient in this direction is expected to be very small due

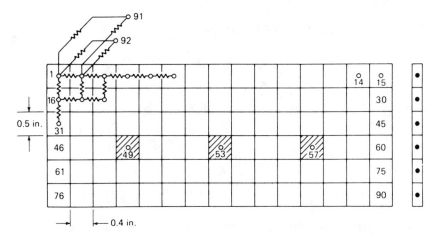

Fig. 8-29. Thermal network model of aluminum panel.

DATA SET

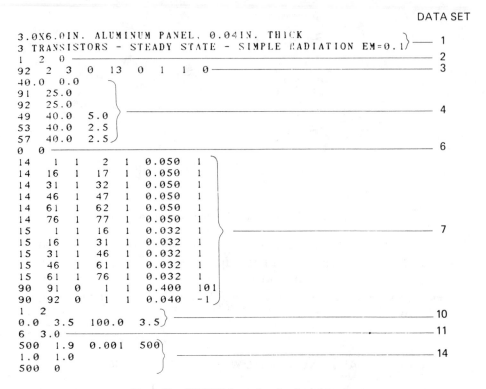

```
3.0X6.0IN.  ALUMINUM PANEL,  0.04IN.  THICK                          ⎫— 1
3 TRANSISTORS  -  STEADY STATE  -  SIMPLE RADIATION  EM=0.1⎠
1    2    0  ————————————————————————————————————————————————————— 2
92    2    3    0   13    0    1    1    0 ———————————————————————— 3
40.0   0.0
91    25.0
92    25.0
49    40.0   5.0                                                  ⎫— 4
53    40.0   2.5
57    40.0   2.5 
0    0  ——————————————————————————————————————————————————————————— 6
14    1    1     2    1    0.050    1   ⎫
14    16   1    17    1    0.050    1
14    31   1    32    1    0.050    1
14    46   1    47    1    0.050    1
14    61   1    62    1    0.050    1
14    76   1    77    1    0.050    1
15    1    1    16    1    0.032    1  ⎬ ————————————————————————— 7
15    16   1    31    1    0.032    1
15    31   1    46    1    0.032    1
15    46   1    61    1    0.032    1
15    61   1    76    1    0.032    1
90    91   0     1    1    0.400    101⎫
90    92   0     1    1    0.040    -1 ⎭
1    2  ——————————————————————————————————————————————————————————— 10
0.0   3.5   100.0   3.5⎠
6    3.0  ———————————————————————————————————————————————————•————— 11
500    1.9   0.001   500⎫
1.0    1.0                ⎬ ————————————————————————————————————— 14
500    0
```

Fig. 8-30. TNETFA input for aluminum panel.

to the rather large thermal conductivity. Ambient nodes 91 and 92 are used to collect convective and radiative heat transfer. Conductive losses from the panel into supporting structures are not included in this model.

The TNETFA input listing in Fig. 8-30 indicates that 13 lines are used to interconnect the nodes with the appropriately specified conductances. The first six conductor lines specify the conductive cross-sectional area/path length ratio for the horizontal direction:

$$A_k/L = (0.50 \text{ in.})(0.040 \text{ in.})/(0.40 \text{ in.})$$

$$= 0.050 \text{ in.}$$

The equivalent ratio for the vertical direction is:

$$A_k/L = (0.40 \text{ in.})(0.040 \text{ in.})/(0.5 \text{ in.})$$

$$= 0.032 \text{ in.}$$

The thermal conductivity is specified via two data pairs in array 1. This method is used to permit very easy modification of k by chang-

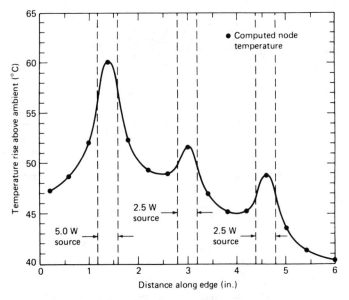

Fig. 8-31. Computed temperature profile taken through nodes 46–60 of aluminum panel.

ing only two input constants. Note that a temperature-dependent k is not required to use the array feature.

Natural convection from both surfaces of each node is accounted for by three lines of input. Line 23 connects 90 nodes to node 91. The total surface area is $(0.5 \text{ in.})(0.4 \text{ in.})(2) = 0.4 \text{ in.}^2$. The fourth from last line specifies natural convection from a small device with a dimensional parameter $P = H = 3.0$ in.

Simple radiation from each node to node 92 requires quantitative input of $\epsilon A_s = (0.1)(0.5 \text{ in.})(0.4 \text{ in.})(2)$.

Iteration is started at 40.0°C for all non-constant-temperature nodes.

The computed temperature rise above ambient for nodes 46–60 is plotted in Fig. 8-31. Temperatures between nodes can only be estimated, and in this case a graphical method is more than adequate. A more easily obtained solution would have resulted with TAMS although the network method has the advantage of easy incorporation of nonlinearities such as radiation and natural convection.

8.8 METAL ENCLOSURE WITH TEN CIRCUIT CARDS

Three variations of the cabinet illustrated in Fig. 8-32 are considered: (1) totally sealed; (2) inlet area = 11 in.², exit area = 9 in.²; (3) inlet area = 11 in.², exit area = 20 in.². The hole pattern in the top panel is shown for case (3). The top panel hole pattern somewhat resembles the bottom panel for case (2). The specified ventilation areas are free areas available for air flow. The circuit cards are 4.0 in. high and vertically oriented. The heat-dissipating components are uniformly distributed and have a very low profile.

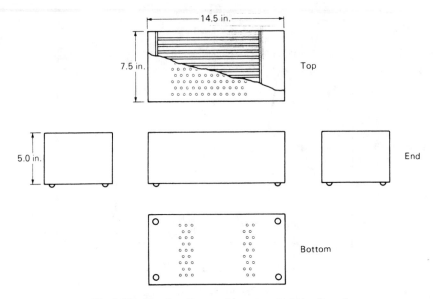

Fig. 8-32.   Metal enclosure with ten vertical circuit cards.

The thermal network model for the metal enclosure is shown in Fig. 8-33. Temperature gradients through the cabinet walls are so small that elements for wall conduction are omitted. Including the relatively large wall conductances could lead to very slow convergence of the iteration process, as discussed in Section 8.4.

The internal air temperature is obtained from node 1, which convects (natural) to wall nodes 4–9. The wall nodes in turn convect to ambient node 10 and radiate to ambient node 11. The exterior cabi-

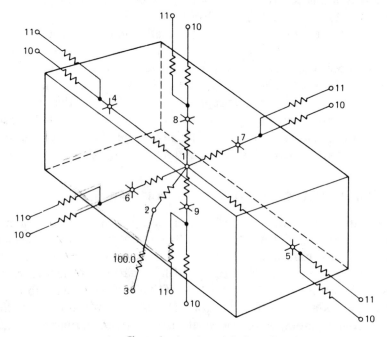

Fig. 8-33.   Thermal network model of metal enclosure.

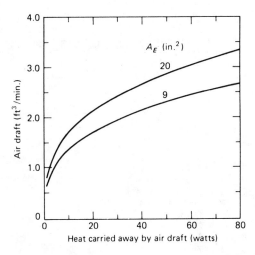

Fig. 8-34. Computed air draft for metal enclosure with ten circuit cards.

net surfaces have an emissivity of 0.7. No compensation is made to the surface areas for the reduction by ventilation perforations. The air draft is simulated by the one-way element between nodes 1 and 2 for airflow. Node 2 is connected by an arbitrarily large conductance (100.0) to ambient node 3 in the same fashion as discussed in Section 8.4.

The natural ventilation air draft is computed with the iteration air-draft formula in the manner used in Section 8.4. A dissipation height $d = 3.5$ in. is used. The circuit cards have a negligible airflow resistance compared to the inlet/exit resistance. The total airflow resistance in case (2) is:

$$R = 2.4 \times 10^{-3} \left[\frac{1}{A_I^2} + \frac{1}{A_E^2} \right]$$

$$= 2.4 \times 10^{-3} \left[\frac{1}{(11)^2} + \frac{1}{(9)^2} \right]$$

$$= 4.95 \times 10^{-5} \text{ in. } H_2O/(ft^3/min.)^2$$

In case (3):

$$R = 2.4 \times 10^{-3} \left[\frac{1}{(11)^2} + \frac{1}{(20)^2} \right]$$

$$= 2.58 \times 10^{-5} \text{ in. } H_2O/(ft^3/min.)^2$$

The computed air draft vs. heat into the airstream is plotted in Fig. 8-34 for cases (2) and (3).

DATA SET

```
PORTABLE METAL ENCLOSURE WITH 10 CIRCUIT CARDS⎫── 1
VENTED. EMISSIVITY=0.7                         ⎭
 1   2   0 ─────────────────────────────────────────── 2
11   3   1   0   9   2   0   3   0 ──────────────────── 3
30.0    0.0        ⎫
 3   20.0          ⎪
10   20.0          ⎬ ────────────────────────────────── 4
11   20.0          ⎪
 1   30.0  100.0 ──⎭
 0   0 ──────────────────────────────────────────────── 6
 2  10   0   4   1   37.5   101 ⎫
 2   1   0   4   1   37.5   101 ⎪
 2  10   0   6   1   72.5   101 ⎪
 2   1   0   6   1   72.5   101 ⎪
 2   1   7   8   2  108.8   102 ⎬ ────────────────────── 7
 2   1   8   9   1  108.8   103 ⎪
 2  11   0   4   1   26.3   - 1 ⎪
 2  11   0   6   1   50.8   - 1 ⎪
 2  11   0   8   1   76.1   - 1 ⎭
 2   3  100.0   0 ⎫
 1   2   3.2   301 ⎭ ────────────────────────────────── 8
 1   5.0 ⎫
 2   2.47⎬ ───────────────────────────────────────────11
 3   2.47⎭
50   1.0   0.001   50 ⎫
1.0   1.0             ⎬ ──────────────────────────────14
50   1                ⎭
```

Fig. 8-35. TNETFA input for metal enclosure.

A sample TNETFA input is reproduced in Fig. 8-35 for a total heat dissipation of 100 watts in case (3). The appropriate air draft between nodes 1 and 2 was obtained by computing the problem a few times and changing the input air draft until the airflow and heat transfer (computed and indicated in node details in the printout for node 1) are consistent with Fig. 8-34. This same method is described in Section 8.4. A draft of 3.2 ft³/min. was the final airflow for case (3) with a total of 100 watts dissipation. The corresponding output is reproduced in Fig. 8-36.

Computed and measured air temperature rises are plotted in Fig. 8-37 for several values of total heat dissipation. The empirical temperatures are the average of nine thermocouple readings, three for each of three cards in one half of the box. The thermocouple beads were distributed across the top of the circuit cards at distances from each card equivalent to an approximate component height. The standard deviation of each set of nine measurements is indicated next to each data point.

The empirical–theoretical correlation is consistent with other studies not discussed here; in summary, vertical–horizontal flat plate theoretical models for natural convection and radiation cooled systems predict air temperatures consistent with an "average maximum" obtained from the warmer air in the upper portion of the cabinet, or chamber in the case of complex systems.

```
                    NETWORK THERMAL ANALYSIS

PORTABLE METAL ENCLOSURE WITH 10 CIRCUIT CARDS
VENTED. EMISSIVITY=0.7

UNITS=2

NUMBER OF NODES=  11    NUMBER OF CONDUCTORS=  40

NLOOP=  50   TPRINT=  50   NPRINT=  1   LOOPEN=  50   ALDT= .1000E-02   BETA= 1.00

NATURAL CONVECTION PARAMETER

  1   VERTICAL FLAT PLATE OR CYLINDER                                              P= .5000E+01
  2   HORIZONTAL FLAT PLATE OR CYLINDER, HEATED SIDE FACING UP OR COOLED SIDE FACING DOWN    P= .2470E+01
  3   HORIZONTAL FLAT PLATE OR CYLINDER, HEATED SIDE FACING DOWN OR COOLED SIDE FACING UP     P= .2470E+01

                           LOOPCT=  0         MAXDT=  -1

                                TEMPERATURES

T(  1)= .3000E+02   T(  2)= .3000E+02   T(  3)= .2000E+02   T(  4)= .3000E+02   T(  5)= .3000E+02   T(  6)= .3000E+02
T(  7)= .3000E+02   T(  8)= .3000E+02   T(  9)= .3000E+02   T( 10)= .2000E+02   T( 11)= .2000E+02

                           LOOPCT=  7         MAXDT= .8824E-03

                                TEMPERATURES

T(  1)= .5803E+02   T(  2)= .2000E+02   T(  3)= .2000E+02   T(  4)= .3418E+02   T(  5)= .3418E+02   T(  6)= .3418E+02
T(  7)= .3418E+02   T(  8)= .3448E+02   T(  9)= .3181E+02   T( 10)= .2000E+02   T( 11)= .2000E+02

                           LOOPCT=  7         MAXDT= .8824E-03
```

Fig. 8-36. TNETFA output for vented metal enclosure.

DETAIL OF NODE 1

TEMPERATURE= .5803E+02   POWER= .1000E+03   STABILITY CONSTANT =   -1   CAP= .1000E-19

NODE	CTYPE	CMODE	C	CONDUCTANCE	FLUX	REYN NO.	BNDRY LAYER	HYDR DIA	HT TRANS COEF
2	301		.3200E+01	.1734E+01	.6595E+02				
9	103	3	.1088E+03	.2153E+00	.5644E+01				.1979E-02
8	102	2	.1088E+03	.4186E+00	.9856E+01				.3847E-02
7	101	1	.7250E+02	.2564E+00	.6113E+01				.3536E-02
6	101	1	.7250E+02	.2564E+00	.6113E+01				.3536E-02
5	101	1	.3750E+02	.1326E+00	.3163E+01				.3536E-02
4	101	1	.3750E+02	.1326E+00	.3163E+01				.3536E-02

NET TOTAL = .1000E+03

DETAIL OF NODE 2

TEMPERATURE= .2000E+02   POWER=0.   STABILITY CONSTANT =   -1   CAP= .1000E-19

NODE	CTYPE	CMODE	C	CONDUCTANCE	FLUX	REYN NO.	BNDRY LAYER	HYDR DIA	HT TRANS COEF
3	0		.1000E+03	.1000E+03	0.				

NET TOTAL. =0.

DETAIL OF NODE -3

TEMPERATURE= .2000E+02   POWER=0.   STABILITY CONSTANT =   -1   CAP= .1000E-19

THIS IS A CONSTANT TEMPERATURE NODE

NODE	CTYPE	CMODE	C	CONDUCTANCE	FLUX	REYN NO.	BNDRY LAYER	HYDR DIA	HT TRANS COEF
2	0		.1000E+03	.1000E+03	0.				

NET TOTAL =0.

DETAIL OF NODE 4

TEMPERATURE= .3411E+02   POWER=0.   STABILITY CONSTANT =   -1   CAP= .1000E-19

NODE	CTYPE	CMODE	C	CONDUCTANCE	FLUX	REYN NO.	BNDRY LAYER	HYDR DIA	HT TRANS COEF
11	-1		.2630E+02	.1040E+00	.1475E+01				.3955E-02
1	101	1	.3750E+02	.1326E+00	-.3163E+01				.3536E-02
10	101	1	.3750E+02	.1191E+00	.1688E+01				.3175E-02

NET TOTAL = .1997E-05

DETAIL OF NODE 5

TEMPERATURE= .3418E+02   POWER=0.   STABILITY CONSTANT =   -1   CAP= .1000E-19

NODE	CTYPE	CMODE	C	CONDUCTANCE	FLUX	REYN NO.	BNDRY LAYER	HYDR DIA	HT TRANS COEF
11	-1		.2630E+02	.1040E+00	.1475E+01				.3955E-02
1	101	1	.3750E+02	.1326E+00	-.3163E+01				.3536E-02
10	101	1	.3750E+02	.1191E+00	.1688E+01				.3175E-02

NET TOTAL = .1997E-05

Fig. 8-36. (Continued)

DETAIL OF NODE 6   TEMPERATURE= .3418E+02   POWER=0.   STABILITY CONSTANT =   -1   CAP= .1000E-19

NODE	CTYPE	CMODE	C	CONDUCTANCE	FLUX	REYN NO.	BNDRY LAYER	HYDR DIA	HT TRANS COEF
11	-1		.5080E+02	.2009E+00	.2849E+01				.3955E-02
1	101	1	.7250E+02	.2564E+00	-.6113E+01				.3536E-02
10	101	1	.7250E+02	.2302E+00	.3264E+01				.3175E-02
				NET TOTAL =	.4003E-05				

DETAIL OF NODE 7   TEMPERATURE= .3418E+02   POWER=0.   STABILITY CONSTANT =   -1   CAP= .1000E-19

NODE	CTYPE	CMODE	C	CONDUCTANCE	FLUX	REYN NO.	BNDRY LAYER	HYDR DIA	HT TRANS COEF
11	-1		.5080E+02	.2009E+00	.2849E+01				.3955E-02
10	101	1	.7250E+02	.2564E+00	-.6113E+01				.3536E-02
1	101	1	.7250E+02	.2302E+00	.3264E+01				.3175E-02
				NET TOTAL =	.4003E-05				

DETAIL OF NODE 8   TEMPERATURE= .3448E+02   POWER=0.   STABILITY CONSTANT =   -1   CAP= .1000E-19

NODE	CTYPE	CMODE	C	CONDUCTANCE	FLUX	REYN NO.	BNDRY LAYER	HYDR DIA	HT TRANS COEF
11	-1		.7610E+02	.3015E+00	.4366E+01				.3961E-02
10	102	2	.1088E+03	.3791E+00	.5490E+01				.3484E-02
1	102	2	.1088E+03	.4186E+00	-.9856E+01				.3847E-02
				NET TOTAL =	.3093E-04				

DETAIL OF NODE 9   TEMPERATURE= .3181E+02   POWER=0.   STABILITY CONSTANT =   -1   CAP= .1000E-19

NODE	CTYPE	CMODE	C	CONDUCTANCE	FLUX	REYN NO.	BNDRY LAYER	HYDR DIA	HT TRANS COEF
11	-1		.7610E+02	.2974E+00	.3513E+01				.3908E-02
10	103	3	.1088E+03	.1804E+00	.2131E+01				.1658E-02
1	103	3	.1088E+03	.2153E+00	-.5644E+01				.1979E-02
				NET TOTAL =	-.3352E-04				

Fig. 8-36. (Continued)

DETAIL OF NODE -10   TEMPERATURE= .2000E+02   POWER=0.   STABILITY CONSTANT = -1   CAP= .1000E-19

THIS IS A CONSTANT TEMPERATURE NODE

NODE	CTYPE	CMODE	C	CONDUCTANCE	FLUX	REYN NO.	BNDRY LAYER	HYDR DIA	HT TRANS COEF
9	103	3	.1088E+03	.1804E+00	-.2131E+01				.1658E-02
8	102	2	.1088E+03	.3791E+00	-.5490E+01				.3484E-02
7	101	1	.7250E+02	.2302E+00	-.3264E+01				.3175E-02
6	101	1	.7250E+02	.2302E+00	-.3264E+01				.3175E-02
5	101	1	.3750E+02	.1191E+00	-.1688E+01				.3175E-02
4	101	1	.3750E+02	.1191E+00	-.1688E+01				.3175E-02
				NET TOTAL =-.1753E+02					

DETAIL OF NODE -11   TEMPERATURE= .2000E+02   POWER=0.   STABILITY CONSTANT = -1   CAP= .1000E-19

THIS IS A CONSTANT TEMPERATURE NODE

NODE	CTYPE	CMODE	C	CONDUCTANCE	FLUX	REYN NO.	BNDRY LAYER	HYDR DIA	HT TRANS COEF
9	-1		.7610E+02	.2974E+00	-.3513E+01				.3908E-02
8	-1		.7610E+02	.3015E+00	-.4366E+01				.3961E-02
7	-1		.5080E+02	.2009E+00	-.2849E+01				.3955E-02
6	-1		.5080E+02	.2009E+00	-.2849E+01				.3955E-02
5	-1		.2630E+02	.1040E+00	-.1475E+01				.3955E-02
4	-1		.2630E+02	.1040E+00	-.1475E+01				.3955E-02
				NET TOTAL =-.1653E+02					

ENERGY BALANCE =-.2206E-03

Fig. 8-36. (Continued)

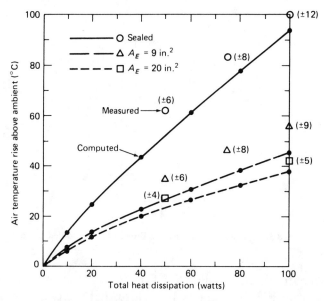

Fig. 8-37. Computed and measured internal air temperature rise for metal enclosure.

8.9 PROBLEM—FORCED AIR COOLED CABINET

The forced air cooled cabinet in Fig. 8-38 is selected as an example of a pressure-airflow system that is soluble only by using an iterative method.

The corresponding airflow resistance network, Fig. 8-39, simulates the main card cage (R_c), two air inlets (R_a, R_b), and perforations in a longitudinal bulkhead (R_d, R_e). A simple wire finger guard over the rear panel mounted fan is presumed to have negligible resistance. The chamber opposing the card cage has a sufficiently low packaging density to offer negligible airflow resistance. The quanti-

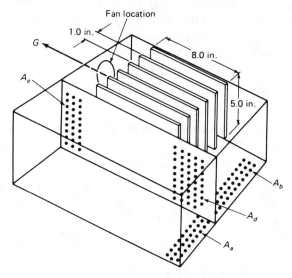

Fig. 8-38. Forced air cooled cabinet.

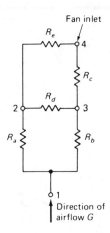

Fig. 8-39. Airflow network model for forced air cooled cabinet.

tative estimates listed in Table 8-4 for the five resistance elements are obtained from Figs. 6-9 and 6-10. Note that the circuit in Fig. 8-39 cannot be reduced to a single equivalent resistance using the standard formulae for combining series and parallel elements.

A system resistance curve (pressure vs. total airflow) may be constructed by computing the pressures (sum of losses) at node 4 for several different values of total airflow. Since the cabinet is negatively pressurized, node 4 may be considered as a negative source of airflow. Node 1 is set as a fixed pressure of 0.0 in. H_2O.

The TNETFA input for a computation at the intersection of the fan and resistance curves, Fig. 8-41, is reproduced in Fig. 8-40. The value CTYPE = 402 indicates a turbulent flow or $\Delta h = RG^2$ element definition. Forced airflow modeling is the one occasion in which TNETFA requires a resistance rather than conductance input. The output, Fig. 8-42, corresponding to this input contains the computed pressures in the temperature printout and the airflow distribution for a total of 11.5 ft³/min. in the detailed node printout under the column labeled flux. TNETFA uses a pressure/airflow analogy corresponding to temperature/heat rate flow. The airflow results are tabulated in Table 8-5. The various air temperatures are easily com-

Table 8-4. Resistance values for forced air cooled cabinet.

Resistance	Value (in. $H_2O/(ft^3/min.)^2$)
R_a	$2.4 \times 10^{-3}/[(3.0)(1.0)(0.33)]^2 = 2.45 \times 10^{-3}$
R_b	$2.4 \times 10^{-3}/[(4.0)(1.0)(0.33)]^2 = 1.38 \times 10^{-3}$
R_c	$3.08 \times 10^{-4}(8.0)/[(6)(5.0)(1.0)]^2 = 2.74 \times 10^{-6}$
R_d	$2.4 \times 10^{-3}/[(2.0)(3.0)(0.33)]^2 = 6.12 \times 10^{-4}$
R_e	$2.4 \times 10^{-3}/[(3.0)(1.0)(0.33)]^2 = 2.45 \times 10^{-3}$

DATA SET

```
FORCED AIR COOLED CABINET
TOTAL AIRFLOW = 11.5 CFM              ⎫——— 1
1    0    0 ——————————————————————————— 2
4    1    1    0    0    5    0    0    0 ——— 3
-0.05    0.0                       ⎫
1    0.0                           ⎬——————— 4
4    -0.05    -11.5                ⎭
0    0 ——————————————————————————————— 6
1    2    2.45E-3    402           ⎫
1    3    1.38E-3    402           ⎪
2    3    6.12E-4    402           ⎬——————— 8
2    4    2.45E-3    402           ⎪
3    4    2.74E-6    402           ⎭
100    1.9    1.0E-5    100        ⎫
1.0    1.0                         ⎬————14
100    1                           ⎭
```

Fig. 8-40. TNETFA input for forced air cooled cabinet.

puted for any given heat dissipation within the enclosure. Average air velocities may also be estimated for the two chambers and used in conjunction with component thermal resistance curves to obtain device case and junction temperatures.

8.10 PROBLEM—TWO DIMENSIONAL IDEAL FLUID ANALYSIS OF A CIRCUIT CARD

The similarity between the equations for a thermal network and an ideal fluid makes it possible to compute velocity potentials and streamlines with TNETFA. In actual practice, the laminar streamlines and othogonal velocity potentials are an idealization that seldom exists in electronic equipment. The concept is sometimes useful, however, in making preliminary judgments concerning component placement on circuit cards. Figure 8-43 illustrates such an application.

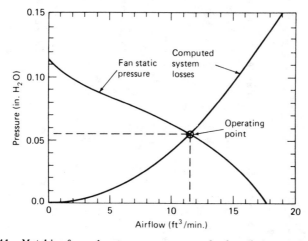

Fig. 8-41. Matching fan and system pressure curves for forced air cooled cabinet.

NETWORK THERMAL ANALYSIS

FORCED AIR COOLED CABINET
TOTAL AIRFLOW = 11.5 CFM

UNITS=0

NUMBER OF NODES= 4   NUMBER OF CONDUCTORS= 10

NLOOP= 100   TPRINT= 100   NPRINT= 1   LOOPEN= 100   ALDT= .1000E-04   BETA= 1.90

LOOPCT= 0   MAXDT= -1

TEMPERATURES

T(1)=0.   T(2)=-.5000E-01   T(3)=-.5000E-01   T(4)=-.5000E-01

LOOPCT= 100   MAXDT= .3522E-04

TEMPERATURES

T(1)=0.   T(2)=-.4892E-01   T(3)=-.5430E-01   T(4)=-.5457E-01

ENERGY BALANCE = .7588E+00

LOOPCT= 100   MAXDT= .3522E-04

Fig. 8-42. TNETFA output for forced air cooled cabinet.

DETAIL OF NODE -1   TEMPERATURE=0.   POWER=0.   STABILITY CONSTANT =   -I   CAP= .1000E-19

THIS IS A CONSTANT TEMPERATURE NODE

NODE	CTYPE	CMODE	C	CONDUCTANCE	FLUX	REYN NO.	BNDRY LAYER	HYDR DIA	HT TRANS COEF
3	402		.1380E-02	.1155E+03	.6273E+01				
2	402		.2450E-02	.9134E+02	.4468E+01				
				NET TOTAL =	.1074E+02				

DETAIL OF NODE 2   TEMPERATURE=-.4892E-01   POWER=0.   STABILITY CONSTANT =   -I   CAP= .1000E-19

NODE	CTYPE	CMODE	C	CONDUCTANCE	FLUX	REYN NO.	BNDRY LAYER	HYDR DIA	HT TRANS C(EF
4	402		.2450E-02	.2688E+03	.1519E+01				
3	402		.6120E-03	.5511E+03	.2965E+01				
1	402		.2450E-02	.9134E+02	-.4468E+01				
				NET TOTAL =	.1520E-01				

DETAIL OF NODE 3   TEMPERATURE=-.5430E-01   POWER=0.   STABILITY CONSTANT =   -I   CAP= .1000E-19

NODE	CTYPE	CMODE	C	CONDUCTANCE	FLUX	REYN NO.	BNDRY LAYER	HYDR DIA	HT TRANS COEF
4	402		.2740E-05	.3679E+05	.9920E+01				
2	402		.6120E-03	.5511E+03	-.2965E+01				
1	402		.1380E-02	.1155E+03	-.6273E+01				
				NET TOTAL =	.6825E+00				

DETAIL OF NODE 4   TEMPERATURE=-.5457E-01   POWER=-.1150E+02   STABILITY CONSTANT =   -I   CAP= .1000E-19

NODE	CTYPE	CMODE	C	CONDUCTANCE	FLUX	REYN NO.	BNDRY LAYER	HYDR DIA	HT TRANS COEF
3	402		.2740E-05	.3679E+05	-.9920E+01				
2	402		.2450E-02	.2688E+03	-.1519E+01				
				NET TOTAL =	-.1144E+02				

ENERGY BALANCE = .7588E+00

Fig. 8-42. (Continued)

Table 8-5. Airflow components
in various resistance legs of a
forced air cooled cabinet. Total
airflow = 11.5 ft^3/min.

Resistance	Airflow (ft^3/min.)
R_a	4.47
R_b	6.27
R_c	9.92
R_d	2.97
R_e	1.52

Card guides, cabinet walls, and bulkheads often prevent flow across certain card boundaries. In this example, air enters a left bottom slot about 1.0 in. high and exits across the top card edge.

Construction of two slightly different grids is required for the potential and streamline models. The grids are sufficiently fine to maintain adequate accuracy, as the results demonstrate. The same grid subdividing should be used for each problem, but differing boundary conditions lead to two different node numbering schemes. The velocity potential model is shown in Fig. 8-44. Node 85 defines

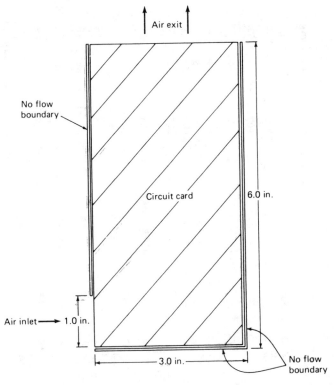

Fig. 8-43. Forced air cooled circuit card geometry applicable to ideal fluid flow analysis.

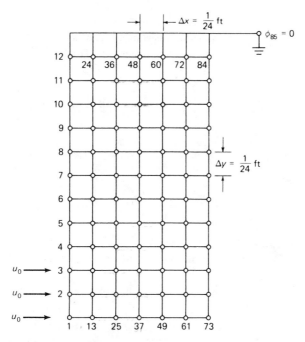

Fig. 8-44. Grid and node identification for two-dimensional velocity potential solution.

an equipotential of value 0.0. Nodes 1, 2, and 3 are, as discussed in Section 6.7, thought of as connected to fictitious nodes just to the left. The inlet flow of 100 ft/min. is distributed between nodes 1, 2, and 3 and quantified as source inputs $u_0/\Delta x$ (100 ft/min.)/(1/24 ft) = 2400.0.

TNETFA input rules require conductance input between nodes defined by:

$$C_{ij} = 1/(\Delta x)^2: \quad x\text{-direction (nodes } 1 \to 13, \text{ etc.)}$$

$$= 1/(1/24)^2 = 576.0$$

$$C_{ij} = 1/(\Delta y)^2: \quad y\text{-direction (nodes } 1 \to 2, \text{ etc.)}$$

$$= 1/(1/24)^2 = 576.0$$

An x- (CTYPE = 0) or y-direction (CTYPE = 1) indicator is also specified. One two-pair data array with a 1.0 multiplicative factor is also required. Note that the third line of input for a velocity potential solution requires MODE = 2 (Fig. 8-45).

A partial output is reproduced in Fig. 8-46 for the velocity potential. The potential at each node is obtained from the temperature printout (T analogous to ϕ). An output feature unique to this problem is the velocity component map specifying v_x, v_y, $v = \sqrt{v_x^2 + v_y^2}$,

DATA SET

```
VELOCITY POTENTIAL SOLUTION ⎫──── 1
                            ⎭
 2    0    0 ────────────────────────── 2
85    1    3    0   14    0    1    0    0 ── 3
 0.0    0.0
85      0.0
 1      0.0    2400.0  ⎫
 2      0.0    2400.0  ⎬──────────── 4
 3      0.0    2400.0  ⎭
 0      0 ──────────────────────────── 6
12      1    1    13    1    576.0    0
12     13    1    25    1    576.0    0
12     25    1    37    1    576.0    0
12     37    1    49    1    576.0    0
12     49    1    61    1    576.0    0
12     61    1    73    1    576.0    0
11      1    1     2    1    576.0    1
11     13    1    14    1    576.0    1  ⎫── 7
11     25    1    26    1    576.0    1
11     37    1    38    1    576.0    1
11     49    1    50    1    576.0    1
11     61    1    62    1    576.0    1
11     73    1    74    1    576.0    1
 7     85    0    12   12    1.0E10    1 ⎭
 1    2
 0.0    1.0    100.0    1.0 ⎫──────── 10
100.0 ──────────────────────────────── 13
100     1.9    0.001    100 ⎫
 1.0    1.0                 ⎬──────── 14
100     1                  ⎭
```

Fig. 8-45. TNETFA input for two-dimensional velocity potential.

and theta, the angle between v and v_x where the x-direction was specified by CTYPE = 0 in the input.

The node and grid patterns for the streamline problem are shown in Fig. 8-47. The nonflow boundaries are specified by constant stream function values $\psi_{63} = 0$ and ψ_1. The value of ψ_1 is computed from the inlet velocity u_0 in the x-direction and the grid spacing in the y-direction using E6.23:

$$\psi_1 = \psi_{63} + u_0 h$$

$$= 0.0 + (100)(2/24)$$

$$= 8.33$$

Source magnitude definitions are not required in stream function problems. The equivalent conductances are $C_{ij} = 1/(\Delta x)^2 = 1/(\Delta y)^2 = 1/(24)^2 = 576$, i.e., identical to the potential problem. The direct analogy between stream function and temperature fields permits MODE = 1 input (Fig. 8-48) with no special x-y directional indicators required; thus all CTYPE = 0. The problem output is identical to temperature output and thus is not reproduced here. It is sufficient to indicate that the stream function values are contained in the temperature printout.

NETWORK THERMAL ANALYSIS

VELOCITY POTENTIAL SOLUTION

UNITS=0

NUMBER OF NODES= 85   NUMBER OF CONDUCTORS= 312

NLOOP= 100   TPRINT= 100   NPRINT= 1   LOOPEN= 100   ALDT= .1000E-02   BETA= 1.90

ARRAY DATA

ARRAY 1   2 X-Y PAIRS

0.   ,   .1000E+01   .1000E+03,   .1000E+01

LOOPCT=   0   MAXDT=   -I

TEMPERATURES

T(1)=0.	T(2)=0.	T(3)=0.	T(4)=0.	T(5)=0.	T(6)=0.
T(7)=0.	T(8)=0.	T(9)=0.	T(10)=0.	T(11)=0.	T(12)=0.
T(13)=0.	T(14)=0.	T(15)=0.	T(16)=0.	T(17)=0.	T(18)=0.
T(19)=0.	T(20)=0.	T(21)=0.	T(22)=0.	T(23)=0.	T(24)=0.
T(25)=0.	T(26)=0.	T(27)=0.	T(28)=0.	T(29)=0.	T(30)=0.
T(31)=0.	T(32)=0.	T(33)=0.	T(34)=0.	T(35)=0.	T(36)=0.
T(37)=0.	T(38)=0.	T(39)=0.	T(40)=0.	T(41)=0.	T(42)=0.
T(43)=0.	T(44)=0.	T(45)=0.	T(46)=0.	T(47)=0.	T(48)=0.
T(49)=0.	T(50)=0.	T(51)=0.	T(52)=0.	T(53)=0.	T(54)=0.
T(55)=0.	T(56)=0.	T(57)=0.	T(58)=0.	T(59)=0.	T(60)=0.
T(61)=0.	T(62)=0.	T(63)=0.	T(64)=0.	T(65)=0.	T(66)=0.
T(67)=0.	T(68)=0.	T(69)=0.	T(70)=0.	T(71)=0.	T(72)=0.
T(73)=0.	T(74)=0.	T(75)=0.	T(76)=0.	T(77)=0.	T(78)=0.
T(79)=0.	T(80)=0.	T(81)=0.	T(82)=0.	T(83)=0.	T(84)=0.
T(85)=0.				

LOOPCT=   83   MAXDT= .9393E-03

Fig. 8-46. TNETFA output for velocity potential.

TEMPERATURES

```
T(  1)= .2396E+02   T(  2)= .2295E+02   T(  3)= .2077E+02   T(  4)= .1691E+02   T(  5)= .1404E+02   T(  6)= .1165E+02
T(  7)= .9510E+01   T(  8)= .7503E+01   T(  9)= .5576E+01   T( 10)= .3696E+01   T( 11)= .1842E+01   T( 12)= .1062E-06
T( 13)= .2079E+02   T( 14)= .1996E+02   T( 15)= .1830E+02   T( 16)= .1590E+02   T( 17)= .1357E+02   T( 18)= .1141E+02
T( 19)= .9372E+01   T( 20)= .7423E+01   T( 21)= .5529E+01   T( 22)= .3670E+01   T( 23)= .1830E+01   T( 24)= .1055E-06
T( 25)= .1846E+02   T( 26)= .1781E+02   T( 27)= .1655E+02   T( 28)= .1482E+02   T( 29)= .1294E+02   T( 30)= .1103E+02
T( 31)= .9147E+01   T( 32)= .7287E+01   T( 33)= .5448E+01   T( 34)= .3624E+01   T( 35)= .1810E+01   T( 36)= .1043E-06
T( 37)= .1678E+02   T( 38)= .1626E+02   T( 39)= .1527E+02   T( 40)= .1391E+02   T( 41)= .1233E+02   T( 42)= .1064E+02
T( 43)= .8896E+01   T( 44)= .7129E+01   T( 45)= .5351E+01   T( 46)= .3569E+01   T( 47)= .1785E+01   T( 48)= .1028E-06
T( 49)= .1561E+02   T( 50)= .1518E+02   T( 51)= .1436E+02   T( 52)= .1321E+02   T( 53)= .1183E+02   T( 54)= .1030E+02
T( 55)= .8669E+01   T( 56)= .6982E+01   T( 57)= .5260E+01   T( 58)= .3516E+01   T( 59)= .1761E+01   T( 60)= .1015E-06
T( 61)= .1488E+02   T( 62)= .1450E+02   T( 63)= .1377E+02   T( 64)= .1274E+02   T( 65)= .1148E+02   T( 66)= .1005E+02
T( 67)= .8500E+01   T( 68)= .6870E+01   T( 69)= .5189E+01   T( 70)= .3475E+01   T( 71)= .1742E+01   T( 72)= .1003E-06
T( 73)= .1453E+02   T( 74)= .1417E+02   T( 75)= .1349E+02   T( 76)= .1251E+02   T( 77)= .1131E+02   T( 78)= .9923E+01
T( 79)= .8411E+01   T( 80)= .6811E+01   T( 81)= .5151E+01   T( 82)= .3452E+01   T( 83)= .1731E+01   T( 84)= .9981E-07
T( 85)= 0.
```

LOOPCT=   83   MAXDT= .9393E-03

POTENTIAL FLOW CALCULATION - ANGLES IN DEGREES

V0 = .1000E+03

```
NODE   1    VX = .7592E+02    VY = .2407E+02    V = .7964E+02    THETA =   17.59
NODE   2    VX = .7177E+02    VY = .5229E+02    V = .8880E+02    THETA =   36.08
NODE   3    VX = .5946E+02    VY = .9282E+02    V = .1102E+03    THETA =   57.36
NODE   4    VX = .2415E+02    VY = .6868E+02    V = .7280E+02    THETA =   70.63
NODE   5    VX = .1131E+02    VY = .5737E+02    V = .5848E+02    THETA =   78.85
NODE   6    VX = .5902E+01    VY = .5147E+02    V = .5181E+02    THETA =   83.46
```

Fig. 8-46. (Continued)

```
NODE    7    VX = .3303E+01    VY = .4817E+02    V = .4828E+02    THETA = 86.08
NODE    8    VX = .1921E+01    VY = .4624E+02    V = .4628E+02    THETA = 87.62
NODE    9    VX = .1124E+01    VY = .4512E+02    V = .4513E+02    THETA = 88.57
NODE   10    VX = .6284E+00    VY = .4449E+02    V = .4450E+02    THETA = 89.19
NODE   11    VX = .2795E+00    VY = .4421E+02    V = .4421E+02    THETA = 89.64
NODE   12    VX = .1708E-07    VY = .1062E-01    V = .1062E-01    THETA = 90.00
NODE   13    VX = .5600E+02    VY = .1992E+02    V = .5943E+02    THETA = 19.58
NODE   14    VX = .5172E+02    VY = .3998E+02    V = .6537E+02    THETA = 37.70
NODE   15    VX = .4193E+02    VY = .5751E+02    V = .7117E+02    THETA = 53.90
NODE   16    VX = .2582E+02    VY = .5585E+02    V = .6153E+02    THETA = 65.19
NODE   17    VX = .1520E+02    VY = .5196E+02    V = .5414E+02    THETA = 73.70
NODE   18    VX = .8990E+01    VY = .4887E+02    V = .4969E+02    THETA = 79.58
NODE   19    VX = .5395E+01    VY = .4678E+02    V = .4709E+02    THETA = 83.42
NODE   20    VX = .3264E+01    VY = .4544E+02    V = .4556E+02    THETA = 85.89
NODE   21    VX = .1952E+01    VY = .4462E+02    V = .4466E+02    THETA = 87.49
NODE   22    VX = .1099E+01    VY = .4415E+02    V = .4416E+02    THETA = 88.57
NODE   23    VX = .4942E+00    VY = .4393E+02    V = .4393E+02    THETA = 89.36
NODE   24    VX = .2783E-07    VY = .1055E-01    V = .1055E-01    THETA = 90.00
NODE   25    VX = .4037E+02    VY = .1564E+02    V = .4329E+02    THETA = 21.17
NODE   26    VX = .3716E+02    VY = .3019E+02    V = .4788E+02    THETA = 39.09
```

Fig. 8-46. *(Continued)*

POTENTIAL FLOW CALCULATION - ANGLES IN DEGREES

V0 = .1000E+03

NODE	27	VX =	.3073E+02	VY =	.4139E+02	V =	.5156E+02	THETA =	53.41
NODE	28	VX =	.2199E+02	VY =	.4522E+02	V =	.5029E+02	THETA =	64.07
NODE	29	VX =	.1466E+02	VY =	.4575E+02	V =	.4805E+02	THETA =	72.24
NODE	30	VX =	.9470E+01	VY =	.4528E+02	V =	.4626E+02	THETA =	78.19
NODE	31	VX =	.6027E+01	VY =	.4465E+02	V =	.4506E+02	THETA =	82.31
NODE	32	VX =	.3783E+01	VY =	.4413E+02	V =	.4429E+02	THETA =	85.10
NODE	33	VX =	.2314E+01	VY =	.4377E+02	V =	.4383E+02	THETA =	86.97
NODE	34	VX =	.1322E+01	VY =	.4354E+02	V =	.4356E+02	THETA =	88.26
NODE	35	VX =	.5996E+00	VY =	.4344E+02	V =	.4344E+02	THETA =	89.21
NODE	36	VX =	.3616E-07	VY =	.1043E-01	V =	.1043E-01	THETA =	90.00
NODE	37	VX =	.2793E+02	VY =	.1243E-01	V =	.3057E+02	THETA =	23.99
NODE	38	VX =	.2582E+02	VY =	.2376E+02	V =	.3509E+02	THETA =	42.63
NODE	39	VX =	.2184E+02	VY =	.3265E+02	V =	.3928E+02	THETA =	56.22
NODE	40	VX =	.1675E+02	VY =	.3789E+02	V =	.4143E+02	THETA =	66.16
NODE	41	VX =	.1198E+02	VY =	.4057E+02	V =	.4230E+02	THETA =	73.54
NODE	42	VX =	.8204E+01	VY =	.4183E+02	V =	.4263E+02	THETA =	78.90
NODE	43	VX =	.5449E+01	VY =	.4241E+02	V =	.4276E+02	THETA =	82.68
NODE	44	VX =	.3527E+01	VY =	.4266E+02	V =	.4281E+02	THETA =	85.27
NODE	45	VX =	.2202E+01	VY =	.4277E+02	V =	.4283E+02	THETA =	87.05

Fig. 8-46. (Continued)

NODE 46	VX = .1272E+01	VY = .4282E+02	V = .4284E+02	THETA = 88.30
NODE 47	VX = .5805E+00	VY = .4284E+02	V = .4284E+02	THETA = 89.22
NODE 48	VX = .3175E-07	VY = .1028E-01	V = .1028E-01	THETA = 90.00
NODE 49	VX = .1759E+02	VY = .1032E+02	V = .2040E+02	THETA = 30.40
NODE 50	VX = .1635E+02	VY = .1979E+02	V = .2567E+02	THETA = 50.43
NODE 51	VX = .1407E+02	VY = .2756E+02	V = .3094E+02	THETA = 62.96
NODE 52	VX = .1118E+02	VY = .3313E+02	V = .3497E+02	THETA = 71.35
NODE 53	VX = .8332E+01	VY = .3679E+02	V = .3772E+02	THETA = 77.24
NODE 54	VX = .5915E+01	VY = .3908E+02	V = .3952E+02	THETA = 81.39
NODE 55	VX = .4046E+01	VY = .4049E+02	V = .4069E+02	THETA = 84.29
NODE 56	VX = .2676E+01	VY = .4134E+02	V = .4142E+02	THETA = 86.30
NODE 57	VX = .1696E+01	VY = .4184E+02	V = .4188E+02	THETA = 87.68
NODE 58	VX = .9911E+00	VY = .4213E+02	V = .4214E+02	THETA = 88.65
NODE 59	VX = .4572E+00	VY = .4226E+02	V = .4226E+02	THETA = 89.38
NODE 60	VX = .2914E-07	VY = .1015E-01	V = .1015E-01	THETA = 90.00
NODE 61	VX = .8560E+01	VY = .9081E+01	V = .1244E+02	THETA = 46.86
NODE 62	VX = .7939E+01	VY = .1759E+02	V = .1922E+02	THETA = 65.60
NODE 63	VX = .6896E+01	VY = .2467E+02	V = .2562E+02	THETA = 74.38
NODE 64	VX = .5581E+01	VY = .3028E+02	V = .3079E+02	THETA = 79.56
NODE 65	VX = .4249E+01	VY = .3437E+02	V = .3463E+02	THETA = 82.95
NODE 66	VX = .3076E+01	VY = .3721E+02	V = .3734E+02	THETA = 85.27

Fig. 8-46. (Continued)

POTENTIAL FLOW CALCULATION - ANGLES IN DEGREES

V0 = .1000E+03

NODE	VX	VY	V	THETA
NODE 67	VX = .2138E+01	VY = .3912E+02	V = .3918E+02	THETA = 86.87
NODE 68	VX = .1432E+01	VY = .4036E+02	V = .4038E+02	THETA = 87.97
NODE 69	VX = .9154E+00	VY = .4114E+02	V = .4115E+02	THETA = 88.72
NODE 70	VX = .5392E+00	VY = .4159E+02	V = .4160E+02	THETA = 89.26
NODE 71	VX = .2476E+00	VY = .4180E+02	V = .4180E+02	THETA = 89.66
NODE 72	VX = .1160E-07	VY = .1003E-01	V = .1003E-01	THETA = 90.00
NODE 73	VX = 0.	VY = .8511E+01	V = .8511E+01	THETA = 90.00
NODE 74	VX = 0.	VY = .1646E+02	V = .1646E+02	THETA = 90.00
NODE 75	VX = 0.	VY = .2336E+02	V = .2336E+02	THETA = 90.00
NODE 76	VX = 0.	VY = .2895E+02	V = .2895E+02	THETA = 90.00
NODE 77	VX = 0.	VY = .3320E+02	V = .3320E+02	THETA = 90.00
NODE 78	VX = 0.	VY = .3627E+02	V = .3627E+02	THETA = 90.00
NODE 79	VX = 0.	VY = .3841E+02	V = .3841E+02	THETA = 90.00
NODE 80	VX = 0.	VY = .3984E+02	V = .3984E+02	THETA = 90.00
NODE 81	VX = 0.	VY = .4076E+02	V = .4076E+02	THETA = 90.00
NODE 82	VX = 0.	VY = .4130E+02	V = .4130E+02	THETA = 90.00
NODE 83	VX = 0.	VY = .4155E+02	V = .4155E+02	THETA = 90.00
NODE 84	VX = 0.	VY = .9981E-02	V = .9981E-02	THETA = 90.00
NODE -85	VX = 0.	VY = 0.	V = 0.	

Fig. 8-46. (Continued)

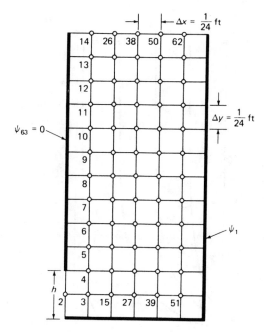

Fig. 8-47. Grid and node identification for two-dimensional streamline solution.

The equipotentials ϕ and equistream function values ψ are sketched in Fig. 8-49. Intra-node values of ϕ and ψ were computed from the TNETFA output using simple linear interpolation along each grid line. Note that the functions ϕ and ψ are orthogonal at all points of intersection.

```
                                                        DATA SET
STREAM FUNCTION SOLUTION  }———— 1
 1    0   0 ——————————————————————————— 2
63    2   0   0   14   0   0   0   0 —— 3
 4.0    0.0
 1    8.333  }——————————————————————— 4
63    0.0
 0    0 ——————————————————————————————— 6
12     3   1   15   1   576.0    0
12    15   1   27   1   576.0    0
12    27   1   39   1   576.0    0
12    39   1   51   1   576.0    0
12    51   1    1   0   576.0    0
11     4   1   63   0   576.0    0
 1     2   0    3   0   576.0    0
11     3   1    4   1   576.0    0  }—— 7
11    15   1   16   1   576.0    0
11    27   1   28   1   576.0    0
11    39   1   40   1   576.0    0
11    51   1   52   1   576.0    0
 5     3  12    1   0   576.0    0
 2     2   0    1  62   576.0    0
100    1.9   0.001    0
 1.0    1.0                     }——— 14
100    1
```

Fig. 8-48. TNETFA input for two-dimensional stream function.

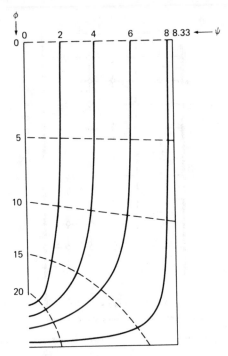

Fig. 8-49. Potentials and streamlines for two-dimensional solution.

8.11 PROBLEM–NATURAL CONVECTION COOLED METAL CABINET WITH THREE INTERNAL CHAMBERS

This example illustrates the basics of a method for modeling cabinets more complex than a single-chamber system. Figure 8-50 shows a three-chamber cabinet with natural ventilation air draft holes in both the bottom and side panels. The left and right halves of the cabinet interact via convection to the interior, front–rear bulkhead. The two left chambers are separated by a slotted horizontal panel that permits a ventilating air draft to pass through. The horizontal panel is the only internal air draft resistance because of adequate circuit card spacing (circuit cards are not shown in drawing). No components are heat sunk to internal panels or outer walls. The cabinet exterior transfers heat to ambient air at 20°C via natural convection and simple radiation ($\epsilon = 0.8$).

The three chambers are individually shown in Figs. 8-51 through 8-53 with the corresponding elements of the thermal network model. The exterior surfaces show simple radiation and flat plate natural convection elements in parallel connecting wall surface nodes with ambient (node 1).

One node is used to simulate air (nodes 2, 14, 20) within each chamber. Flat plate natural convection transfers heat between the interior air and wall surfaces. Note that Figs. 8-51 through 8-53 indicate common nodes at walls connecting adjacent chambers.

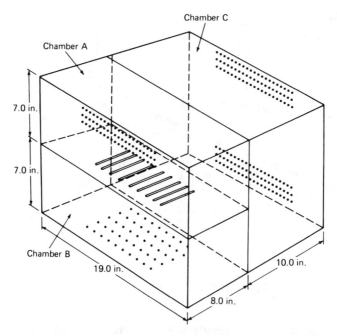

Fig. 8-50. Natural convection cooled cabinet with three internal chambers.

The ambient air (node 1) is connected to inlet air nodes 21 and 22 by a large $(C = 100.0)$ conductance to be consistent with the TNETFA idiosyncrasy of not permitting direct connection of an air node and an ambient with a one-way conductance—an intermediate constant conductance is required.

The air draft vs. heat dissipated into airstream is computed for chamber C and for chambers A, B. The latter two chambers have a

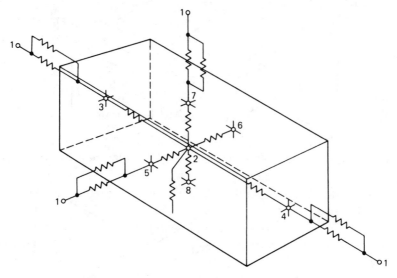

Fig. 8-51. Cabinet chamber A—node detail.

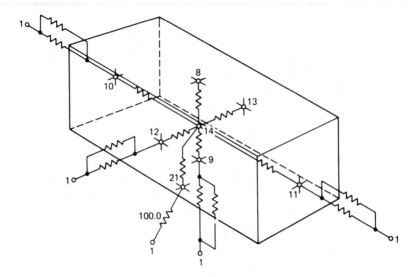

Fig. 8-52. Cabinet chamber B−node detail.

common air draft. The iteration method is used to compute the air draft in a manner identical to that used for the ventilated cabinet described earlier in this chapter. The airflow resistance for chamber C is due to free inlet and exit areas:

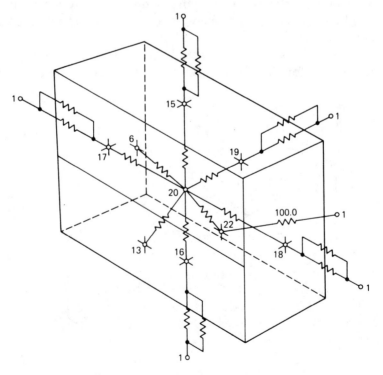

Fig. 8-53. Cabinet chamber C−node detail.

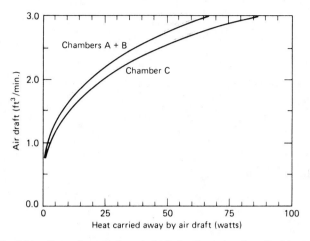

Fig. 8-54. Natural ventilation air draft for three-chambered cabinet.

$$A_I = A_E = (10.0 \text{ in.})(2.0 \text{ in.})(0.33) = 6.6 \text{ in.}^2, 33\% \text{ perforated}$$

$$R = [2.4 \times 10^{-3}/(6.6)^2](2) = 1.1 \times 10^{-4}$$

The horizontal panel for chambers A and B has a free area $A = (10.0$ in.$)(5.0$ in.$)(0.2) = 10.0$ in.^{2}, 20% slotted. The inlet area for chamber B is $A = (10.0 \text{ in.})(6.0 \text{ in.})(0.33) = 19.8 \text{ in.}^2$, 33% perforated. The free exit area for chamber A is 6.6 in.^{2}. Thus the total combined airflow resistance for chambers A and B is:

$$R = 2.4 \times 10^{-3}[(1/19.8)^2 + (1/10.0)^2 + (1/6.6)^2]$$

$$= 8.5 \times 10^{-5}$$

The air draft vs. heat input calculations are plotted in Fig. 8-54.

The TNETFA input is reproduced in Fig. 8-55. Chambers A, B, and C each contain 75.0, 25.0, and 100.0 watts, respectively, dissipated by components. These values are presented as sources at air nodes 2, 14, and 20, respectively. The conductance/node connection input contains flat plate surface areas and emissivity–area products for convection and radiation, respectively.

The final three conductance inputs (CTYPE = 301) contain the air drafts in ft^3/min. Clearly the draft between nodes 2 and 14 and nodes 14 and 21 must be identical because it is the same airstream. The first run of TNETFA used an arbitrary guess of the draft for all three one-way conductances, e.g., 1.0 ft^3/min. The detailed node output for nodes 2, 14, and 20 was examined. The flux (watts) for node 2 to 14 plus node 14 to 21 was summed and a better draft estimate determined from Fig. 8-54. The flux (watts) from node 20 to

```
                                                                  DATA SET
NATURAL CONVECTION COOLED CABINET, EM=0.8 ⎫——— 1
THREE INTERNAL CHAMBERS                    ⎬
1   2   0 —————————————————————————————————————————————— 2
22  1   3   0   31   5   0   8   0 ——————————————————————— 3
30.0  0.0
1    20.0                           ⎫
2    40.0    75.0                   ⎬——————————————————————— 4
14   40.0    25.0                   ⎭
20   40.0    100.0
0   0 ——————————————————————————————————————————————————————— 6
2   1   0   3   1     56.0    101   ⎫
1   1   0   5   0    133.0    101   ⎪
1   1   0   7   0    152.0    102   ⎪
2   1   0   3   1     44.8    -1    ⎪
1   1   0   5   0    106.4    -1    ⎪
1   1   0   7   0    121.6    -1    ⎪
2   2   0   3   1     56.0    103   ⎪
2   2   0   5   1    133.0    103   ⎪
1   2   0   7   0    152.0    104   ⎪
1   2   0   8   0    152.0    105   ⎪
2   1   0   10  1     56.0    101   ⎪
1   1   0   12  0    133.0    101   ⎪
1   1   0   9   0    152.0    106   ⎪
2   1   0   10  1     44.0    -1    ⎪
1   1   0   12  0    106.0    -1    ⎪
1   1   0   9   0    121.6    -1    ⎬——————————————————————— 7
2   14  0   10  1    56.0    103    ⎪
2   14  0   12  1   133.0    103    ⎪
2   14  0   8   1   152.0    105    ⎪
2   1   0   17  1   140.0    101    ⎪
1   1   0   19  0   266.0    101    ⎪
1   1   0   15  0   190.0    102    ⎪
1   1   0   16  0   190.0    106    ⎪
2   1   0   17  1   112.0    -1     ⎪
1   1   0   19  0   212.8    -1     ⎪
2   1   0   15  1   152.0    -1     ⎪
2   20  0   17  1    40.0    101    ⎪
2   20  0   6   7   133.0    101    ⎪
1   20  0   19  0   266.0    101    ⎪
1   20  0   15  0   190.0    107    ⎪
1   20  0   16  0   190.0    108    ⎭
1   21  100.0   0                  ⎫
1   22  100.0   0                  ⎪
2   14  2.90  301                  ⎬——————————————————————— 8
14  21  2.90  301                  ⎪
20  22  2.50  301                  ⎭
1   14.0                           ⎫
2   4.6                            ⎪
1   7.0                            ⎪
2   2.8                            ⎬——————————————————————— 11
3   2.8                            ⎪
3   4.6                            ⎪
2   3.3                            ⎪
3   3.3                            ⎭
100   1.0   0.001   100            ⎫
1.0   1.0                          ⎬——————————————————————— 14
100   1                            ⎭
```

Fig. 8-55. TNETFA input for three-chamber, natural convection, radiation cooled cabinet.

NETWORK THERMAL ANALYSIS

NATURAL CONVECTION COOLED CABINET, EM=0.8
THREE INTERNAL CHAMBERS

UNITS=2

NUMBER OF NODES= 22   NUMBER OF CONDUCTORS= 100

NLOOP= 100   TPRINT= 100   NPRINT= 100   LOOPEN= 100   ALDT= .1000E-02   BETA= 1.00

NATURAL CONVECTION PARAMETER

```
1   VERTICAL FLAT PLATE OR CYLINDER                                                      P= .1400E+02
2   HORIZONTAL FLAT PLATE OR CYLINDER, HEATED SIDE FACING UP OR COOLED SIDE FACING DOWN  P= .4600E+01
3   VERTICAL FLAT PLATE OR CYLINDER                                                      P= .7000E+01
4   HORIZONTAL FLAT PLATE OR CYLINDER, HEATED SIDE FACING UP OR COOLED SIDE FACING DOWN  P= .2800E+01
5   HORIZONTAL FLAT PLATE OR CYLINDER, HEATED SIDE FACING DOWN OR COOLED SIDE FACING UP  P= .2800E+01
6   HORIZONTAL FLAT PLATE OR CYLINDER, HEATED SIDE FACING DOWN OR COOLED SIDE FACING UP  P= .4600E+01
7   HORIZONTAL FLAT PLATE OR CYLINDER, HEATED SIDE FACING UP OR COOLED SIDE FACING DOWN  P= .3300E+01
8   HORIZONTAL FLAT PLATE OR CYLINDER, HEATED SIDE FACING DOWN OR COOLED SIDE FACING UP  P= .3300E+01
```

LOOPCT=   0   MAXDT=   -1

TEMPERATURES

```
T( 1)= .2000E+02   T( 2)= .4000E+02   T( 3)= .3000E+02   T( 4)= .3000E+02   T( 5)= .3000E+02   T( 6)= .3000E+02
T( 7)= .3000E+02   T( 8)= .3000E+02   T( 9)= .3000E+02   T(10)= .3000E+02   T(11)= .3000E+02   T(12)= .3000E+02
T(13)= .3000E+02   T(14)= .4000E+02   T(15)= .3000E+02   T(16)= .3000E+02   T(17)= .3000E+02   T(18)= .3000E+02
T(19)= .3000E+02   T(20)= .4000E+02   T(21)= .3000E+02   T(22)= .3000E+02
```

LOOPCT=   9   MAXDT=   .6139E-03

TEMPERATURES

```
T( 1)= .2000E+02   T( 2)= .5898E+02   T( 3)= .3434E+02   T( 4)= .3434E+02   T( 5)= .3434E+02   T( 6)= .5742E+02
T( 7)= .3474E+02   T( 8)= .4651E+02   T( 9)= .2390E+02   T(10)= .2496E+02   T(11)= .2496E+02   T(12)= .2494E+02
T(13)= .4412E+02   T(14)= .3419E+02   T(15)= .3315E+02   T(16)= .3034E+02   T(17)= .2516E+02   T(18)= .2516E+02
T(19)= .3188E+02   T(20)= .5564E+02   T(21)= .2000E+02   T(22)= .2000E+02
```

Fig. 8-56. TNETFA output for three-chambered cabinet.

TEMPERATURES

LOOPCT= 9   MAXDT= .6139E-03

DETAIL OF NODE -1   TEMPERATURE= .2000E+02   POWER=0.   STABILITY CONSTANT =   -1   CAP= .1000E-19

THIS IS A CONSTANT TEMPERATURE NODE

NODE	CTYPE	CMODE	C	CONDUCTANCE	FLUX	REYN NO.	BNDRY LAYER	HYDR DIA	HT TRANS COEF.
22	0		.1000E+03	.1000E+03	0.				
21	0		.1000E+03	.1000E+03	0.				
16	-1		.1520E+03	.5896E+00	-.6094E+01				.3879E-02
15	-1		.1520E+03	.5981E+00	-.7867E+01				.3935E-02
19	-1		.2128E+03	.8319E+00	-.9881E+01				.3909E-02
18	-1		.1120E+03	.4231E+00	-.2183E+01				.3778E-02
17	-1		.1120E+03	.4231E+00	-.2183E+01				.3778E-02
16	106	3	.1900E+03	.2611E+00	-.2699E+01				.1374E-02
15	102	2	.1900E+03	.5537E+00	-.7283E+01				.2914E-02
19	101	1	.2660E+03	.6256E+00	-.7330E+01				.2352E-02
18	101	1	.1400E+03	.2684E+00	-.1385E+01				.1917E-02
17	101	1	.1400E+03	.2684E+00	-.1365E+01				.1917E-02
9	-1		.1216E+03	.4564E+00	-.1779E+01				.3754E-02
12	-1		.1060E+03	.4000E+00	-.1974E+01				.3774E-02
11	-1		.4400E+02	.1661E+00	-.8236E+00				.3774E-02
10	-1		.4400E+02	.1661E+00	-.8236E+00				.3774E-02
9	106	3	.1520E+03	.1643E+00	-.6405E+00				.1081E-02
12	101	1	.1330E+03	.2522E+00	-.1245E+01				.1856E-02
11	101	1	.5600E+02	.1063E+00	-.5273E+00				.1899E-02
10	101	1	.5600E+02	.1063E+00	-.5273E+00				.1899E-02
7	-1		.1216E+03	.4823E+00	-.7111E+01				.3967E-02
5	-1		.1064E+03	.4212E+00	-.6039E+01				.3959E-02
4	-1		.4480E+02	.1773E+00	-.2543E+01				.3959E-02
3	-1		.4480E+02	.1773E+00	-.2543E+01				.3959E-02
7	102	2	.1520E+03	.4553E+00	-.6713E+01				.2996E-02
5	101	1	.1330E+03	.3274E+00	-.4694E+01				.2461E-02
4	101	1	.5600E+02	.1378E+00	-.1976E+01				.2461E-02
3	101	1	.5600E+02	.1378E+00	-.1976E+01				.2461E-02

NET TOTAL =-.9033E+02

DETAIL OF NODE 2   TEMPERATURE= .5898E+02   POWER= .7500E+02   STABILITY CONSTANT =   -1   CAP= .1000E-19

Fig. 8-56. (Continued)

NODE	CTYPE	CMODE	C	CONDUCTANCE	FLUX	REYN NO.	BNDRY LAYER	HYDR DIA	HT TRANS COEF
14									
8	301	3	.2900E+01	.1536E+01	.3808E+02				.1579E-02
7	105	2	.1520E+03	.2400E+00	.2991E+01				.3753E-02
6	104		.1520E+03	.5705E+00	.1382E+02				.1620E-02
5	103	1	.1330E+03	.2155E+00	.3346E+00				.3275E-02
4	103	1	.1330E+03	.4356E+00	.1073E+02				.3275E-02
3	103	1	.5600E+02	.1834E+00	.4519E+01				.3275E-02
3	103	1	.5600E+02	.1834E+00	.4519E+01				

NET TOTAL = .7500E+02

DETAIL OF NODE 3   TEMPERATURE= .3434E+02   POWER=0.   STABILITY CONSTANT = −1   CAP = .1000E−19

NODE	CTYPE	CMODE	C	CONDUCTANCE	FLUX	REYN NO.	BNDRY LAYER	HYDR DIA	HT TRANS COEF
2	103	1	.5600E+02	.1834E+00	−.4519E+01				.3275E-02
1	−1		.4480E+02	.1773E+00	.2543E+01				.3959E-02
1	101	1	.5600E+02	.1378E+00	.1976E+01				.2461E-02

NET TOTAL =−.1838E−04

DETAIL OF NODE 4   TEMPERATURE= .3434E+02   POWER=0.   STABILITY CONSTANT = −1   CAP = .1000E−19

NODE	CTYPE	CMODE	C	CONDUCTANCE	FLUX	REYN NO.	BNDRY LAYER	HYDR DIA	HT TRANS COEF
2	103	1	.5600E+02	.1834E+00	−.4519E+01				.3275E-02
1	−1		.4480E+02	.1773E+00	.2543E+01				.3959E-02
1	101	1	.5600E+02	.1378E+00	.1976E+01				.2461E-02

NET TOTAL =−.1838E−04

DETAIL OF NODE 5   TEMPERATURE= .3434E+02   POWER=0.   STABILITY CONSTANT = −1   CAP = .1000E−19

NODE	CTYPE	CMODE	C	CONDUCTANCE	FLUX	REYN NO.	BNDRY LAYER	HYDR DIA	HT TRANS COEF
2	103	1	.1330E+03	.4356E+00	−.1073E+02				.3275E-02
1	−1		.1064E+03	.4212E+00	.6039E+01				.3959E-02
1	101	1	.1330E+03	.3274E+00	.4694E+01				.2461E-02

NET TOTAL =−.4365E−04

DETAIL OF NODE 6   TEMPERATURE= .5742E+02   POWER=0.   STABILITY CONSTANT = −1   CAP = .1000E−19

NODE	CTYPE	CMODE	C	CONDUCTANCE	FLUX	REYN NO.	BNDRY LAYER	HYDR DIA	HT TRANS COEF
20	101	1	.1330E+03	.1879E+00	.3347E+00				.1412E-02
2	103	1	.1330E+03	.2155E+00	−.3346E+00				.1620E-02

NET TOTAL = .6395E−04

Fig. 8-56.   (Continued)

DETAIL OF NODE 7   TEMPERATURE= .3474E+02   POWER=0.   STABILITY CONSTANT =   -I   CAP = .1000E-19

NODE	CTYPE	CMODE	C	CONDUCTANCE	FLUX	REYN NO.	BNDRY LAYER	HYDR DIA	HT TRANS COEF
2	104	2	.1520E+03	.5705E+00	-.1382E+02				.3753E-02
1	-1		.1216E+03	.4823E+00	.7111E+01				.3967E-02
1	102	2	.1520E+03	.4553E+00	.6713E+01				.2996E-02
				NET TOTAL =-.6124E-04					

DETAIL OF NODE 8   TEMPERATURE= .4651E+02   POWER=0.   STABILITY CONSTANT =   -I   CAP = .1000E-19

NODE	CTYPE	CMODE	C	CONDUCTANCE	FLUX	REYN NO.	BNDRY LAYER	HYDR DIA	HT TRANS COEF
14	105	3	.1520E+03	.2427E+00	.2991E+01				.1597E-02
2	105	3	.1520E+03	.2400E+00	-.2991E+01				.1579E-02
				NET TOTAL = .1040E-04					

DETAIL OF NODE 9   TEMPERATURE= .2390E+02   POWER=0.   STABILITY CONSTANT =   -I   CAP = .1000E-19

NODE	CTYPE	CMODE	C	CONDUCTANCE	FLUX	REYN NO.	BNDRY LAYER	HYDR DIA	HT TRANS COEF
14	105	3	.1520E+03	.2351E+00	-.2419E+01				.1547E-02
1	-1		.1216E+03	.4564E+00	.1779E+01				.3754E-02
1	106	3	.1520E+03	.1643E+00	.6405E+00				.1081E-02
				NET TOTAL = .1420E-04					

DETAIL OF NODE 10   TEMPERATURE= .2496E+02   POWER=0.   STABILITY CONSTANT =   -I   CAP = .1000E-19

NODE	CTYPE	CMODE	C	CONDUCTANCE	FLUX	REYN NO.	BNDRY LAYER	HYDR DIA	HT TRANS COEF
14	103	1	.5600E+02	.1464E+00	-.1351E+01				.2614E-02
1	-1		.4400E+02	.1661E+00	.8236E+00				.3774E-02
1	101	1	.5600E+02	.1063E+00	.5273E+00				.1899E-02
				NET TOTAL = .1132E-04					

DETAIL OF NODE 11   TEMPERATURE= .2496E+02   POWER=0.   STABILITY CONSTANT =   -I   CAP = .1000E-19

NODE	CTYPE	CMODE	C	CONDUCTANCE	FLUX	REYN NO.	BNDRY LAYER	HYDR DIA	HT TRANS COEF
14	103	1	.5600E+02	.1464E+00	-.1351E+01				.2614E-02
1	-1		.4400E+02	.1661E+00	.8236E+00				.3774E-02
1	101	1	.5600E+02	.1063E+00	.5273E+00				.1899E-02
				NET TOTAL = .1132E-04					

DETAIL OF NODE 12   TEMPERATURE= .2494E+02   POWER=0.   STABILITY CONSTANT =   -I   CAP = .1000E-19

NODE	CTYPE	CMODE	C	CONDUCTANCE	FLUX	REYN NO.	BNDRY LAYER	HYDR DIA	HT TRANS COEF
14	103	1	.1330E+03	.3479E+00	-.3219E+01				.2616E-02

Fig. 8-56. (Continued)

```
1   -1    1    .1060E+03   .4000E+00   .1974E+01   .3774E-02
1   101   1    .1330E+03   .2522E+00   .1245E+01   .1896E-02
                           NET TOTAL = .2670E-04
```

DETAIL OF NODE 13   TEMPERATURE= .4412E+02   POWER=0.   STABILITY CONSTANT =   -1   CAP= .1000E-19

NODE	CTYPE	CMODE	C	CONDUCTANCE	REYN NO.	FLUX	BNDRY LAYER	HYDR DIA	HT TRANS COEF
20	101	1	.1330E+03	.3018E+00		-.3478E+01			.2269E-02
14	103	1	.1330E+03	.3501E+00		.3478E+01			.2633E-02
				NET TOTAL =-.4084E-04					

DETAIL OF NODE 14   TEMPERATURE= .3419E+02   POWER= .2500E+02   STABILITY CONSTANT =   -1   CAP= .1000E-19

NODE	CTYPE	CMODE	C	CONDUCTANCE	REYN NO.	FLUX	BNDRY LAYER	HYDR DIA	HT TRANS COEF
21	301		.2900E+01	.1630E+01		.2313E+02			.1547E-02
9	105	3	.1520E+03	.2351E+00		.2419E+01			.1597E-02
8	105	3	.1520E+03	.2427E+00		.2991E+01			.2633E-02
13	103	1	.1330E+03	.3501E+00		-.3478E+01			.2616E-02
12	103	1	.1330E+03	.3479E+00		.3219E+01			.2614E-02
11	103	1	.5600E+02	.1464E+00		.1351E+01			.2614E-02
10	103	1	.5600E+02	.1464E+00		.1351E+01			
				NET TOTAL = .2500E+02					

DETAIL OF NODE 15   TEMPERATURE= .3315E+02   POWER=0.   STABILITY CONSTANT =   -1   CAP= .1000E-19

NODE	CTYPE	CMODE	C	CONDUCTANCE	REYN NO.	FLUX	BNDRY LAYER	HYDR DIA	HT TRANS COEF
20	107	2	.1900E+03	.6736E+00		-.1515E+02			.3545E-02
1	-1		.1520E+03	.5981E+00		.7867E+01			.3935E-02
1	102	2	.1900E+03	.5537E+00		.7283E+01			.2914E-02
				NET TOTAL =-.2048E-05					

DETAIL OF NODE 16   TEMPERATURE= .3034E+02   POWER=0.   STABILITY CONSTANT =   -1   CAP= .1000E-19

NODE	CTYPE	CMODE	C	CONDUCTANCE	REYN NO.	FLUX	BNDRY LAYER	HYDR DIA	HT TRANS COEF
20	108	3	.1900E+03	.3475E+00		-.8793E+01			.1829E-02
1	-1	3	.1520E+03	.5896E+00		.6094E+01			.3879E-02
1	106	3	.1900E+03	.2611E+00		.2699E+01			.1374E-02
				NET TOTAL = .1720E-05					

DETAIL OF NODE 17   TEMPERATURE= .2516E+02   POWER=0.   STABILITY CONSTANT =   -1   CAP= .1000E-19

NODE	CTYPE	CMODE	C	CONDUCTANCE	REYN NO.	FLUX	BNDRY LAYER	HYDR DIA	HT TRANS COEF
20	101	1	.4000E+02	.1170E+00		-.3568E+01			.2926E-02
1	-1		.1120E+03	.4231E+00		.2183E+01			.3778E-02

Fig. 8-56. (Continued)

```
NODE  CTYPE  CMODE       C      CONDUCTANCE      FLUX      REYN NO.  BNDRY LAYER   HYDR DIA   HT TRANS COEF
  1    101    1       .1400E+03   .2684E+00    .1385E+01                             -1        .1917E-02
                                  NET TOTAL = .6365E-06

DETAIL OF NODE  18    TEMPERATURE= .2516E+02    POWER=0.    STABILITY CONSTANT =          CAP= .1000E-19


NODE  CTYPE  CMODE       C      CONDUCTANCE      FLUX      REYN NO.  BNDRY LAYER   HYDR DIA   HT TRANS COEF
 20    101    1       .4000E+02   .1170E+00   -.3568E+01                             -1        .2926E-02
 -1    -1     1       .1120E+03   .4231E+00    .2183E+01                                       .3778E-02
 -1    101    1       .1400E+03   .2684E+00    .1385E+01                                       .1917E-02
                                  NET TOTAL = .6365E-06

DETAIL OF NODE  19    TEMPERATURE= .3188E+02    POWER=0.    STABILITY CONSTANT =          CAP= .1000E-19


NODE  CTYPE  CMODE       C      CONDUCTANCE      FLUX      REYN NO.  BNDRY LAYER   HYDR DIA   HT TRANS COEF
 20    101    1       .2660E+03   .7284E+00   -.1731E+02                             -1        .2739E-02
 -1    -1     1       .2128E+03   .8319E+00    .9881E+01                                       .3909E-02
 -1    101    1       .2660E+03   .6256E+00    .7430E+01                                       .2352E-02
                                  NET TOTAL = .5284E-06

DETAIL OF NODE  20    TEMPERATURE= .5564E+02    POWER= .1000E+03    STABILITY CONSTANT =       CAP= .1000E-19


NODE  CTYPE  CMODE       C      CONDUCTANCE      FLUX      REYN NO.  BNDRY LAYER   HYDR DIA   HT TRANS COEF
 22    301    1       .2500E+01   .1360E+01    .4847E+02                             -1        .1829E-02
 16    108    3       .1900E+03   .3475E+00    .8793E+01                                       .3545E-02
 15    107    2       .1900E+03   .6736E+00    .1515E+02                                       .2739E-02
 19    101    1       .2660E+03   .7284E+00    .1731E+02                                       .2269E-02
 13    101    1       .1330E+03   .3018E+00    .3478E+01                                       .1412E-02
  6    101    1       .1330E+03   .1879E+00   -.3347E+00                                       .2926E-02
 18    101    1       .4000E+02   .1170E+00    .3568E+01                                       .2926E-02
 17    101    1       .4000E+02   .1170E+00    .3568E+01
                                  NET TOTAL = .1000E+03

DETAIL OF NODE  21    TEMPERATURE= .2000E+02    POWER=0.    STABILITY CONSTANT =          CAP= .1000E-19


NODE  CTYPE  CMODE       C      CONDUCTANCE      FLUX      REYN NO.  BNDRY LAYER   HYDR DIA   HT TRANS COEF
  1    0      0       .1000E+03   .1000E+03    0.                                    -1
                                  NET TOTAL =0.

DETAIL OF NODE  22    TEMPERATURE= .2000E+02    POWER=0.    STABILITY CONSTANT =          CAP= .1000E-19


NODE  CTYPE  CMODE       C      CONDUCTANCE      FLUX      REYN NO.  BNDRY LAYER   HYDR DIA   HT TRANS COEF
  1    0      0       .1000E+03   .1000E+03    0.                                    -1
                                  NET TOTAL =0.

                              ENERGY BALANCE = .1128E-03
```

8-56(4)

Fig. 8-56. (Continued)

22 was also used to obtain a better airflow estimate from Fig. 8-54. TNETFA was rerun with the revised air drafts and even better draft estimates obtained from the node to node heat flux output. This process seldom requires more than about three runs of TNETFA to obtain a sufficient consistency between air draft and heat dissipated into the air. The input of Fig. 8-55 and the corresponding output, Fig. 8-56, are the last runs made for this problem.

Chapter 9
Digital Computer Program Instruction—TAMS*

9.1 HISTORICAL BACKGROUND

A family of programs has evolved in the last few years (1969–1980) that is an exception to the limited use of analytical methods. Probably the earliest contribution to significant practical application was TASIC, [34]. While the use of classical Fourier-series techniques has been of obvious importance in heat transfer problems for many decades, the author of TASIC developed a particularly clever method of incorporating a lumped parameter representation of a finite number of thermal resistances.

A series of improvements in the Fourier-series method has resulted in a multifold increase in program applicability. The first step was reported in 1973, in [35], when the formalism was provided for extending the Fourier-series solution to a multilayer configuration with the results for two layers given in detail. In 1976 the solution and a significant application were reported for an early four-layer version of the program now identified as TAMS, [10].

R. F. David reported an interesting numerical method that avoids the almost overwhelming algebraic requirements for calculating the Fourier-series coefficients, [36]. Because of this, his program will accommodate up to ten layers, but, as reported, it is limited to an isothermal base and an adiabatic top boundary, and excludes lead or edge connector losses.

Finally, all previously unpublished features of TAMS were documented at the 1978 meeting of the International Conference on Hybrid Microelectronics, [37]. The latest improvements were achieved primarily by adapting the Fourier-series solution to an anisotropic thermal conductivity for each layer, subsurface heat sources, and lumped parameter thermal resistances intermixed with both the top and bottom surfaces. Each resistance terminates away from the substrate at a unique sink temperature. Figure 9-1 illustrates most of these features.

---

*Portions of this chapter are abstracted from "TAMS—Thermal Analyzer for Multilayer Structures," courtesy Tektronix, Inc.

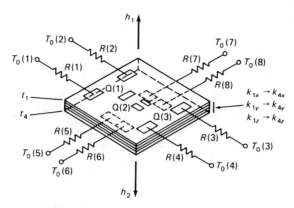

Fig. 9-1. Principal features of TAMS illustrating multilayered anisotropic thermal conductivity and lumped parameter thermal resistances at both surfaces. Reprinted from the Proceedings of the 1978 International Microelectronics Symposium, [37], by permission of the International Society for Hybrid Microelectronics, P.O. Box 3255, Montgomery, AL 36109.

9.2 PROGRAM FEATURES

TAMS is written in FORTRAN IV in a manner that permits the program to be run on most computing systems with at least 30,000 (decimal) words of memory. The equations derived for sources and resistances at various levels within the multilayer structure have been included with preset values of depth that meet the requirements of most practical problems.

Application of TAMS requires the user to select from one of six possible cases, any of which may use anisotropic thermal conductivity. These options are illustrated in Fig. 9-2 using an edge view that

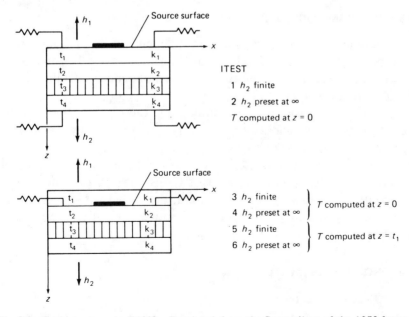

Fig. 9-2. Problem cases in TAMS. Reprinted from the Proceedings of the 1978 International Microelectronics Symposium, [37], by permission of the International Society for Hybrid Microelectronics, P.O. Box 3255, Montgomery, AL 36109.

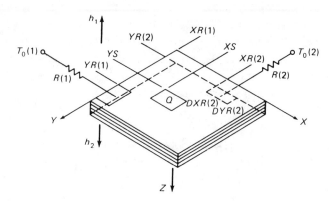

Fig. 9-3. Coordinate system and nomenclature applicable to TAMS. Reprinted from the Proceedings of the 1978 International Microelectronics Symposium, [37], by permission of the International Society for Hybrid Microelectronics, P.O. Box 3255, Montgomery, AL 36109.

emphasizes the multilayer aspect. Cases 1, 3, and 5 are concerned with Newton's law cooling from both surfaces, whereas cases 2, 4, and 6 assume an isothermal surface (T_A) at the base. In all cases except 3 and 4, the temperatures are calculated at the source centers and in the same plane as the sources. The two exceptions, cases 3 and 4, still provide the temperature calculations at the x, y centers of the source, but displaced in the z-direction to the top surface ($z = 0$). The use of resistances attached to the bottom surface ($z = C_4$) is limited to case 1, sufficient for most problems. The appropriate x, y, z coordinate system is shown in Fig. 9-3, with emphasis on the definition of the source and resistance coordinates.

9.3 PROGRAM ACCURACY

The accuracy of a particular solution is dependent upon at least three factors. The first is the extent to which the physical problem and the TAMS model have a one-to-one correspondence. Certainly this is difficult to assess quantitatively.

Second, the basic physical parameters, thermal conductivity and heat transfer coefficients (particularly the latter), are very often known to no better than ±20%.

A third element addressed is the error obtained by truncating the Fourier series to a finite number of terms. TAMS does not have a self-contained series truncation system, but rather the user resorts to a program option that outputs a single-source plot of temperature vs. number of terms. This option is discussed in Section 9.4.2.

9.4 PROGRAM OPERATION

9.4.1 Required Data Input

1. A number that defines the problem type (ITEST as indicated in Fig. 9-2).

2. The maximum number of Fourier-series terms.
3. Substrate length, width, and layer thicknesses.
4. Heat transfer coefficients, thermal conductivities, and ambient temperature.
5. The number of sources and resistors.
6. Source coordinates and power.
7. Source locations at which temperature calculations are required.
8. Coordinates of resistor attachment location, thermal resistances, and sink temperatures.

Items (3), (6), (7), and (8) are defined within the constraints of Fig. 9-3.

9.4.2 Fourier-Series Convergence Test

A line printer plot of a single source temperature vs. LMAX = MMAX is obtained as follows:

1. Set up standard, one-source problem with no lumped resistance. Use the source or resistance with the smallest DX, DY for the source in this case.
2. Set LMAX = MMAX = 70.
3. Add 10 to selected ITEST, e.g., ITEST = 1 becomes ITEST = 11, etc.

From the output plot, determine the maximum number of terms and then run the desired multi-sourced problem.

9.4.3 File Handling for Input, Output*

Data enter TAMS via a free formatted file named DIN, and output via D1OUT, D2OUT, and D3OUT. File D1OUT is the standard problem output. D2OUT and D3OUT are single-source temperature vs. LMAX = MMAX. File D2OUT is a nonformatted table, whereas D3OUT is formatted to produce a line printer graph.

9.4.4 Definition of Variables in Data Input File

TAMS input requirements are most easily understood by recognizing the existence of ten different data sets, some of which contain only one line. Table 9-1 outlines the definitions, followed by Table 9-2 which lists the order in which the input must appear.

---

*The first line in the source defines these files in a CDC-FORTRAN deck. Use of other computing systems may require modification of the TAMS READ/WRITE statements.

Table 9-1. Definition of variables in data sets.

Data-Set	Variable Name	Variable Description
1		Title line
2	ITEST	Determines type of problem—see Fig. 9-2
	MODE	Determines run to be for isotropic k (MODE = 0) or run to be for anisotropic k (MODE = 1)
3	LMAX	Maximum number terms (Fourier series) for X component
	MMAX	Maximum number terms (Fourier series) for Y component
4	A	X dimension of device
	B	Y dimension of device
	t_1	Thickness of layer 1
	t_2	Thickness of layer 2
	t_3	Thickness of layer 3
	t_4	Thickness of layer 4
5	h_1	Heat transfer coefficient for $Z = 0$ surface
	h_2	Heat transfer coefficient for $Z = C_4$ surface $(C_4 = t_1 + t_2 + t_3 + t_4)$
	T_A	Ambient temperature seen by surfaces $Z = 0, C_4$
6	k_1	Thermal conductivity of layer 1 ⎤
	k_2	Thermal conductivity of layer 2 ⎪ MODE = 0
	k_3	Thermal conductivity of layer 3 ⎬
	k_4	Thermal conductivity of layer 4 ⎦
	k_{1x}	Thermal conductivity of layer 1 in X direction ⎤
	k_{1y}	Thermal conductivity of layer 1 in Y direction ⎪
	k_{1z}	Thermal conductivity of layer 1 in Z direction ⎪
	k_{2x}	Thermal conductivity of layer 2 in X direction ⎬ MODE = 1
	$\vdots$	$\vdots$ ⎪
	k_{4z}	⎦
7	NS	Total number of sources (Maximum value = 20)
	$NR1$	Number of leads or thermal resistors at $Z = 0$
	$NR2$	Number of leads or thermal resistors at $Z = C_4$ ($NR2$ must be zero for all problems except ITEST = 1) (Maximum value of $NR1 + NR2 = 20$)
8	XS	Minimum X coordinate of a source
	DXS	Source dimension in X direction
	YS	Minimum Y coordinate of a source
	DYS	Source dimension in Y direction
	Q	Source power
9		Every Ith source that requires a temp. calculation requires the nonzero integer I. Input zero for all others.
10	XR	Minimum X coordinate of a resistor pad
	DXR	Resistor pad dimension in X-direction
	YR	Minimum Y coordinate of a resistor pad
	DYR	Resistor pad dimension in Y-direction
	R	Thermal resistance
	T_0	Sink temperature of resistor at resistor end opposite substrate end

Table 9-2. Data set input requirements–list directed input. An I superscript indicates integer input format required.

Data Set	Variables					
1	Title line–all 80 columns may be used					
2	$ITEST^I$ $MODE^I$					
3	$LMAX^I$ $MMAX^I$					
4	A B t_1 t_2 t_3 t_4					
5	h_1 h_2 T_A					
6	k_1 k_2 k_3 k_4 } If MODE = 0					
	$\left.\begin{array}{ccc} k_{1x} & k_{1y} & k_{1z} \\ k_{2x} & k_{2y} & k_{2z} \\ k_{3x} & k_{3y} & k_{3z} \\ k_{4x} & k_{4y} & k_{4z} \end{array}\right\}$ If MODE = 1					
7	NS^I $NR1^I$ $NR2^I$					
	$XS(1)$	$DXS(1)$	$YS(1)$	$DYS(1)$	$Q(1)$	
	$\vdots$	$\vdots$	$\vdots$	$\vdots$	$\vdots$	
	$XS(NS)$	$DXS(NS)$	$YS(NS)$	$DYS(NS)$	$Q(NS)$	
9	1^I	2^I	3^I			
	$XR(1)$	$DXR(1)$	$YR(1)$	$DYR(1)$	$R(1)$	$T_0(1)$
	$\vdots$	$\vdots$	$\vdots$	$\vdots$	$\vdots$	
10	$XR(NR1)$	$DXR(NR1)$	$YR(NR1)$	$DYR(NR1)$	$R(NR1)$	$T_0(NR1)$
	$\vdots$	$\vdots$	$\vdots$	$\vdots$	$\vdots$	
	$XR(NR)$	$DXR(NR)$	$YR(NR)$	$DYR(NR)$	$R(NR)$	$T_0(NR)$

List directed input is used. Input variables must be separated by blanks. Very large or very small quantities such as 3.25E-5 may be used. Once the user has become familiar with the program, Table 9-2 and Fig. 9-2 are usually the only material required to set up a TAMS problem.

Chapter 10
Digital Computer Program Instruction—TNETFA

10.1 INTRODUCTION

TNETFA is a general-purpose steady state and transient thermal analysis program that may be used to predict temperature, pressure, and airflow. Although the program is not difficult to use, the engineer/designer will have to spend some time to develop familiarity with the basic input rules and features peculiar to this program.

A thermal network analysis requires a theoretical model that provides a one-to-one correspondence between the physical and mathematical models. As in all simulation techniques, the theoretical model is in reality an approximation of the physical model. A successful thermal analysis requires the correct identification of significant aspects of the physical system that must be individually incorporated into the theoretical model as conductance elements. The geometry of some heat transfer and airflow problems is sufficiently complex that the analyst may be able to include only the most significant flow paths.

Each element of a thermal model must be quantitatively specified. In a conduction problem, a path length, cross-sectional area, and thermal conductivity are required to completely specify a single element. A convectively cooled surface must be described by a surface element area and a heat transfer coefficient. Accurate values of heat transfer coefficients are particularly difficult to obtain because complex and irregular surface shapes give rise to airflow and heat transfer characteristics that are often quite different from what is predicted using classical heat transfer criteria. The thermal designer must be aware of this and consider supplementary experimental data to either support or replace the classical textbook data.

A few comments are in order concerning airflow predictions. Two techniques are indicated in this book. The first uses classical methods to predict pressure and airflow distribution within an electronic enclosure. The theory was developed in Chapter 6 and applied to some simple problems. TNETFA permits solution of pressure network problems too complex for scientific calculator solution. Occasionally it is necessary to have both laminar and turbulent flow resistances in a single network, and TNETFA will accommodate this kind of problem.

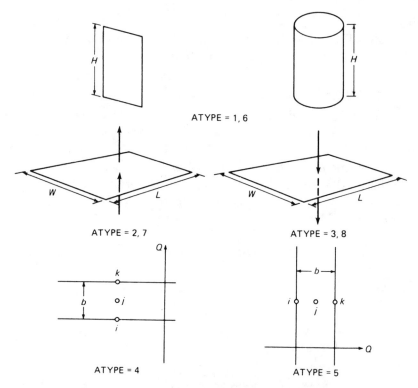

Fig. 10-1. Free convection geometry.

A second analytical tool available for airflow characterization uses two-dimensional laminar flow theory. Air velocity and temperature rise may be computed from streamline computations. This method is most valid for forced airflow, and the accuracy of the results is dependent upon the extent to which the actual airflow is of a streamline nature. There are very few data at present to indicate the accuracy of this method other than qualitative-type indications of potential trouble spots.

10.2 COMMENTS CONCERNING RESTRICTIONS ON MODEL SIZE

Three major limitations to model size are:

1. The number of nodes is limited to a maximum of 650.
2. The total number of elements connecting nodes is limited to a maximum of 1999. TNETFA, however, considers a node and its connecting neighbors as having a total count of twice the actual number of elements in a schematic representation of the circuit. Therefore the count (NUMBER OF CONDUCTORS) printed at the beginning of every printout must be less than exactly 3999.
3. The third limitation on model size is concerned with the multi-surface radiation option (CTYPE = −2 in Data Sets 7, 8). A maximum of 50 nodes may utilize this feature.

Most of the remaining limitations pertain to line count in the various data sets. Table 10-6 lists these limits. Quantities not listed in the table (other than those indicated in 1 and 2) have no limit.

10.3 INSTRUCTIONS

Input for TNETFA consists of specifying values for variables contained in a maximum of 14 data sets. The variable definitions are listed in Table 10-1. The order of each data set and arrangement of variables within data sets are contained in Table 10-2.

Tables 10-3, 10-4, and 10-5 provide information required to specify element types and values using the TNETFA conductance element library.

Table 10-1. Description of data sets and variables.

Data Set Number	Data Sets and Variables
1	TITLE LINES Use for problem identification. Two lines required.
2	PROBLEM TYPE IDENTIFIER A single line specifying noniterate or iterate, units, parameter run requests. *MODE:* 0–suppress iteration, print node connections based on input. 1–attempt steady state solution. 2–attempt velocity potential solution. 3–attempt time-dependent thermal solution. *UNITS:* 0–X, L, D–ft V–ft/sec Q–Btu/hr T–deg F C–Btu/(hr · deg F) G–ft^3/min. ρC_p–Btu/(ft^3 · deg F) CAP–Btu/deg F (E1.19) Time–hr 1–X, L, D–cm V–cm/sec Q–watts T–deg C C–watts/deg C G–cm^3/sec ρC_p–cal/(cm^3 · deg C) CAP–joules/deg C Time–sec 2–X, L, D–in. V–ft/min. Q–watts T–deg C C–watts/deg C G–ft^3/min. ρC_p–cal/(cm^3 · deg C) CAP–joules/deg C Time–sec

Table 10-1. (*Continued*)

Data Set Number	Data Sets and Variables

ICSE: Number of additional problem runs using repeats of Data Sets 9–14 with parameter variations within those sets. Additional sets (9–14) are added to input following last line of set 14 for first run. Set ICSE = $\emptyset$ for no parameter variations.

3   **BASIC INPUT QUANTITIES**
A single line specifying number of nodes, and number of lines required for most data sets. Exceptions, etc., are appropriately identified as required.

NN: total number of nodes.

NCT: number of fixed-temperature nodes specified in Data Set 4, e.g., an ambient temperature node.

NZS: number of nodes requiring individual starting temperature and power input in Data Set 4.

NQCRV: number of time-dependent curves in Data Set 5 specifying time dependence of heat rate (power) dissipation of any heat source node.

NCBLC: number of lines in Data Set 7 for automatic construction of conductor strings.

NCS: number of lines in Data Set 8 specifying single conductor input.

NCRV: number of curves in Data Set 10 used to specify temperature-dependent thermal conductivity or any multiplying factor of variable C in Data Sets 7, 8.

NNCNV: number of lines in Data Set 11 specifying natural convection parameters.

NFCNV: number of lines in Data Set 12 specifying forced convection parameters.

4   **TEMPERATURE, POWER INPUT**
A multiple-line data set specifying starting temperatures for steady state and time-dependent problems, fixed-temperature nodes, and constant heat rates.

TSET, QSET: starting temperature and heat rate for all nodes. Exceptions accounted for in remainder of this data set. One input line used.

NC, TC: node number and temperature for fixed-temperature node, e.g., ambient. Number of lines indicated by NCT in Data Set 3. A minimum of one data line required for all problems.

N, T, Q: node number, starting temperature, and constant heat rate. Number of lines indicated by NZS in Data Set 3. No minimum required number of data lines.

5   **TIME-DEPENDENT HEAT RATE CURVES**
A multiple-line data set specifying time-dependent heat rate dissipation for any non-fixed-temperature node.

N: node number

NQPRS: number of time, heat rate data pairs to specify complete $Q(t)$ dependence.

TIME, Q: time, heat rate values. A minimum of two pairs per curve required. Multiple data pairs per line acceptable. Data pairs must be input in order of increasing time. Heat rates at times outside limits specified by first and last data pair are set at first and last specified heat rates, respectively.

Table 10-1. (Continued)

Data Set Number	Data Sets and Variables

<table>
<tr><td>6</td><td>

CAPACITANCE INPUT

A multiple-line data set specifying nodal thermal capacitance for time-dependent problems. These data may be input for a steady state problem in anticipation of later use in a time-dependent analysis of the same network model.

SINCAP, STRCAP: a header line required by all problems. Both values are zero if no additional data in this set.

SINCAP: number of input lines specifying single node capacitance input via data line N, CAP, CURVE. Use zero if no input lines of this type.

N, CAP, CURVE: node number, capacitance, and curve number (in Data Set 10) for temperature-dependent capacitance. If no curve required, input zero for CURVE. Actual capacitance determined by product of *CAP* in Data Set 6 and *k* in Data Set 10; therefore any combination of *CAP*, *k* that gives correct value of *CAP*k* may be used. Value of *k* returned from curve in Data Set 10 determined by temperature of node N.

STRCAP: number of input lines specifying identical values of CAP, CURVE for consecutive nodes NA through NB in data line NA, NB, CAP, CURVE. Use zero if no input lines of this type.

NA, NB, CAP, CURVE: consecutive node numbers from and including NA through NB for nodes with capacitance *CAP* and using curve CURVE. All preceding comments regarding CAP, CURVE in this data set apply.

</td></tr>
</table>

| 7 | AUTOMATIC CONDUCTOR STRING GENERATOR |

A data set used to simultaneously specify several conductances with identical values of C, CTYPE (see definition to follow), and the appropriate node connections. The number of input lines is indicated by NCBLC in Data Set 3. Conductor input not accommodated by the string generator is specified in Data Set 8.

NBLD: number of conductors generated by this line of data.

NA1: first node number in string.

NAS: node NA1 incrementor or decrementor—may be set at zero if required.

NB1: second node number in string.

NBS: node NB1 incrementor or decrementor—may be set at zero if required.

C: element value as specified by Table 10-3.

CTYPE: term that specifies element type corresponding to C (see Table 10-3).

| 8 | SINGLE CONDUCTOR INPUT |

A data set used to specify single conductances and the appropriate node connections that are not accommodated by Data Set 7. The number of input lines is indicated by NCS in Data Set 3.

Table 10-1. (*Continued*)

Data Set Number	Data Sets and Variables

NA: first node number.

NB: second node number.

C: element value as specified by Table 10-3.

CTYPE: term that specifies element type corresponding to C (see Table 10-3).

9   AREA, EMISSIVITY INPUT FOR MULTI-SURFACE RADIATION EXCHANGE

A multiple-line data set specifying the area and emissivity of every surface participating in multi-surface radiation exchange. This set is required when and only when a CTYPE value of "–2" is specified in any input line in Data Sets 7 and/or 8. Data Set 3 does not specify any input relevant here.

SINAE, STRAE: a header line required when this data set is used. Either value, but not both values may be zero.

SINAE: number of input lines specifying a single surface area and emissivity input via data line N, AR, EM. Use zero if no input lines of this type.

N, AR, EM: node number, area, and emissivity.

STRAE: number of input lines specifying identical values of AR, EM for a string of consecutive node numbers NA through NB in data line NA, NB, AR, EM. Use zero if no input lines of this type.

NA, NB AR, EM: consecutive node numbers from and including NA through NB for nodes with area AR and emissivity EM.

10   TEMPERATURE-DEPENDENT MULTIPLICATIVE FACTORS

A multi-line data set used to provide a temperature-dependent multiplying factor to the C values specified in Data Sets 7, 8 with associated CTYPE values of 1–100. This may be used to input temperature-dependent quantities such as thermal conductivity, heat transfer coefficients not within the TNETFA library of elements, etc.

Data Set 6 temperature-dependent nodal capacitance will also refer to this data set via the value CURVE in Set 6.

Conductance or capacitance common to several nodes may be varied by altering a curve value here rather than several input lines in Data Sets 6, 7, or 8.

The number of curves is indicated by NCRV in Data Set 3.

CURVE, NPAIRS: a header line preceding each curve set.

CURVE: an integer corresponding directly to the value indicated by CTYPE in Data Sets 7, 8. CURVE values must start with the value "1" and be numbered consecutively.

NPAIRS: The number of T, k data pairs for the respective CURVE. NPAIRS must specify at least two data pairs.

T, k: The X, Y coordinates appropriately specifying the curve. A curve indicating a capacitance factor for Data Set 6 linearly interpolated will return a k value based on the temperature T of the node. CAP in Data Set 6 is multiplied by k.

A CTYPE indicating a CURVE for Data Sets 7, 8 will return a linearly interpolated k value based

Table 10-1. *(Continued)*

Data Set Number	Data Sets and Variables

10   TEMPERATURE-DEPENDENT MULTIPLICATIVE FACTORS *(Cont.)* on a T equal to an average temperature of the two nodes interconnecting the conductance. C in Data Sets 7, 8 is multiplied by k.

11   NATURAL CONVECTION PARAMETERS

Natural convection TNETFA library elements are indicated by CTYPE = 101–200 in Data Sets 7, 8, for which the appropriate C input is nodal surface area. Data Set 11 is used to input both the type (ATYPE) of natural convection and the significant dimensional parameter (AA1).

The input sequence in which ATYPE, AA1 appear is precisely CTYPE – 100; e.g., if CTYPE = 109, the corresponding ATYPE, AA1 is the ninth data line in Data Set 11.

ATYPE: term that specifies device orientation and direction of heat transfer.

AA1: significant dimensional parameter (see Table 10-4 for details).

12   FORCED CONVECTION PARAMETERS

Forced convection TNETFA library elements are indicated by CYTPE = 201–300 in Data Sets 7, 8, for which the appropriate C input is nodal surface area. Data Set 12 is used to input the type (BTYPE) of forced convection, dimensional parameters, airflow rate, etc.

BTYPE: specifies type of forced airflow.

BB1–BB4: required parameters (see Table 10-5).

13   VELOCITY POTENTIAL

This is a one-line data set used only when MODE = 2, the velocity potential case.

VO: inlet air velocity.

14   RUN CONTROL STATEMENTS

This three-line data set controls iterations (steady state), time steps (time-dependent), and print intervals,

NLOOP: maximum number of *steady state* iterations.

BETA: *steady state* overrelaxation constant β; $1.0 \leqslant \beta < 2.0$ (E1.17).

ALDT: maximum allowed temperature change between any two successive *steady state* iterations. Iteration is terminated by a maximum temperature change per iteration, MAXDT* $\leqslant$ ALDT.

LOOPEN: number of *steady state* iterations between printouts of total system energy balance.

DELT: time step for *time-dependent* computations. $\Delta t \leqslant CAP/\Sigma$ conductances (E1.21).

MAXT: time limit on *time-dependent* mode.

TPRINT: number of steady state iterations or time steps (TPRINT = time printout interval/DELT) between temperature printouts.

NPRINT: indicator used to suppress (NPRINT = 0) or print (NPRINT = 1) details of node connections.

---

*Ignore MAXDT output value for zero iterations (LOOPCT = 0).

Table 10-2. Arrangement of variables in order of input.

Data Set	Input Variables									Required (1)
1	TITLE LINE 1									X
	TITLE LINE 2									X
2	MODE^I	UNITS^I	ICSE^I							X
3	NN^I	NCT^I	NZS^I	NQCRV^I	NCBLC^I	NCS^I	NCRV^I	NNCNV^I	NFCNV^I	X
4	TSET	QSET								X
	NC^I	TC								X
	N^I	T	Q							
5	N^I	NQPRS^I								
	TIME	Q	TIME	Q	⋯					
6	SINCAP^I	STRCAP^I								X
	N^I	CAP	CAP							
	NA^I	NB^I	CURVE^I	CURVE^I						
7	NBLD^I	NA1^I	NAS^I	NB1^I	NBS^I	C	CTYPE^I			
8	NA^I	NB^I	C	CTYPE^I						
9	SINAE^I	STRAE^I								
	N^I	AR	EM							
	NA^I	NB^I	AR	EM						
10	CURVE^I	NPAIRS^I								
	T_1	k_1	T_2	k_2	⋯					
11	ATYPE^I	AA1								
12	BTYPE^I	BB1	BB2	BB3	BB4^I					
13	VO									
14	NLOOP^I	BETA	LOOPEN^I							X
	DELT	MAXT	ALDT							X
	TPRINT^I	NPRINT^I								X

(1) Identifies required input for *all* problems. ^{I} Indicates integer input.

Table 10-3. Element descriptions for Data Sets 7, 8

MODE	CTYPE	C Specification	
		Description	Formula
0, 1, 3	−1	Simple radiation library element.	$\mathcal{F}A_s$
	−2	Multi-surface radiation library element.	FA_s
	0	Nonspecific. C used as conductance.	$kA_k/L, hA_s, 1/(\Delta x)^2, 1/(\Delta y)^2$, etc.
	1–100	Temperature-dependent multiplicative factor.	$A_k/L, \rho$, etc.
	101–200	Natural convection library element.	A_s
	201–300	Forced convection library element.	A_s
	301	Volumetric airflow rate.	G
	311–400	Volumetric non-airflow rate. ρC_p must be input in Data Set 10.	G
	401	Airflow resistance, laminar.	R
	402	Airflow resistance, turbulent.	R
2	0	Velocity potential problem, x-direction.	$1/(\Delta x)^2$
	1	Velocity potential problem, y-direction.	$1/(\Delta y)^2$

Table 10-4. Additional detail on Data Set 11—natural convection.

Mode of Heat Transfer	ATYPE	AA1	Reference
Vertical plate or cylinder	1	Plate height H, cylinder height H	E2.7, Table 2-2
Horizontal rectangular plate Convection from *upper* surface to air $(T_s > T_{AIR})$ or convection from air to *lower* surface $(T_{AIR} > T_s)$.	2	$WL/[(W+L)2]$	E2.7, Table 2-2
Horizontal rectangular plate Convection from air to *upper* surface $(T_{AIR} > T_s)$ or convection from *lower* surface to air $(T_s > T_{AIR})$.	3	$WL/[(W+L)2]$	E2.7, Table 2-2
Horizontal air space Heat transfer in upward direction. b = air space thickness. One node required at midspace.	4	b	[7]
Vertical air space Heat transfer in horizontal direction. b = air space thickness. One node required at midspace.	5	b	[7]
Small rectangular plate Vertical orientation, $H \lesssim 6$ in. Heat transfer to or from either surface.	6	H	E2.14
Horizontal orientation $H, L \leqslant 6$ in. Convection from *upper* surface to air or convection from air to *lower* surface.	7	$WL/[2(W+L)]$	E2.14
Convection from air to *upper* surface or convection from *lower* surface to air.	8	$WL/[2(W+L)]$	E2.14

Table 10-5. Additional detail on Data Set 12—forced convection.

Mode of Heat Transfer	BTYPE	BB1	BB2	BB3	BB4	Reference
Duct, laminar flow $(Re_D < 2100)$	1	V	D_H	L	A	E2.34
Duct, turbulent flow $(Re_D > 10,000)$	2	V	D_H	L	A	E2.37, E2.38, E2.39
Flat plate, laminar flow ave h $(Re_L < 5 \times 10^5)$	3	V	A	L	A	E2.20, E2.21, E2.22
Flat plate, turbulent flow ave h $(Re_L > 5 \times 10^5)$	4	V	A	L	A	[2], Eqn. 6-67
Flat plate, laminar flow local h $(Re_x < 5 \times 10^5)$	5	V	A	ΔX	I	E2.20
Flat plate, turbulent flow local h $(Re_x > 5 \times 10^5)$	6	V	A	ΔX	I	[2], Eqn. 6-66

A: Arbitrary, but required numeric input. V: Flow velocity.

ΔX: Length of node in string of equal length nodes. D, D_H: Hydraulic diameter.

I: number of node at leading edge of string. Nodes must be numbered in ascending order starting with node I. Missing node numbers not allowed in any given string. Use integer format.

L: Length of plate or duct.

Table 10-6. Restrictions on input code.

Data Set	Variable	Limit
4	NC, TC	50 lines
	N, T, Q	650 lines
5	N, NQPRS, TIME, Q	20 curves–total of 100 pairs for all curves in this data set
6	N, CAP, CURVE	650 lines, 20 different curves
	NA, NB, CAP, CURVE	20 different curves
7	NBLD, NA1, NAS, NB1, C, CTYPE	300 lines
8	NA, NB, C, CTYPE	1999 lines
9	AR, EM	50 different values
10	CURVE, NPAIRS	20 curves, total of 100 pairs for all curves in this data set
11	ATYPE, AA1	50 lines
12	BTYPE, BB1–BB4	50 lines

Glossary

The most common symbols used herein have been tabulated as an aid to the reader. Occasionally a symbol has more than one meaning. However, little confusion should result when the context of the associated discussion is considered. The reader should remain flexible with regard to subscripts, although some of these are included here. Dimensionless quantities are indicated, but no attempt is made to specify the units because more than one system is used within the text.

A_k	Conduction path cross-sectional area
A_s	Surface area
a	Thermal diffusivity
C_p	Specific heat at constant pressure
C	Thermal conductance
CAP_i	Capacitance of node i
D	Diameter
E	Gray, diffuse surface radiation emissive power
E_b	Black body radiation emissive power
$E_{\lambda b}$	Black body monochromatic radiation flux per unit wavelength interval
F	Radiation angle factor, dimensionless
$\mathcal{F}$	Gray body radiation exchange factor, dimensionless
f	Heat transfer coefficient correction factor, dimensionless
Gr_p	Grashof number, dimensionless
G	Volumetric flow rate
g	Acceleration due to gravity
H_b	Buoyancy pressure head
H_L	Total pressure head loss
H	Height
H	Irradiance
h_c	Convective heat transfer coefficient
h_r	Radiative heat transfer coefficient
h	Combined (convection, radiation) heat transfer coefficient
I	Radiation field intensity
I_0	Modified Bessel function of first kind, zero order
J	Radiosity
K_c	Contraction loss coefficient, dimensionless
K_e	Expansion loss coefficient, dimensionless
k	Thermal conductivity
L	Length
$\dot{m}$	Mass flow rate
Nu_p	Nusselt number, dimensionless
N	Number of fins, dimensionless

Pr	Prandtl number, dimensionless
P_w	Wetted perimeter
P	Convection parameter
p	Pressure
Q_c	Heat transfer from a surface by convection
Q_k	Heat transfer by conduction
Q_r	Heat transfer from a surface by radiation
Q_V	Heat dissipation per unit volume
q	Heat transfer per unit area
Ra^*	Modified Rayleigh number, dimensionless
Re_p	Reynold's number, dimensionless
R	Airflow resistance
R	Thermal resistance
r_c	Thermal interface resistance
r_i	Residual heat energy for node i
S	Spacing
$STAB_i$	Stability constant for node i
T	Temperature
T'	Temperature in absolute scale
t	Thickness
t	Time
u	Velocity component
V	Velocity
v	Velocity component
W	Width
x	Length

Greek symbols:

β	Coefficient of thermal expansion of air
β	$hA_s/\dot{m}C_p$, dimensionless
β	Overrelaxation constant, dimensionless
δ	Convective boundary layer thickness
ϵ_λ	Monochromatic emissivity, dimensionless
ϵ	Total emissivity, dimensionless
η	Fin efficiency, dimensionless
λ	Wavelength
μ	Absolute viscosity
ν	Kinematic viscosity
π	3.1416
ρ	Density
ρ	Total reflectivity, dimensionless
σ	Stefan-Boltzmann constant
ω	Solid angle

APPENDICES

Appendix I
References

1. Barzelay, M. E., Kin Nee Tong, and Holloway, G., "Effect of Pressure on Thermal Conductance of Contact Joints," National Advisory Committee for Aeronautics, Technical Note 3295, May 1955.
2. Kreith, Frank, *Principles of Heat Transfer*, 3rd Edition, Harper & Row, Publishers, New York, 1973.
3. *Handbook of Heat Transfer*, Editors, Rohsenow, W. M., and Hartnett, J. P., McGraw-Hill Book Co., New York, 1973.
4. McAdams, W. H., *Heat Transmission*, 3rd Edition, McGraw-Hill Book Co., New York, 1954.
5. Goldstein, R. J., Sparrow, E. M., and Jones, D. C., "Natural Convection Mass Transfer Adjacent to Horizontal Plates," *International Journal of Heat and Mass Transfer*, vol. 16, no. 5, May 1973, pp. 1025–1034.
6. Lloyd, J. R., and Moran, W. R., "Natural Convection Adjacent to Horizontal Surface of Various Planforms," *Journal of Heat Transfer*, Nov. 1974, pp. 443–447.
7. Jakob, M., *Heat Transfer*, vols. I and II, John Wiley & Sons, Inc., New York, 1957.
8. Nusselt, W., *Zeitschr. d. Ver deutsch. Ing.*, vol. 73, 1929, p. 1475.
9. King, W. J., *Mech. Eng.*, 1932, p. 347.
10. Ellison, G. N., "The Thermal Design of an LSI Single Chip Package," *IEEE Trans. Parts, Hybrids, and Packaging*, vol. PHP-12, Dec. 1976, pp. 371–378.
11. Schmidt, E., and Beckman, W., "Das Temperatur und Geschwindigkeitsfeld vor einer warmeabgebgebended senkrechten Platte bei naturlicher Konvection," *Tech. Mech. u. Thermodynamic*, Bd. 1, no. 10, Oct. 1930, pp. 341–349; cont. Bd. 1, no. 11, Nov. 1930, pp. 391–406.
12. Kays, W. M., and Crawford, M. E., *Convective Heat and Mass Transfer*, 2nd Edition, McGraw Hill Book Co., New York, 1980.
13. Kays, W. M., "Numerical Solutions for Laminar-Flow Heat Transfer in Circular Tubes," *Trans. ASME*, vol. 77, 1955, pp. 1265–1274.
14. Boelter, L. M. K., Young, G., and Iversen, H. W., NACA TN 1451 (now NASA), Washington, D.C., July 1948.
15. Bolvin, Robert, "Thermal Characteristics of ICs Gain Importance," *Electronics*, Oct. 31, 1974, pp. 87–90.
16. Dunkle, R. V., "Thermal Radiation Tables and Applications," *Trans. ASME*, May 1954, pp. 549–552.
17. Bevans, J. T., Gier, J. T., and Dunkle, R. V., "Comparison of Total Emittances with Values Computed from Spectral Measurements," *Trans. ASME*, vol. 80, 1958, p. 80.
18. Ellison, G. N., "Theoretical Calculation of the Thermal Resistance of a Conducting and Convecting Surface," *IEEE Trans. Parts, Hybrids, and Packaging*, vol. PHP-12, no. 3, Sept. 1976, pp. 265–266.
19. Carslow, H. S., and Jaeger, J. C., *Conduction of Heat in Solids*, Oxford at the Clarendon Press, p. 216, 1959.
20. Joy, C. R., and Schlig, E. S., "Thermal Properties of Very Fast Transistors," *IEEE Trans. Electron Devices*, vol. ED-17, Aug. 1970, pp. 586–593.

21. Ellison, G. N., "Generalized Computations of the Gray Body Shape Factor for Thermal Radiation from a Rectangular U-Channel," *IEEE Trans. Components, Hybrids, and Manfg. Tech.*, CHMT-2, no. 4, Dec. 1979, pp. 517–522.

22. Rea, S. N., and West, S. E., "Thermal Radiation from Finned Heat Sinks," *IEEE Trans. Parts, Hybrids, and Packaging*, vol. PHP-12, no. 2, June 1976.

23. Elenbaas, W., "Dissipation of Heat by Free Convection," Parts I and II, *Philips Research Report*, vol. 3, N. V. Philips, Gloeilampenfabrieken, Eindhoven, Netherlands, 1948, pp. 338–360, 450–465.

24. Van de Pol, D. W., and Tierney, J. K., "Free Convection Heat Transfer from Vertical Fin Arrays," *IEEE Trans. Parts, Hybrids, and Packaging*, vol. PHP-10, no. 4, Dec. 1974, pp. 267–271.

25. Thermalloy Inc., Bulletin EHS-81, P.O. Box 34829, 2021 Valley View Lane, Dallas, Texas.

26. Osborne, W. C., *Fans*, 2nd Edition, Pergamon Press Ltd., New York, 1977.

27. American Society of Heating, Refrigerating, and Air Conditioning Engineers, Standard 51-75.

28. Hay, Donald, "Cooling Card-Mounted Solid-State Component Circuits," McLean Engineering, a division of Zero Corp., Princeton Junction, New Jersey.

29. Han, L. S., *J. Appl. Mech., Trans. ASME*, vol. 27, 1960, p. 403.

30. Lundgren, T. S., Sparrow, E. M., and Starr, J. S., *J. Bas. Eng., Trans. ASME*, vol. 86, 1964, p. 620.

31. Deissler, R. G., NACA TN 3016, 1953.

32. Heat Transfer and Fluid Data Books, General Electric Co., Corporate Research and Development, Schenectady, New York.

33. Streeter, V. L., and Benjamin, W. E., *Fluid Mechanics*, 7th Edition, McGraw-Hill Book Co., New York, 1979.

34. Hein, V. L., and Lenzi, V. K., "Thermal Analysis of Substrates and Integrated Circuits," presented at the 1969 Electronic Components Conference.

35. Ellison, G. N., "The Effect of Some Composite Structures on the Thermal Resistance of Substrates and Integrated Circuit Chips," *IEEE Trans. Electron Devices*, vol. ED-20, Mar. 1973, pp. 233–238.

36. David, R. F., "Computerized Thermal Analysis of Hybrid Circuits," presented at the 1977 Electronic Components Conference, Arlington, Virginia.

37. Ellison, G. N., "A Thermal Analysis Computer Program Applicable to a Variety of Microelectronic Devices," presented at the 1978 International Conference on Hybrid Microelectronics, Minneapolis, Minnesota.

Appendix II
Theoretical Basis of TAMS

II.1 SPECIAL NOMENCLATURE

A	Substrate dimension in the x-direction of a three-dimensional coordinate system
B	Substrate dimension in the y-direction of a three-dimensional coordinate system
C_i	Total thickness of all layers up to and including layer i
h_1	Heat transfer coefficient of surface one at $z = 0$
h_2	Heat transfer coefficient of surface two at $z = c_4$
k_{xi}	Thermal conductivity in x-direction of layer i
k_{yi}	Thermal conductivity in y-direction of layer i
k_{zi}	Thermal conductivity in z-direction of layer i
$Q(r)$	Source heat dissipation per unit volume
$q(r)$	Source heat dissipation per unit area
R_i	Lumped thermal resistance
$r(x, y, z)$	Magnitude of a radius vector
$T(r)$	Temperature at field point r
$T_0(i)$	Sink temperature at end of R_i
T_A	Ambient temperature
T_{AR}	Ambient temperature at $z = 0$ side when $T_{AR} \neq T_A$
t_i	Thickness of layer i

II.2 BASIC PROBLEM

The theoretical foundations of TAMS are only summarized here. Readers with interest in more than a cursory mathematical description are referred to [10], [35], and [37].

The rectangular coordinate system and nomenclature representative of the physical model are illustrated in Fig. 9-3 for the complete four-layer system. Thermal conductivity components are specified for each of the three coordinate system axes, a total of 12 different values. Cases where fewer than four layers are required are accommodated by merely specifying the same conductivity for as many layers as necessary. Buried heat sources are provided by a solution for sources located at the boundary between the first and second layers. Nonburied sources are located at the $z = 0$ boundary.

The equation requiring a solution is the differential equation for steady state heat conduction with an anisotropic thermal conductivity k_{xi}, k_{yi}, k_{zi} independent of x, y, z, respectively, within each layer. From E1.2:

$$k_{xi}(\partial^2 T/\partial x^2) + k_{yi}(\partial^2 T/\partial y^2) + k_{zi}(\partial^2 T/\partial z^2) = -Q$$

II.3 BOUNDARY CONDITIONS

The most applicable edge boundary conditions on $T(r)$ assume negligible heat loss across these usually small areas:

$$\partial T/\partial x = 0; \; x = 0, A \qquad \partial T/\partial y = 0; \; y = 0, B \qquad \text{EII.1}$$

Boundary conditions at edges

Radiation and convection from either or both of the substrate surfaces are accounted for by:

$$k_z \partial T/\partial z - h_1 T = 0; \qquad k_z \partial T/\partial z + h_2 T = 0; \qquad \text{EII.2}$$
$$z = 0 \qquad\qquad z = C_4$$

Boundary conditions at surfaces

where h_1 and h_2 are the appropriate heat transfer coefficients. An isothermal at $z = C_4$ is obtained in the final solution by letting h_2 approach infinity. Note that the surface flux density is written as having a linear temperature dependence.

II.4 SOLUTION OF DIFFERENTIAL EQUATION FOR ONE SOURCE

A solution $T(r)$ where $r = r(x, y, z)$ is anticipated as an expansion in a set of orthogonal functions of x and y that satisfy the boundary conditions at the edges. Clearly, this expansion is a double Fourier cosine series.

The first objective is to obtain the expansion of the source function, $Q(r)$:

$$Q(r) = \sum_{\alpha} \sum_{\beta} \epsilon_\alpha \epsilon_\beta \phi_{\alpha\beta}(z) \cos \alpha x \cos \beta y$$

where $\alpha = (l\pi/A)\sqrt{k_x/k_z}$, $\beta = (m\pi/B)\sqrt{k_y/k_z}$ for $l = 0, 1, 2, \ldots, \infty$; $m = 0, 1, 2, \ldots \infty$ and:

$$\epsilon_\alpha = \begin{cases} 1/2, \; \alpha = 0 \\ 1, \; \alpha \neq 0 \end{cases} \qquad \epsilon_\beta = \begin{cases} 1/2, \; \beta = 0 \\ 1, \; \beta \neq 0 \end{cases}$$

The coefficients $\phi_{\alpha\beta}(z)$ are:

$$\phi_{\alpha\beta}(z) = (4/AB) \int_{x=0}^{A} \int_{y=0}^{B} Q(r) \cos \alpha x \cos \beta y \, dxdy; \, \alpha \neq 0, \beta \neq 0$$

$$\phi_{\alpha 0}(z) = (4/AB) \int_{x=0}^{A} \int_{y=0}^{B} Q(r) \cos \alpha x \, dxdy; \, \alpha \neq 0, \beta = 0$$

$$\phi_{0\beta}(z) = (4/AB) \int_{x=0}^{A} \int_{y=0}^{B} Q(r) \cos \beta y \, dxdy; \, \alpha = 0, \beta \neq 0$$

The temperature $T(r)$ is similarly expanded:

$$T(r) = \sum_\alpha \sum_\beta \epsilon_\alpha \epsilon_\beta \psi_{\alpha\beta}(z) \cos \alpha x \cos \beta y$$

The source Q is assumed to be a surface source at $z = z_0$ of zero thickness; hence it may conveniently be expressed as $Q(r) = q(x, y)\delta(z - z_0)$ where $\delta(z - z_0)$ is the Dirac delta function. Note that $q(x, y)$ is a flux density (power/area) because:

$$\int Q \, dr = \int q(x, y)\delta(z - z_0) \, dx dy dz$$

$$= \int q(x, y) \, dx dy$$

The flux density is assumed to be uniform over the region of the x and y integrations.

The expansions for $T(r)$ and $Q(r)$ are substituted into the partial differential equation for heat conduction, E1.2. The problem then becomes one of solving a second-order, inhomogeneous differential equation for $\psi_{\alpha\beta}$ with one variable (z) with appropriate matching to boundary conditions at intermediate surfaces that define the various layers:

$$\frac{d^2\psi_{\alpha\beta}}{dz^2} - \gamma^2_{\alpha\beta} \, \psi_{\alpha\beta} = -\frac{1}{k_{zi}} \phi_{\alpha\beta},$$

$$\gamma^2_{\alpha\beta} = \left(\frac{l}{\pi A}\right)^2 \left(\frac{k_{xi}}{k_{zi}}\right) + \left(\frac{m}{\pi B}\right)^2 \left(\frac{k_{yi}}{k_{zi}}\right)$$

The solution is rather complex and will not be discussed here.

II.5 SOLUTION FOR MULTI-SOURCES

The theoretical considerations of extending the single-source solution to the significantly more practical problem of a multiple-source, multiple-lumped-resistance application are accomplished by superposition of the single-source solutions:

$$T(r_i) = \sum_{j=1}^{NS} Q_j A_{ij}(r_i, r_j) - \sum_{k=1}^{NR} F_k A_{ik}(r_i, r_k) + T_A \qquad \text{EII.3}$$

Multi-source, multi-resistance temperature

The A_{ij} and A_{ik} are solutions for single sources of unit magnitude; i.e., A_{ij} is the temperature $T(r_i)$ at r_i due to a unit source ($Q = 1$ watt) at r_j for zero ambient temperature. Q_j is the heat into the system at r_j, and F_k is the heat out of the system at r_k and through a lumped parameter thermal resistance R_k. The reader is referred to the references for details on the method of calculating A_{ij}, A_{ik}, and F_k.

Appendix III
TAMS Boundary Conditions for Radiation and Convection with Two Different Ambient Temperatures

The original mathematical formulation of TAMS assumed convection from two surfaces to the same ambient temperature. This is occasionally inconvenient when the two surfaces of a rectangular plate "see" different ambients. The difficulty may be surmounted by an appropriate modification of one of the heat transfer coefficients.

Suppose that at $z = C_4$:

$$k_{z4}(\partial T/\partial z) = -h_2(T - T_A)$$

i.e., the ambient is T_A for the surface temperature T at the $z = C_4$ side. At $z = 0$, suppose that:

$$k_{z1}(\partial T/\partial z) = h_1(T - T_{AR})$$

or:

$$k_{z1}(\partial T/\partial z) = h'_1(T - T_A)$$

as required. A little algebra shows that:

$$\boxed{h'_1 = h_1\left(1 - \frac{T_{AR} - T_A}{T - T_A}\right)}$$

EIII.1
h_1 modified for elevated ambient

Appendix IV
Thermal Conductivity of a
Nonhomogeneous Medium

The analysis of a device may be complicated by a nonhomogeneous thermal conductivity. An example is a layer of ceramic with a large number of thick film circuit lines. Very often the circuit lines are approximately parallel or perpendicular. Such configurations should be considered to consist of parallel or series resistances in directions parallel or perpendicular, respectively, to the circuit lines.

The formulae for the effective thermal conductivity k_e of two materials, k_1 and k_2, are derived below, where:

$$l = \text{length of resistor element}$$

$$A = \text{cross-sectional area of resistor element}$$

$$k = \text{thermal conductivity of resistor element}$$

For parallel elements the applicable equations are:

$$\frac{k_e A}{l} = \frac{k_1 A_1}{l} + \frac{k_2 A_2}{l}$$

$$k_e = (k_1 A_1/A) + (k_2 A_2/A)$$

Since $A = A_1 + A_2$, we can substitute $A - A_2$ for A_1:

$$k_e = k_1[1 - (A_2/A)] + k_2(A_2/A)$$

Setting $\Gamma_2 = A_2/A$:

$$\boxed{k_e = k_1(1 - \Gamma_2) + k_2\Gamma_2}$$

EIV.1
Conductivity for
parallel elements

For series elements the applicable equations are:

$$l/(k_e A) = l_1/(k_1 A) + l_2/(k_2 A)$$

$$1/k_e = [l_1/(lk_1)] + [l_2/(lk_2)]$$

Since $l = l_1 + l_2$, we can substitute $l - l_2$ for l_1. Setting $\Gamma_2 = l_2/l$:

$$k_e = k_1 k_2 / [(1 - \Gamma_2)k_2 + \Gamma_2 k_1]$$

$$\boxed{k_e = k_1 k_2 / [(1 - \Gamma_2)k_2 + \Gamma_2 k_1]}$$

EIV.2
Conductivity for
series elements

In practice, neither the ratio A_2/A nor l_2/l is known. Instead, it is recommended that the volumetric fraction of element 2 be used.

Appendix V
TAMS Source Listing

```
        PROGRAM TAMSIV (INPUT,OUTPUT,DIN,D1OUT,D2OUT,D3OUT,TAPE1=DIN,TAPE2   A    10
       1=D1OUT,TAPE3=D2OUT,TAPE4=D3OUT)                                      A    20
C          THE TAMS PROGRAM AND ALL RELATED MATERIAL ARE SUPPLIED            A    30
C          AS - IS AND WITHOUT WARRANTY OR REPRESENTATION OF ANY             A    40
C          KIND. THE PROGRAM AUTHOR AND HIS EMPLOYER AND PUBLISHER,          A    50
C          PAST, PRESENT, OR FUTURE MAKE NO REPRESENTATIONS RESPECT-         A    60
C          ING THE PROGRAM AND RELATED MATERIAL, AND EXPRESSLY DIS-          A    70
C          CLAIM ANY LIABILITY FOR DAMAGES WHETHER DIRECT, CONSEQUEN-        A    80
C          TIAL OR OTHERWISE, ARISING FROM THE USE OF THE PROGRAM OR         A    90
C          RELATED MATERIAL, OR ANY PART THEREOF.                           A   100
C       TAMSIV - THERMAL ANALYZER FOR MULTILAYER SYSTEMS -                   A   110
C       ANISOTROPIC K FOR FOUR LAYER SYSTEM,                                 A   120
C       RESISTANCES AT Z=0,C4 FOR ITEST=1,                                   A   130
C       SINGLE SOURCE CONVERGENCE PLOT INCLUDED.                             A   140
C       AUTHOR - GORDON N. ELLISON                                           A   150
C       AUGUST 10, 1983 FORTRAN IV VERSION.                                  A   160
C       HYPERBOLIC COSINES HAVE BEEN FACTORED AS MUCH AS POSSIBLE TO         A   170
C       MINIMIZE THE POSSIBILITY OF AN INFINITE OPERAND. TOTAL FACTORING     A   180
C       HAS NOT BEEN POSSIBLE FOR ALL POSSIBLE CASES,                        A   185
C       I.E. SOME COSH'S STILL REMAIN FOR ITEST=1 WITH RESISTORS ON          A   190
C       BACKSIDE (NL2 NON-ZERO IN CODE, NR2 NON-ZERO IN MANUAL);             A   200
C       ITEST=3, 4.                                                          A   210
C       THE FOLLOWING DIMENSIONS MUST BE EQUAL TO OR GREATER THAN THE        A   220
C       NUMBER OF SOURCES                                                    A   230
        DIMENSION XS(20), DXS(20), YS(20), DYS(20), XSC(20), YSC(20), Q(20   A   240
       1), LOC(20), TEMPS(20), TEMPL(20), TEMP(20), X2S(20), Y2S(20)         A   250
C       THE FOLLOWING DIMENSIONS MUST BE EQUAL TO OR GREATER THAN THE        A   260
C       NUMBER OF LEADS                                                      A   270
        DIMENSION XL(20), DXL(20), YL(20), DYL(20), XLC(20), YLC(20), R(20   A   280
       1), T0(20), SW1(20), F(20), TL(20), X2L(20), Y2L(20)                  A   290
C       THE FOLLOWING DIMENSION C(NL,NL+1) MUST CONTAIN NL EQUAL TO OR       A   300
C       GREATER THAN THE NUMBER OF LEADS                                     A   310
        DIMENSION C(20,21)                                                   A   320
C       THE FOLLOWING DIMENSIONS MUST BE EQUAL TO OR GREATER THAN THE        A   330
C       INPUT DATA FOR LMAX,MMAX                                             A   340
        DIMENSION AL(70), AM(70), ALM(70,70), SPSUM(70),                     A   350
       1          BL(70), BM(70), BLM(70,70),                                A   360
       2    AIL(70),AIM(70),AILM(70,70),                                     A   370
       3    BIL(70),BIM(70),BILM(70,70)                                      A   380
        DIMENSION JC(20), VV(2)                                              A   390
        DIMENSION TITLE(12)                                                  A   400
        REAL K1,K2,K3,K4,K21,K32,K43,K12,K23,K34                             A   410
        DIMENSION LINE(50)                                                   A   420
        REAL K1X,K1Y,K2X,K2Y,K3X,K3Y,K4X,K4Y                                 A   430
        REAL KFX1,KFY1,KFX2,KFY2,KFX3,KFY3,KFX4,KFY4                         A   440
C       FCN DEF STATEMENTS                                                   A   450
        GAM(V,W,ZX,ZY) = SQRT(((((V - 1.0) * AP) ** 2) * ZX + (((W           A   460
       1    - 1.0) * BP) ** 2) * ZY)                                         A   470
        GAML(V,ZX) = (V - 1.0) * AP * SQRT(ZX)                               A   480
        GAMM(V,ZY) = (V - 1.0) * BP * SQRT(ZY)                               A   490
        R4L(V) = K43 * (TH4 + (H2K4 / GAML(V,KFX4))) * GL4 / GL3             A   500
        S4L(V) = 1.0 + (H2K4 / GAML(V,KFX4)) * TH4                           A   510
        UL(V) = TH1 + (H1K1 / GAML(V,KFX1))                                  A   520
        VL(V) = 1.0 + (H1K1 / GAML(V,KFX1)) * TH1                            A   530
        RSCL(V) = R4L(V) * TH3 + S4L(V)                                      A   540
        RCSL(V) = (R4L(V) + S4L(V) * TH3)                                    A   550
        R4M(V) = K43 * (TH4 + (H2K4 / GAMM(V,KFY4))) * GM4 / GM3             A   560
        S4M(V) = 1.0 + (H2K4 / GAMM(V,KFY4)) * TH4                           A   570
        UM(V) = TH1 + (H1K1 / GAMM(V,KFY1))                                  A   580
        VM(V) = 1.0 + (H1K1 / GAMM(V,KFY1)) * TH1                            A   590
        RSCM(V) = R4M(V) * TH3 + S4M(V)                                      A   600
        RCSM(V) = R4M(V) + S4M(V) * TH3                                      A   610
        R4LM(V,W) = K43 * (TH4 + (H2K4 / GAM(V,W,KFX4                        A   620
       1    ,KFY4))) * GLM4 / GLM3                                           A   630
```

351

```
      S4LM(V,W) = 1.0 + (H2K4 / GAM(V,W,KFX4,KFY4)) * TH4      A  640
      ULM(V,W) = TH1 + (H1K1 / GAM(V,W,KFX1,KFY1))             A  650
      VLM(V,W) = 1.0 + (H1K1 / GAM(V,W,KFX1,KFY1)) * TH1       A  660
      RSC(V,W) = R4LM(V,W) * TH3 + S4LM(V,W)                   A  670
      RCS(V,W) = (R4LM(V,W) + S4LM(V,W) * TH3)                 A  680
      PI = 3.1415926535                                        A  690
C     INPUT DATA                                               A  700
C     TITLE CARD                                               A  710
10    READ (1,710) (TITLE(I),I=1,8)                            A  720
      IF (EOF(1) .NE. 0) CALL EXIT                             A  730
C     TEST FOR PROBLEM TYPE                                    A  740
C     ITEST=1 - CALCULATE TEMPERATURE AT Z=0, LEADS            A  750
C     AT Z=0,  T1+T2+T3+T4                                     A  760
C     Z=0,     NEWTON'S LAW COOLING AT Z=T1 + T2 + T3 + T4     A  770
C     ITEST=2 - CALCULATE TEMPERATURE AT Z=0 FOR SOURCES AND LEADS AT  A  780
C     Z=0,     ISOTHERMAL TA AT Z=T1 + T2 + T3 +T4             A  790
C     ITEST=3 - CALCULATE TEMPERATURE AT Z=0 FOR SOURCES AND LEADS AT  A  800
C     Z=T1,    NEWTON'S LAW COOLING AT Z=T1 +T2 +T3 +T4        A  810
C     ITEST=4 - CALCULATE TEMPERATURE AT Z=0 FOR SOURCES AND LEADS AT  A  820
C     Z=T1,    ISOTHERMAL TA AT Z=T1 + T2 + T3 +T4             A  830
C     ITEST=5 - CALCULATE TEMPERATURE AT Z=T1 FOR SOURCES AND LEADS AT  A  840
C     Z=T1,    NEWTON'S LAW COOLING AT Z=T1 + T2 + T3 + T4     A  850
C     ITEST=6 - CALCULATE TEMPERATURE AT Z=T1 FOR SOURCES AND LEADS AT  A  860
C     Z=T1,    ISOTHERMAL TA AT Z=T1 + T2 +T3 + T4             A  870
C     MODE=0 FOR ISOTROPIC K IN ALL LAYERS                     A '880
C     MODE=1 FOR NON-ISOTROPIC K IN ANY LAYER                  A  890
      READ (1, *)ITEST,MODE                                    A  900
C     LIMITS TO FOURIER SERIES, TERMINATION TEST CONSTANTS     A  910
      READ (1, *)LMAX,MMAX                                     A  920
C     SUBSTRATE DIMENSIONS                                     A  930
      READ (1, *)A,B,T1,T2,T3,T4                               A  940
C     PHYSICAL CONSTANTS                                       A  950
C     FOR ISOTHERMAL (ITEST=2,4,6), H2 CAN BE ANYTHING         A  960
      READ (1, *)H1,H2,TA                                      A  970
C     K1,K2,K3,K4 ARE Z-DIRECTION COMPONENTS                   A  980
C     K1X,K2X,K3X,K4X ARE X-DIRECTION COMPONENTS               A  990
C     K1Y,K2Y,K3Y,K4Y ARE Y-DIRECTION COMPONENTS               A 1000
C     IF K ISOTROPIC IN ALL LAYERS, NEED TO READ IN ONLY K1,K2,K3,K4  A 1010
      IF (MODE .EQ. 0) GO TO 20                                A 1020
      READ (1, *)K1X,K1Y,K1                                    A 1030
      READ (1, *)K2X,K2Y,K2                                    A 1040
      READ (1, *)K3X,K3Y,K3                                    A 1050
      READ (1, *)K4X,K4Y,K4                                    A 1060
      GO TO 30                                                 A 1070
20    READ (1, *)K1,K2,K3,K4                                   A 1080
C     NUMBER OF SOURCES, NUMBER OF LEADS AT Z=0,               A 1090
C     NO. OF LEADS AT Z=C4                                     A 1100
30    READ (1, *)NS,NL1,NL2                                    A 1110
C     SOURCE LOCATIONS AND STRENGTHS                           A 1120
C     UNROTATED SOURCES            - X AND Y ARE MINIMUM VALUES  A 1130
C     IN ALL CASES, DX AND DY ARE THE LITERAL SOURCE DIMENSIONS  A 1140
      READ (1, *)(XS(I),DXS(I),YS(I),DYS(I),Q(I),I=1,NS)       A 1150
C     FOR EVERY REQUIRED CALC., INPUT A LOC(I)=I, OTHERWISE SET LOC(I)=0  A 1160
      READ (1, *)(LOC(I),I=1,NS)                               A 1170
C     CHECK FOR SUBSTRATE LEADS                                A 1180
      NL = NL1 + NL2                                           A 1190
      IF (NL .EQ. 0) GO TO 40                                  A 1200
C     IF NL=0, THIS DATA CAN BE ANYTHING                       A 1210
C     LEAD LOCATIONS, THERMAL RESISTANCES, SINK TEMPERATURES   A 1220
C     FIRST NL1 LEADS AT Z=0, LAST NL2 LEADS AT Z=C4           A 1230
      READ (1, *)(XL(I),DXL(I),YL(I),DYL(I),R(I),T0(I),I=1,NL) A 1240
C     CALCULATION OF INTERMEDIATE VARIABLES                    A 1250
40    AP = PI / A                                              A 1260
      BP = PI / B                                              A 1270
```

```
         K21 = K2 / K1                                                A 1280
         K32 = K3 / K2                                                A 1290
         K43 = K4 / K3                                                A 1300
         K12 = K1 / K2                                                A 1310
         K23 = K2 / K3                                                A 1320
         K34 = K3 / K4                                                A 1330
         H1K1 = H1 / K1                                               A 1340
         H2K4 = H2 / K4                                               A 1350
C        CALCULATION OF INTERMEDIATE K-VARIABLES                      A 1360
         IF (MODE .EQ. 0) GO TO 50                                    A 1370
         KFX1 = K1X / K1                                              A 1380
         KFX2 = K2X / K2                                              A 1390
         KFX3 = K3X / K3                                              A 1400
         KFX4 = K4X / K4                                              A 1410
         KFY1 = K1Y / K1                                              A 1420
         KFY2 = K2Y / K2                                              A 1430
         KFY3 = K3Y / K3                                              A 1440
         KFY4 = K4Y / K4                                              A 1450
         GO TO 60                                                     A 1460
50       KFX1 = 1.0                                                   A 1470
         KFX2 = 1.0                                                   A 1480
         KFX3 = 1.0                                                   A 1490
         KFX4 = 1.0                                                   A 1500
         KFY1 = 1.0                                                   A 1510
         KFY2 = 1.0                                                   A 1520
         KFY3 = 1.0                                                   A 1530
         KFY4 = 1.0                                                   A 1540
60       DO 70 I = 1,NS                                               A 1550
         X2S(I) = XS(I) + DXS(I)                                      A 1560
         Y2S(I) = YS(I) + DYS(I)                                      A 1570
         XSC(I) = XS(I) + 0.5 * DXS(I)                                A 1580
         YSC(I) = YS(I) + 0.5 * DYS(I)                                A 1590
70       CONTINUE                                                     A 1600
         IF (ITEST .LT. 10) GO TO 80                                  A 1610
         ITEST2 = ITEST                                               A 1620
         ITEST = ITEST - 10                                           A 1630
80       GO TO (90,130,170,210,170,210),ITEST                        A 1640
C        CALC. OF CASE 1 SERIES COEFFICIENTS AND TERM0                A 1650
C        CALC. OF AL(L)                                               A 1660
90       DO 100 L = 2,LMAX                                            A 1670
         RL = L                                                       A 1680
         GL1 = GAML(RL,KFX1)                                          A 1690
         GL2 = GAML(RL,KFX2)                                          A 1700
         GL3 = GAML(RL,KFX3)                                          A 1710
         GL4 = GAML(RL,KFX4)                                          A 1720
         TH1 = TANH(T1 * GL1)                                         A 1730
         TH2 = TANH(T2 * GL2)                                         A 1740
         TH3 = TANH(T3 * GL3)                                         A 1750
         TH4 = TANH(T4 * GL4)                                         A 1760
         CH1 = COSH(T1 * GL1)                                         A 1770
         CH2 = COSH(T2 * GL2)                                         A 1780
         CH3 = COSH(T3 * GL3)                                         A 1790
         CH4 = COSH(T4 * GL4)                                         A 1800
         RS = RSCL(RL)                                                A 1810
         RC = RCSL(RL)                                                A 1820
         UD = UL(RL)                                                  A 1830
         VD = VL(RL)                                                  A 1840
         G32 = K32 * GL3 / GL2                                        A 1850
         G21 = K21 * GL2 / GL1                                        A 1860
         ADENOM = UD * (RS + G32 * RC * TH2) + G21 * VD * (RS * TH2   A 1870
     1        + G32 * RC)                                             A 1880
         AL(L) = (G21 * TH1 * (RS * TH2 + G32 * RC) + (RS            A 1890
     1        + G32 * RC * TH2)) / (ADENOM * K1 * GL1)                A 1900
         IF (NL2 .EQ. 0) GO TO 100                                    A 1910
```

```
      G23 = K23 * GL2 / GL3                                    A 1920
      G12 = K12 * GL1 / GL2                                    A 1930
      G34 = K34 * GL3 / GL4                                    A 1940
      R4 = R4L(RL)                                             A 1950
      S4 = S4L(RL)                                             A 1960
      AIDENO = G34 * (R4 * ((G12 * UD * TH2 + VD) + G23 * (G12 * UD  A 1970
     1     + VD * TH2) * TH3) + S4 * ((G12 * UD * TH2 + VD) * TH3    A 1980
     2     + G23 * (G12 * UD + VD * TH2)))                    A 1990
      AIL(L) = (G34 * TH4 * ((G12 * UD * TH2 + VD) * TH3      A 2000
     1     + G23 * (G12 * UD + VD * TH2)) + ((G12 * UD * TH2 + VD)   A 2010
     2     * G23 * (G12 * UD + VD * TH2) * TH3)) / (AIDENO * K4 * GL4) A 2020
      BL(L) = 1.0 / (ADENOM * CH1 * CH2 * CH3 * CH4 * K1 * GL1)  A 2030
      BIL(L) = 1.0 / (AIDENO * CH1 * CH2 * CH3 * CH4 * K4 * GL4)  A 2040
100   CONTINUE                                                 A 2050
C     CALC. OF AM(M)                                           A 2060
      DO 110 M = 2,MMAX                                        A 2070
      RM = M                                                   A 2080
      GM1 = GAMM(RM,KFY1)                                      A 2090
      GM2 = GAMM(RM,KFY2)                                      A 2100
      GM3 = GAMM(RM,KFY3)                                      A 2110
      GM4 = GAMM(RM,KFY4)                                      A 2120
      TH1 = TANH(T1 * GM1)                                     A 2130
      TH2 = TANH(T2 * GM2)                                     A 2140
      TH3 = TANH(T3 * GM3)                                     A 2150
      TH4 = TANH(T4 * GM4)                                     A 2160
      CH1 = COSH(T1 * GM1)                                     A 2170
      CH2 = COSH(T2 * GM2)                                     A 2180
      CH3 = COSH(T3 * GM3)                                     A 2190
      CH4 = COSH(T4 * GM4)                                     A 2200
      RS = RSCM(RM)                                            A 2210
      RC = RCSM(RM)                                            A 2220
      UD = UM(RM)                                              A 2230
      VD = VM(RM)                                              A 2240
      G32 = K32 * GM3 / GM2                                    A 2250
      G21 = K21 * GM2 / GM1                                    A 2260
      ADENOM = UD * (RS + G32 * RC * TH2) + G21 * VD * (RS * TH2  A 2270
     1     + G32 * RC)                                         A 2280
      AM(M) = (G21 * TH1 * (RS * TH2 + G32 * RC) + (RS        A 2290
     1     + G32 * RC * TH2)) / (ADENOM * K1 * GM1)            A 2300
      IF (NL2 .EQ. 0) GO TO 110                                A 2310
      G23 = K23 * GM2 / GM3                                    A 2320
      G12 = K12 * GM1 / GM2                                    A 2330
      G34 = K34 * GM3 / GM4                                    A 2340
      R4 = R4M(RM)                                             A 2350
      S4 = S4M(RM)                                             A 2360
      AIDENO = G34 * (R4 * ((G12 * UD * TH2 + VD) + G23 * (G12 * UD  A 2370
     1     + VD * TH2) * TH3) + S4 * ((G12 * UD * TH2 + VD) * TH3    A 2380
     2     + G23 * (G12 * UD + VD * TH2)))                    A 2390
      AIM(M) = (G34 * TH4 * ((G12 * UD * TH2 + VD) * TH3      A 2400
     1     + G23 * (G12 * UD + VD * TH2)) + ((G12 * UD * TH2 + VD)   A 2410
     2     * G23 * (G12 * UD + VD * TH2) * TH3)) / (AIDENO * K4 * GM4) A 2420
      BM(M) = 1.0 / (ADENOM * CH1 * CH2 * CH3 * CH4 * K1 * GM1)  A 2430
      BIM(M) = 1.0 / (AIDENO * CH1 * CH2 * CH3 * CH4 * K4 * GM4) A 2440
110   CONTINUE                                                 A 2450
C     CALC. OF ALM(L,M)                                        A 2460
      DO 120 L = 2,LMAX                                        A 2470
      DO 120 M = 2,MMAX                                        A 2480
      RL = L                                                   A 2490
      RM = M                                                   A 2500
      GLM1 = GAM(RL,RM,KFX1,KFY1)                              A 2510
      GLM2 = GAM(RL,RM,KFX2,KFY2)                              A 2520
      GLM3 = GAM(RL,RM,KFX3,KFY3)                              A 2530
      GLM4 = GAM(RL,RM,KFX4,KFY4)                              A 2540
      TH1 = TANH(T1 * GLM1)                                    A 2550
```

```
        TH2 = TANH(T2 * GLM2)                                           A 2560
        TH3 = TANH(T3 * GLM3)                                           A 2570
        TH4 = TANH(T4 * GLM4)                                           A 2580
        CH1 = COSH(T1 * GLM1)                                           A 2590
        CH2 = COSH(T2 * GLM2)                                           A 2600
        CH3 = COSH(T3 * GLM3)                                           A 2610
        CH4 = COSH(T4 * GLM4)                                           A 2620
        RS = RSC(RL,RM)                                                 A 2630
        RC = RCS(RL,RM)                                                 A 2640
        UD = ULM(RL,RM)                                                 A 2650
        VD = VLM(RL,RM)                                                 A 2660
        G32 = K32 * GLM3 / GLM2                                         A 2670
        G21 = K21 * GLM2 / GLM1                                         A 2680
        ADENOM = UD * (RS + G32 * RC * TH2) + G21 * VD * (RS * TH2      A 2690
     1       + G32 * RC)                                                A 2700
        ALM(L,M) = (G21 * TH1 * (RS * TH2 + G32 * RC) + (RS            A 2710
     1       + G32 * RC * TH2)) / (ADENOM * K1 * GLM1)                  A 2720
        IF (NL2 .EQ. 0) GO TO 120                                       A 2730
        G23 = K23 * GLM2 / GLM3                                         A 2740
        G12 = K12 * GLM1 / GLM2                                         A 2750
        G34 = K34 * GLM3 / GLM4                                         A 2760
        R4 = R4LM(RL,RM)                                                A 2770
        S4 = S4LM(RL,RM)                                                A 2780
        AIDENO = G34 * (R4 * ((G12 * UD * TH2 + VD) + G23 * (G12 * UD   A 2790
     1       + VD * TH2) * TH3 + S4 * ((G12 * UD * TH2 + VD) * TH3      A 2800
     2       + G23 * (G12 * UD + VD * TH2)))                            A 2810
        AILM(L,M) = (G34 * TH4 * ((G12 * UD * TH2 + VD) * TH3           A 2820
     1       + G23 * (G12 * UD + VD * TH2)) + ((G12 * UD * TH2 + VD)    A 2830
     2       + G23 * (G12 * UD + VD * TH2) * TH3)) / (AIDENO * K4 * GLM4) A 2840
        BLM(L,M) = 1.0 / (ADENOM * CH1 * CH2 * CH3 * CH4 * K1 * GLM1)   A 2850
        BILM(L,M) = 1.0 / (AIDENO * CH1 * CH2 * CH3 * CH4 * K4 * GLM4)  A 2860
120     CONTINUE                                                        A 2870
        TERM0 = (T1 + T2 / K21 + T3 * K1 / K3 + T4 * K1 / K4 + K1 / H2) A 2880
     1       / (H1 * (T1 + T2 / K21 + T3 * K1 / K3 + T4 * K1 / K4 + K1  A 2890
     2       * H1 + K1 / H2))                                           A 2900
        IF (NL2 .EQ. 0) GO TO 250                                       A 2910
        TRM0AI = (T4 + K43 * T3 + K4 * T2 / K2 + K4 * T1 / K1 + K4      A 2920
     1       / H1) / (H2 * (T4 + K43 * T3 + K4 * T2 / K2 + K4 * T1 / K1 A 2930
     2       + K4 / H2 + K4 / H1))                                      A 2940
        TRM0B = K1 / ((T1 + K12 * T2 + K1 * T3 / K3 + K1 * T4 / K4 + K1 A 2950
     1       / H1 + K1 / H2) * H1 * H2)                                 A 2960
        TRM0BI = K4 / ((T4 + K43 * T3 + K4 * T2 / K2 + K4 * T1 / K1     A 2970
     1       + K4 / H2 + K4 / H1) * H1 * H2)                            A 2980
        GO TO 250                                                       A 2990
C     CALC. OF CASE 2 SERIES COEFFICIENTS                               A 3000
C     CALC. OF AL(L)                                                    A 3010
130     DO 140 L = 2,LMAX                                               A 3020
        RL = L                                                          A 3030
        GL1 = GAML(RL,KFX1)                                             A 3040
        GL2 = GAML(RL,KFX2)                                             A 3050
        GL3 = GAML(RL,KFX3)                                             A 3060
        GL4 = GAML(RL,KFX4)                                             A 3070
        TH1 = TANH(T1 * GL1)                                            A 3080
        TH2 = TANH(T2 * GL2)                                            A 3090
        TH3 = TANH(T3 * GL3)                                            A 3100
        TH4 = TANH(T4 * GL4)                                            A 3110
        UD = UL(RL)                                                     A 3120
        VD = VL(RL)                                                     A 3130
        G32 = K32 * GL3 / GL2                                           A 3140
        G21 = K21 * GL2 / GL1                                           A 3150
        G43 = K43 * GL4 / GL3                                           A 3160
        AL(L) = (G21 * TH1 * ((G43 * TH3 + TH4) * TH2 + G32 * (G43      A 3170
     1       + TH4 * TH3)) + ((G43 * TH3 + TH4) + G32 * (G43            A 3180
     2       + TH4 * TH3) * TH2)) / ((UD * ((G43 * TH3 + TH4)           A 3190
```

```
      3          + G32 • (G43 + TH4 • TH3) • TH2) + G21 • VD • ((G43 • TH3      A 3200
      4          + TH4) • TH2 + G32 • (G43 + TH4 • TH3))) • K1 • GL1)           A 3210
140   CONTINUE                                                                  A 3220
C     CALC. OF AM(M)                                                            A 3230
      DO 150 M = 2,MMAX                                                         A 3240
      RM = M                                                                    A 3250
      GM1 = GAMM(RM,KFY1)                                                       A 3260
      GM2 = GAMM(RM,KFY2)                                                       A 3270
      GM3 = GAMM(RM,KFY3)                                                       A 3280
      GM4 = GAMM(RM,KFY4)                                                       A 3290
      TH1 = TANH(T1 • GM1)                                                      A 3300
      TH2 = TANH(T2 • GM2)                                                      A 3310
      TH3 = TANH(T3 • GM3)                                                      A 3320
      TH4 = TANH(T4 • GM4)                                                      A 3330
      UD = UM(RM)                                                               A 3340
      VD = VM(RM)                                                               A 3350
      G32 = K32 • GM3 / GM2                                                     A 3360
      G21 = K21 • GM2 / GM1                                                     A 3370
      G43 = K43 • GM4 / GM3                                                     A 3380
      AM(M) = (G21 • TH1 • ((G43 • TH3 + TH4) • TH2 + G32 • (G43               A 3390
      1          + TH4 • TH3)) + ((G43 • TH3 + TH4) + G32 • (G43               A 3400
      2          + TH4 • TH3) • TH2)) / ((UD • ((G43 • TH3 + TH4)              A 3410
      3          + G32 • (G43 + TH4 • TH3) • TH2) + G21 • VD • ((G43 • TH3     A 3420
      4          + TH4) • TH2 + G32 • (G43 + TH4 • TH3))) • K1 • GM1)          A 3430
150   CONTINUE                                                                  A 3440
C     CALC. OF ALM(L,M)                                                         A 3450
      DO 160 L = 2,LMAX                                                         A 3460
      DO 160 M = 2,MMAX                                                         A 3470
      RL = L                                                                    A 3480
      RM = M                                                                    A 3490
      GLM1 = GAM(RL,RM,KFX1,KFY1)                                              A 3500
      GLM2 = GAM(RL,RM,KFX2,KFY2)                                              A 3510
      GLM3 = GAM(RL,RM,KFX3,KFY3)                                              A 3520
      GLM4 = GAM(RL,RM,KFX4,KFY4)                                              A 3530
      TH1 = TANH(T1 • GLM1)                                                     A 3540
      TH2 = TANH(T2 • GLM2)                                                     A 3550
      TH3 = TANH(T3 • GLM3)                                                     A 3560
      TH4 = TANH(T4 • GLM4)                                                     A 3570
      UD = ULM(RL,RM)                                                           A 3580
      VD = VLM(RL,RM)                                                           A 3590
      G32 = K32 • GLM3 / GLM2                                                   A 3600
      G21 = K21 • GLM2 / GLM1                                                   A 3610
      G43 = K43 • GLM4 / GLM3                                                   A 3620
      ALM(L,M) = (G21 • TH1 • ((G43 • TH3 + TH4) • TH2 + G32 • (G43           A 3630
      1          + TH4 • TH3)) + ((G43 • TH3 + TH4) + G32 • (G43              A 3640
      2          + TH4 • TH3) • TH2)) / ((UD • ((G43 • TH3 + TH4)             A 3650
      3          + G32 • (G43 + TH4 • TH3) • TH2) + G21 • VD • ((G43 • TH3    A 3660
      4          + TH4) • TH2 + G32 • (G43 + TH4 • TH3))) • K1 • GLM1)        A 3670
160   CONTINUE                                                                  A 3680
      TERM0 = (T1 + T2 / K21 + T3 • K1 / K3 + T4 • K1 / K4)                    A 3690
      1      / (H1 • (T1 + T2 / K21 + T3 • K1 / K3 + T4 • K1 / K4 + K1        A 3700
      2      / H1))                                                            A 3710
      GO TO 250                                                                 A 3720
C     CALC. OF CASES 3 AND 5 SERIES COEFFICIENTS FOR T CALC. AT Z=C1           A 3730
C     CALC. OF AL(L)                                                            A 3740
170   DO 180 L = 2,LMAX                                                         A 3750
      RL = L                                                                    A 3760
      GL1 = GAML(RL,KFX1)                                                       A 3770
      GL2 = GAML(RL,KFX2)                                                       A 3780
      GL3 = GAML(RL,KFX3)                                                       A 3790
      GL4 = GAML(RL,KFX4)                                                       A 3800
      TH1 = TANH(T1 • GL1)                                                      A 3810
      TH2 = TANH(T2 • GL2)                                                      A 3820
      TH3 = TANH(T3 • GL3)                                                      A 3830
```

```
      TH4 = TANH(T4 • GL4)                                     A 3840
      RS = RSCL(RL)                                            A 3850
      RC = RCSL(RL)                                            A 3860
      UD = UL(RL)                                              A 3870
      VD = VL(RL)                                              A 3880
      G32 = K32 • GL3 / GL2                                    A 3890
      G21 = K21 • GL2 / GL1                                    A 3900
      AL(L) = VD • (RS + G32 • RC • TH2) / ((UD • (RS          A 3910
     1      + G32 • RC • TH2) + G21 • VD • (RS • TH2           A 3920
     2      + G32 • RC)) • K1 • GL1)                           A 3930
180   CONTINUE                                          .      A 3940
C     CALC. OF AM(M)                                          A 3950
      DO 190 M = 2,MMAX                                        A 3960
      RM = M                                                   A 3970
      GM1 = GAMM(RM,KFY1)                                      A 3980
      GM2 = GAMM(RM,KFY2)                                      A 3990
      GM3 = GAMM(RM,KFY3)                                      A 4000
      GM4 = GAMM(RM,KFY4)                                      A 4010
      TH1 = TANH(T1 • GM1)                                     A 4020
      TH2 = TANH(T2 • GM2)                                     A 4030
      TH3 = TANH(T3 • GM3)                                     A 4040
      TH4 = TANH(T4 • GM4)                                     A 4050
      RS = RSCM(RM)                                            A 4060
      RC = RCSM(RM)                                            A 4070
      UD = UM(RM)                                              A 4080
      VD = VM(RM)                                              A 4090
      G32 = K32 • GM3 / GM2                                    A 4100
      G21 = K21 • GM2 / GM1                                    A 4110
      AM(M) = VD • (RS + G32 • RC • TH2) / ((UD • (RS          A 4120
     1      + G32 • RC • TH2) + G21 • VD • (RS • TH2           A 4130
     2      + G32 • RC)) • K1 • GM1)                           A 4140
190   CONTINUE                                                 A 4150
C     CALC. OF ALM(L,M)                                        A 4160
      DO 200 L = 2,LMAX                                        A 4170
      DO 200 M = 2,MMAX                                        A 4180
      RL = L                                                   A 4190
      RM = M                                                   A 4200
      GLM1 = GAM(RL,RM,KFX1,KFY1)                              A 4210
      GLM2 = GAM(RL,RM,KFX2,KFY2)                              A 4220
      GLM3 = GAM(RL,RM,KFX3,KFY3)                              A 4230
      GLM4 = GAM(RL,RM,KFX4,KFY4)                              A 4240
      TH1 = TANH(T1 • GLM1)                                    A 4250
      TH2 = TANH(T2 • GLM2)                                    A 4260
      TH3 = TANH(T3 • GLM3)                                    A 4270
      TH4 = TANH(T4 • GLM4)                                    A 4280
      RS = RSC(RL,RM)                                          A 4290
      RC = RCS(RL,RM)                                          A 4300
      UD = ULM(RL,RM)                                          A 4310
      VD = VLM(RL,RM)                                          A 4320
      G32 = K32 • GLM3 / GLM2                                  A 4330
      G21 = K21 • GLM2 / GLM1                                  A 4340
      ALM(L,M) = VD • (RS + G32 • RC • TH2) / ((UD • (RS       A 4350
     1      + G32 • RC • TH2) + G21 • VD • (RS • TH2           A 4360
     2      + G32 • RC)) • K1 • GLM1)                          A 4370
200   CONTINUE                                                 A 4380
      TERM0 = (T2 / K2 + T3 / K3 + T4 / K4 + 1.0 / H2) • (T1 + K1  A 4390
     1      / H1) / (T1 + T2 / K21 + T3 • K1 / K3 + T4 • K1 / K4 + K1  A 4400
     2      / H1 + K1 / H2)                                    A 4410
      GO TO 250                                                A 4420
C     CALC. OF CASES 4 AND 6 SERIES COEFFICIENTS FOR T CALC. AT Z=C1  A 4430
C     CALC. OF AL(L)                                           A 4440
210   DO 220 L = 2,LMAX                                        A 4450
      RL = L                                                   A 4460
      GL1 = GAML(RL,KFX1)                                      A 4470
```

```
      GL2 = GAML(RL,KFX2)                                      A 4480
      GL3 = GAML(RL,KFX3)                                      A 4490
      GL4 = GAML(RL,KFX4)                                      A 4500
      TH1 = TANH(T1 * GL1)                                     A 4510
      TH2 = TANH(T2 * GL2)                                     A 4520
      TH3 = TANH(T3 * GL3)                                     A 4530
      TH4 = TANH(T4 * GL4)                                     A 4540
      UD = UL(RL)                                              A 4550
      VD = VL(RL)                                              A 4560
      G43 = K43 * GL4 / GL3                                    A 4570
      G32 = K32 * GL3 / GL2                                    A 4580
      G21 = K21 * GL2 / GL1                                    A 4590
      AL(L) = VD * ((G43 * TH3 + TH4) + G32 * (G43            A 4600
     1      + TH4 * TH3) * TH2) / ((UD * ((G43 * TH3 + TH4)    A 4610
     2      + G32 * (G43 + TH4 * TH3) * TH2) + G21 * VD * ((G43 * TH3  A 4620
     3      + TH4) * TH2 + G32 * (G43 + TH4 * TH3))) * K1 * GL1)  A 4630
  220 CONTINUE                                                 A 4640
C     CALC. OF AM(M)                                           A 4650
      DO 230 M = 2,MMAX                                        A 4660
      RM = M                                                   A 4670
      GM1 = GAMM(RM,KFY1)                                      A 4680
      GM2 = GAMM(RM,KFY2)                                      A 4690
      GM3 = GAMM(RM,KFY3)                                      A 4700
      GM4 = GAMM(RM,KFY4)                                      A 4710
      TH1 = TANH(T1 * GM1)                                     A 4720
      TH2 = TANH(T2 * GM2)                                     A 4730
      TH3 = TANH(T3 * GM3)                                     A 4740
      TH4 = TANH(T4 * GM4)                                     A 4750
      UD = UM(RM)                                              A 4760
      VD = VM(RM)                                              A 4770
      G43 = K43 * GM4 / GM3                                    A 4780
      G32 = K32 * GM3 / GM2                                    A 4790
      G21 = K21 * GM2 / GM1                                    A 4800
      AM(M) = VD * ((G43 * TH3 + TH4) + G32 * (G43            A 4810
     1      + TH4 * TH3) * TH2) / ((UD * ((G43 * TH3 + TH4)    A 4820
     2      + G32 * (G43 + TH4 * TH3) * TH2) + G21 * VD * ((G43 * TH3  A 4830
     3      + TH4) * TH2 + G32 * (G43 + TH4 * TH3))) * K1 * GM1)  A 4840
  230 CONTINUE                                                 A 4850
C     CALC. OF ALM(L,M)                                        A 4860
      DO 240 L = 2,LMAX                                        A 4870
      DO 240 M = 2,MMAX                                        A 4880
      RL = L                                                   A 4890
      RM = M                                                   A 4900
      GLM1 = GAM(RL,RM,KFX1,KFY1)                              A 4910
      GLM2 = GAM(RL,RM,KFX2,KFY2)                              A 4920
      GLM3 = GAM(RL,RM,KFX3,KFY3)                              A 4930
      GLM4 = GAM(RL,RM,KFX4,KFY4)                              A 4940
      TH1 = TANH(T1 * GLM1).                                   A 4950
      TH2 = TANH(T2 * GLM2)                                    A 4960
      TH3 = TANH(T3 * GLM3)                                    A 4970
      TH4 = TANH(T4 * GLM4)                                    A 4980
      UD = ULM(RL,RM)                                          A 4990
      VD = VLM(RL,RM)                                          A 5000
      G43 = K43 * GLM4 / GLM3                                  A 5010
      G32 = K32 * GLM3 / GLM2                                  A 5020
      G21 = K21 * GLM2 / GLM1                                  A 5030
      ALM(L,M) = VD * ((G43 * TH3 + TH4) + G32 * (G43         A 5040
     1      + TH4 * TH3) * TH2) / ((UD * ((G43 * TH3 + TH4)    A 5050
     2      + G32 * (G43 + TH4 * TH3) * TH2) + G21 * VD * ((G43 * TH3  A 5060
     3      + TH4) * TH2 + G32 * (G43 + TH4 * TH3))) * K1 * GLM1)  A 5070
  240 CONTINUE                                                 A 5080
      TERM0 = (T2 / K21 + T3 * K1 / K3 + T4 * K1 / K4) * (T1 / K1  A 5090
     1      + 1.0 / H1) / (T1 + T2 / K21 + T3 * K1 / K3 + T4 * K1 / K4  A 5100
     2      + K1 / H1)                                         A 5110
```

```
C         CHECK TO SEE IF MATRIX CALC. MAY BE BYPASSED                        A 5120
250       IF (NL .EQ. 0) GO TO 420                                            A 5130
C         CALC. OF COEFFICIENT MATRIX                                         A 5140
C         CALC. OF LOCATIONS OF UNIT TEMP CALC.                               A 5150
          DO 260 I = 1,NL                                                     A 5160
          X2L(I) = XL(I) + DXL(I)                                             A 5170
          Y2L(I) = YL(I) + DYL(I)                                             A 5180
          XLC(I) = XL(I) + 0.5 * DXL(I)                                       A 5190
260       YLC(I) = YL(I) + 0.5 * DYL(I)                                       A 5200
C         CALC. OF TEMP AT LEADS DUE TO UNIT FLUX AT LEADS                    A 5210
          DO 340 I = 1,NL                                                     A 5220
          DO 340 J = 1,NL                                                     A 5230
          INLTS = I - NL1                                                     A 5240
          JNLTS = J - NL1                                                     A 5250
          IF (INLTS) 270,270,300                                             A 5260
270       IF (JNLTS) 280,280,290                                             A 5270
C         TEMP AT Z=0 DUE TO LEAD AT Z=0                                      A 5280
280       CALL SST (A,B,AP,BP,XL(J),X2L(J),YL(J),Y2L(J),XLC(I),YLC(I),TERM0, A 5290
         1AL,AM,ALM,LMAX,MMAX,C(I,J))                                        A 5300
          GO TO 330                                                           A 5310
C         TEMP AT Z=0 DUE TO LEAD AT Z=C4                                     A 5320
290       CALL SST (A,B,AP,BP,XL(J),X2L(J),YL(J),Y2L(J),XLC(I),YLC(I),TRM0B, A 5330
         1BL,BM,BLM,LMAX,MMAX,C(I,J))                                        A 5340
          GO TO 330                                                           A 5350
300       IF (JNLTS) 310,310,320                                             A 5360
C         TEMP AT Z=C4 DUE TO LEAD AT Z=0                                     A 5370
310       CALL SST (A,B,AP,BP,XL(J),X2L(J),YL(J),Y2L(J),XLC(I),YLC(I),       A 5380
         1     TRM0BI,BIL,BIM,BILM,LMAX,MMAX,C(I,J))                          A 5390
          GO TO 330                                                           A 5400
C         TEMP AT Z=C4 DUE TO LEAD AT Z=C4                                    A 5410
320       CALL SST (A,B,AP,BP,XL(J),X2L(J),YL(J),Y2L(J),XLC(I),YLC(I),       A 5420
         1     TRM0AI,AIL,AIM,AILM,LMAX,MMAX,C(I,J))                          A 5430
330       IF (I .NE. J) GO TO 340                                            A 5440
          C(I,J) = C(I,J) + R(I)                                             A 5450
340       CONTINUE                                                            A 5460
C         CALC. OF TEMPS AT LEADS DUE TO SOURCES Q                            A 5470
          DO 390 I = 1,NL                                                     A 5480
          SW1(I) = 0.0                                                        A 5490
          INLTS = I - NL1                                                     A 5500
          DO 380 J = 1,NS                                                     A 5510
          IF (Q(J) .EQ. 0.0) GO TO 380                                       A 5520
          IF (INLTS) 350,350,360                                             A 5530
C         TEMP AT Z=0 DUE TO SOURCE AT Z=0                                    A 5540
350       CALL SST (A,B,AP,BP,XS(J),X2S(J),YS(J),Y2S(J),XLC(I),YLC(I),TERM0, A 5550
         1AL,AM,ALM,LMAX,MMAX,W1)                                            A 5560
          GO TO 370                                                           A 5570
C         TEMP AT Z=C4 DUE TO SOURCE AT Z=0                                   A 5580
360       CALL SST (A,B,AP,BP,XS(J),X2S(J),YS(J),Y2S(J),XLC(I),YLC(I),       A 5590
         1     TRM0BI,BIL,BIM,BILM,LMAX,MMAX,W1)                             A 5600
370       W1 = W1 * Q(J)                                                     A 5610
          SW1(I) = SW1(I) + W1                                               A 5620
          C(I,NL + 1) = SW1(I) + TA - T0(I)                                  A 5630
380       CONTINUE                                                            A 5640
390       CONTINUE                                                            A 5650
          VV(1) = 4.0                                                         A 5660
          LLTEST = 0                                                          A 5670
          CALL GER (C,21,20,NL,NL + 1,LLTEST,JC,VV)                          A 5680
          IF (LLTEST .NE. 1) GO TO 400                                       A 5690
          WRITE (2,720)                                                       A 5700
          STOP                                                                A 5710
C         NAMING OF LEAD FLUX FROM MATRIX SOLN OUTPUT                         A 5720
400       DO 410 I = 1,NL                                                     A 5730
410       F(I) = C(I,NL + 1)                                                 A 5740
C         CALCULATION OF TEMP. AT SELECTED SOURCES DUE TO SOURCES ONLY(TEMPS A 5750
```

```
C        TEST TO CHECK IF NEW SERIES COEF. NEED TO BE          A 5760
C        CALC.(FOR CASES 3 AND 4 ONLY)                         A 5770
420      GO TO (510,510,430,470,510,510),ITEST                 A 5780
C        NEW SERIES COEFF. FOR CASE 3                          A 5790
C        CALC. OF AL(L)                                        A 5800
430      DO 440 L = 2,LMAX                                     A 5810
         RL = L                                                A 5820
         GL1 = GAML(RL,KFX1)                                   A 5830
         GL2 = GAML(RL,KFX2)                                   A 5840
         GL3 = GAML(RL,KFX3)                                   A 5850
         GL4 = GAML(RL,KFX4)                                   A 5860
         TH1 = TANH(T1 * GL1)                                  A 5870
         TH2 = TANH(T2 * GL2)                                  A 5880
         TH3 = TANH(T3 * GL3)                                  A 5890
         TH4 = TANH(T4 * GL4)                                  A 5900
         CH1 = COSH(T1 * GL1)                                  A 5910
         RS = RSCL(RL)                                         A 5920
         RC = RCSL(RL)                                         A 5930
         UD = UL(RL)                                           A 5940
         VD = VL(RL)                                           A 5950
         G32 = K32 * GL3 / GL2                                 A 5960
         G21 = K21 * GL2 / GL1                                 A 5970
         AL(L) = (RS + G32 * RC * TH2) / ((UD * (RS + G32 * RC * TH2)  A 5980
1            + G21 * VD * (RS * TH2 + G32 * RC)) * K1 * CH1 * GL1)     A 5990
440      CONTINUE                                              A 6000
C        CALC. OF AM(M)                                        A 6010
         DO 450 M = 2,MMAX                                     A 6020
         RM = M                                                A 6030
         GM1 = GAMM(RM,KFY1)                                   A 6040
         GM2 = GAMM(RM,KFY2)                                   A 6050
         GM3 = GAMM(RM,KFY3)                                   A 6060
         GM4 = GAMM(RM,KFY4)                                   A 6070
         TH1 = TANH(T1 * GM1)                                  A 6080
         TH2 = TANH(T2 * GM2)                                  A 6090
         TH3 = TANH(T3 * GM3)                                  A 6100
         TH4 = TANH(T4 * GM4)                                  A 6110
         CH1 = COSH(T1 * GM1)                                  A 6120
         RC = RCSM(RM)                                         A 6130
         VD = VM(RM)                                           A 6140
         RS = RSCM(RM)                                         A 6150
         UD = UM(RM)                                           A 6160
         G32 = K32 * GM3 / GM2                                 A 6170
         G21 = K21 * GM2 / GM1                                 A 6180
         AM(M) = (RS + G32 * RC * TH2) / ((UD * (RS + G32 * RC * TH2)  A 6190
1            + G21 * VD * (RS * TH2 + G32 * RC)) * K1 * CH1 * GM1)     A 6200
450      CONTINUE                                              A 6210
C        CALC. OF ALM(L,M)                                     A 6220
         DO 460 L = 2,LMAX                                     A 6230
         DO 460 M = 2,MMAX                                     A 6240
         RL = L                                                A 6250
         RM = M                                                A 6260
         GLM1 = GAM(RL,RM,KFX1,KFY1)                           A 6270
         GLM2 = GAM(RL,RM,KFX2,KFY2)                           A 6280
         GLM3 = GAM(RL,RM,KFX3,KFY3)                           A 6290
         GLM4 = GAM(RL,RM,KFX4,KFY4)                           A 6300
         TH1 = TANH(T1 * GLM1)                                 A 6310
         TH2 = TANH(T2 * GLM2)                                 A 6320
         TH3 = TANH(T3 * GLM3)                                 A 6330
         TH4 = TANH(T4 * GLM4)                                 A 6340
         CH1 = COSH(T1 * GLM1)                                 A 6350
         RS = RSC(RL,RM)                                       A 6360
         RC = RCS(RL,RM)                                       A 6370
         VD = VLM(RL,RM)                                       A 6380
         UD = ULM(RL,RM)                                       A 6390
```

```
           G32 = K32 * GLM3 / GLM2                                          A 6400
           G21 = K21 * GLM2 / GLM1                                          A 6410
           ALM(L,M) = (RS + G32 * RC * TH2) / ((UD * (RS + G32 * RC * TH2)  A 6420
          1    + G21 * VD * (RS * TH2 + G32 * RC)) * K1 * CH1 * GLM1)       A 6430
   460     CONTINUE                                                         A 6440
           TERM0 = (T2 / K21 + T3 * K1 / K3 + T4 * K1 / K4 + K1 / H2)       A 6450
          1    / (H1 * (T1 + T2 / K21 + T3 * K1 / K3 + T4 * K1 / K4 + K1    A 6460
          2    / H1 + K1 / H2))                                             A 6470
           GO TO 510                                                        A 6480
   C       NEW SERIES COEFFICIENTS FOR CASE 4                               A 6490
   C       CALC. OF AL(L)                                                   A 6500
   470     DO 480 L = 2,LMAX                                                A 6510
           RL = L                                                           A 6520
           GL1 = GAML(RL,KFX1)                                              A 6530
           GL2 = GAML(RL,KFX2)                                              A 6540
           GL3 = GAML(RL,KFX3)                                              A 6550
           GL4 = GAML(RL,KFX4)                                              A 6560
           TH1 = TANH(T1 * GL1)                                             A 6570
           TH2 = TANH(T2 * GL2)                                             A 6580
           TH3 = TANH(T3 * GL3)                                             A 6590
           TH4 = TANH(T4 * GL4)                                             A 6600
           CH1 = COSH(T1 * GL1)                                             A 6610
           UD = UL(RL)                                                      A 6620
           VD = VL(RL)                                                      A 6630
           G43 = K43 * GL4 / GL3                                            A 6640
           G32 = K32 * GL3 / GL2                                            A 6650
           G21 = K21 * GL2 / GL1                                            A 6660
           AL(L) = ((G43 * TH3 + TH4) + G32 * (G43 + TH4 * TH3) * TH2)      A 6670
          1    / ((UD * ((G43 * TH3 + TH4) + G32 * (G43                     A 6680
          2    + TH4 * TH3) * TH2) + G21 * VD * ((G43 * TH3 + TH4) * TH2    A 6690
          3    + G32 * (G43 + TH4 * TH3))) * K1 * CH1 * GL1)                A 6700
   480     CONTINUE                                                         A 6710
   C       CALC. OF AM(M)                                                   A 6720
           DO 490 M = 2,MMAX                                                A 6730
           RM = M                                                           A 6740
           GM1 = GAMM(RM,KFY1)                                              A 6750
           GM2 = GAMM(RM,KFY2)                                              A 6760
           GM3 = GAMM(RM,KFY3)                                              A 6770
           GM4 = GAMM(RM,KFY4)                                              A 6780
           TH1 = TANH(T1 * GM1)                                             A 6790
           TH2 = TANH(T2 * GM2)                                             A 6800
           TH3 = TANH(T3 * GM3)                                             A 6810
           TH4 = TANH(T4 * GM4)                                             A 6820
           CH1 = COSH(T1 * GM1)                                             A 6830
           UD = UM(RM)                                                      A 6840
           VD = VM(RM)                                                      A 6850
           G43 = K43 * GM4 / GM3                                            A 6860
           G32 = K32 * GM3 / GM2                                            A 6870
           G21 = K21 * GM2 / GM1                                            A 6880
           AM(M) = ((G43 * TH3 + TH4) + G32 * (G43 + TH4 * TH3) * TH2)      A 6890
          1    / ((UD * ((G43 * TH3 + TH4) + G32 * (G43                     A 6900
          2    + TH4 * TH3) * TH2) + G21 * VD * ((G43 * TH3 + TH4) * TH2    A 6910
          3    + G32 * (G43 + TH4 * TH3))) * K1 * CH1 * GM1)                A 6920
   490     CONTINUE                                                         A 6930
   C       CALC. OF ALM(L,M)                                                A 6940
           DO 500 L = 2,LMAX                                                A 6950
           DO 500 M = 2,MMAX                                                A 6960
           RL = L                                                           A 6970
           RM = M                                                           A 6980
           GLM1 = GAM(RL,RM,KFX1,KFY1)                                      A 6990
           GLM2 = GAM(RL,RM,KFX2,KFY2)                                      A 7000
           GLM3 = GAM(RL,RM,KFX3,KFY3)                                      A 7010
           GLM4 = GAM(RL,RM,KFX4,KFY4)                                      A 7020
           TH1 = TANH(T1 * GLM1)                                            A 7030
```

```
      TH2 = TANH(T2 • GLM2)                                          A 7040
      TH3 = TANH(T3 • GLM3)                                          A 7050
      TH4 = TANH(T4 • GLM4)                                          A 7060
      CHl = COSH(Tl • GLMl)                                          A 7070
      UD = ULM(RL,RM)                                                A 7080
      VD = VLM(RL,RM)                                                A 7090
      G43 = K43 • GLM4 / GLM3                                        A 7100
      G32 = K32 • GLM3 / GLM2                                        A 7110
      G21 = K21 • GLM2 / GLMl                                        A 7120
      ALM(L,M) = ((G43 • TH3 + TH4) + G32 • (G43 + TH4 • TH3) • TH2) A 7130
     1      / ((UD • ((G43 • TH3 + TH4) + G32 • (G43              A 7140
     2      + TH4 • TH3) • TH2) + G21 • VD • ((G43 • TH3 + TH4) • TH2 A 7150
     3      + G32 • (G43 + TH4 • TH3))) • Kl • CHl • GLMl)          A 7160
500   CONTINUE                                                       A 7170
      TERM0 = (T2 / K21 + T3 • Kl / K3 + T4 • Kl / K4) / ((Tl + T2   A 7180
     1      / K21 + T3 • Kl / K3 + T4 • Kl / K4 + Kl / Hl) • Hl)     A 7190
510   DO 580 I = 1,NS                                                A 7200
      J = LOC(I)                                                     A 7210
      IF (J .EQ. 0) GO TO 580                                        A 7220
      TEMPS(J) = 0.0                                                 A 7230
      TEMPL(J) = 0.0                                                 A 7240
      DO 530 L = 1,NS                                                A 7250
      IF (Q(L) .EQ. 0.0) GO TO 530                                   A 7260
      IF (ITEST2 .LT. 11) GO TO 520                                  A 7270
      CALL SST2 (A,B,AP,BP,XS(L),X2S(L),YS(L),Y2S(L),XSC(J),YSC(J),  A 7280
     1      TERM0,AL,AM,ALM,LMAX,MMAX,SPSUM)                         A 7290
      CALL TOUT (Q(L),LMAX,MMAX,SPSUM)                               A 7300
      CALL LPLOT (LMAX,SPSUM)                                        A 7310
520   CALL SST (A,B,AP,BP,XS(L),X2S(L),YS(L),Y2S(L),XSC(J),YSC(J),TERM0, A 7320
     1AL,AM,ALM,LMAX,MMAX,W2)                                        A 7330
      W2 = W2 • Q(L)                                                 A 7340
      TEMPS(J) = TEMPS(J) + W2                                       A 7350
530   CONTINUE                                                       A 7360
C     CHECK TO SEE IF A LEAD CONTRIBUTION IS NECESSARY               A 7370
      IF (NL .EQ. 0) GO TO 570                                       A 7380
C     LEAD CONTRIBUTION TO TEMP.                                     A 7390
      TEMPL(J) = 0.0                                                 A 7400
C     LOC COORD. FOR TEMP CALC. ALREADY DETERMINED                   A 7410
      DO 560 L = 1,NL                                                A 7420
      LNLTS = L - NL1                                                A 7430
      IF (LNLTS .GT. 0) GO TO 540                                    A 7440
C     TEMP AT Z=0 DUE TO LEAD AT Z=0                                 A 7450
      CALL SST (A,B,AP,BP,XL(L),X2L(L),YL(L),Y2L(L),XSC(J),YSC(J),TERM0, A 7460
     1AL,AM,ALM,LMAX,MMAX,W3)                                        A 7470
      GO TO 550                                                      A 7480
C     TEMP AT Z=0 DUE TO LEAD AT Z=C4                                A 7490
540   CALL SST (A,B,AP,BP,XL(L),X2L(L),YL(L),Y2L(L),XSC(J),YSC(J),TRM0B, A 7500
     1BL,BM,BLM,LMAX,MMAX,W3)                                        A 7510
550   W3 = W3 • F(L)                                                 A 7520
560   TEMPL(J) = TEMPL(J) + W3                                       A 7530
570   TEMP(J) = TEMPS(J) - TEMPL(J) + TA                            A 7540
580   CONTINUE                                                       A 7550
      WRITE (2,770)                                                  A 7560
      WRITE (2,780) (TITLE(I),I=1,8)                                 A 7570
      WRITE (2,790)                                                  A 7580
      GO TO (590,600,610,620,630,640),ITEST                         A 7590
590   WRITE (2,800)                                                  A 7600
      GO TO 650                                                      A 7610
600   WRITE (2,810)                                                  A 7620
      GO TO 650                                                      A 7630
610   WRITE (2,820)                                                  A 7640
      GO TO 650                                                      A 7650
620   WRITE (2,830)                                                  A 7660
      GO TO 650                                                      A 7670
```

```
630     WRITE (2,840)                                                        A 7680
        GO TO 650                                                            A 7690
640     WRITE (2,850)                                                        A 7700
650     IF (MODE .EQ. 1) GO TO 660                                           A 7710
        WRITE (2,730) A,B,T1,T2,T3,T4,H1,H2,K1,K2,K3,K4,NS,NL,NL1,NL2,LMAX   A 7720
       1,MMAX,TA,(I,XS(I),DXS(I),YS(I),DYS(I),Q(I),I=1,NS)                   A 7730
        GO TO 670                                                            A 7740
660     WRITE (2,740) A,B,T1,T2,T3,T4,H1,H2,K1X,K2X,K3X,K4X,K1Y,K2Y,K3Y,K4   A 7750
       1Y,K1,K2,K3,K4,NS,NL,NL1,NL2,LMAX,MMAX,TA,(I,XS(I),DXS(I),YS(I),DYS   A 7760
       2(I),Q(I),I=1,NS)                                                     A 7770
670     IF (NL .EQ. 0) GO TO 690                                             A 7780
        WRITE (2,750) (I,XL(I),DXL(I),YL(I),DYL(I),R(I),T0(I),I=1,NL)        A 7790
        DO 680 I = 1,NL                                                      A 7800
680     TL(I) = F(I) * R(I) + T0(I)                                          A 7810
        WRITE (2,760) (I,F(I),TL(I),I=1,NL)                                  A 7820
690     WRITE (2,860)                                                        A 7830
        DO 700 I = 1,NS                                                      A 7840
        IF (LOC(I) .EQ. 0) GO TO 700                                         A 7850
        TEMPS(I) = TEMPS(I) + TA                                             A 7860
        WRITE (2,870) LOC(I),TEMPS(I)                                        A 7870
        IF (NL .EQ. 0) GO TO 700                                             A 7880
        WRITE (2,880) TEMP(I)                                                A 7890
700     CONTINUE                                                             A 7900
        WRITE (2,770)                                                        A 7910
        GO TO 10                                                             A 7920
C                                                                            A 7930
C                                                                            A 7940
710     FORMAT (8A10)                                                        A 7950
711     FORMAT (2I10,2E10.0)                                                 A 7960
712     FORMAT (6F10.0)                                                      A 7970
713     FORMAT(2E10.0,F10.0)                                                 A 7980
714     FORMAT (3E10.0)                                                      A 7990
715     FORMAT(4E10.0)                                                       A 8000
716     FORMAT (3I10)                                                        A 8010
717     FORMAT (5F10.0)                                                      A 8020
718     FORMAT (8I10)                                                        A 8030
720     FORMAT (22H MATRIX SOLUTION ERROR)                                   A 8040
730     FORMAT (1H ,9X,43HSUBSTRATE DIMENSIONS AND PHYSICAL CONSTANTS/1H ,   A 8050
       112X,2HA=,F8.5,4X,2HB=,F8.5,4X,3HT1=,F7.5,4X,3HT2=,F7.5,4X,3HT3=,F7   A 8060
       2.5,4X,3HT4=,F7.5/1H ,12X,3HH1=,E10.4,2X,3HH2=,E10.4/1H ,12X,3HK1=,   A 8070
       3E10.4,2X,3HK2=,E10.4,2X,3HK3=,E10.4,2X,3HK4=,E10.4//1H ,12X,18HNUM   A 8080
       4BER OF SOURCES=,I3,2X,15HNUMBER OF RES.=,I3,2X,                      A 8090
       54HNR1=,I3,2X,4HNR2=,I3//1H ,12X,                                     A 8100
       65HLMAX=,I4,2X,                                                       A 8110
       7  5HMMAX=,I4//1H ,12X,3HTA=,F6.1,1X//1                              A 8120
       8H ,12X,11HSOURCE DATA/1H ,12X,12HSOURCE NO. I,6X,5HXS(I),9X,6HDXS(   A 8130
       9I),10X,5HYS(I),9X,6HDYS(I),16X,4HQ(I)                /(1H0,15X,I3,2X  A 8140
       $,4F15.4,14X,F6.3            ))                                       A 8150
740     FORMAT (1H ,9X,43HSUBSTRATE DIMENSIONS AND PHYSICAL CONSTANTS/1H ,   A 8160
       112X,2HA=,F8.5,4X,2HB=,F8.5,4X,3HT1=,F7.5,4X,3HT2=,F7.5,4X,3HT3=,F7   A 8170
       2.5,4X,3HT4=,F7.5/1H ,12X,3HH1=,E11.4,2X,3HH2=,E11.4/1H ,12X,4HK1X=   A 8180
       3,E10.4,2X,4HK2X=,E10.4,2X,4HK3X=,E10.4,2X,4HK4X=,E10.4/1H ,12X,4HK   A 8190
       41Y=,E10.4,2X,4HK2Y=,E10.4,2X,4HK3Y=,E10.4,2X,4HK4Y=,E10.4/1H ,12X,   A 8200
       54HK1Z=,E10.4,2X,4HK2Z=,E10.4,2X,4HK3Z=,E10.4,2X,4HK4Z=,E10.4//1H ,   A 8210
       612X,18HNUMBER OF SOURCES=,I3,2X,15HNUMBER OF RES.=,I3,2X,            A 8220
       74HNR1=,I3,2X,4HNR2=,I3//1H ,12X,                                5H   A 8230
       8LMAX=,I4,2X,5HMMAX=,I4//1H ,12X,3HTA=,                              A 8240
       9F6.1,1X//1H ,12X,11HSOURCE DATA/1H ,12X,12HSOURCE NO. I,6X,5HXS(I)   A 8250
       $,9X,6HDXS(I),10X,5HYS(I),9X,6HDYS(I),16X,4HQ(I)               /(1H0  A 8260
       $,15X,I3,2X,4F15.4,14X,F6.3            ))                             A 8270
750     FORMAT (1H0,12X,9HRES. DATA/1H ,12X,10HRES. NO. I,8X,5HXR(I),9X,6H   A 8280
       1DXR(I),10X,5HYR(I),9X,6HDYR(I),15X,4HR(I),12X,5HT0(I)/(1H0,15X,I3,   A 8290
       22X,4F15.4,8X,E12.3,11X,F5.1))                                        A 8300
760     FORMAT (1H0,12X,69HTHERMAL FLUX IN RES.  AND TEMPERATURES CALCULAT   A 8310
```

```
       1ED AT RES. PAD CENTERS/1H ,12X,10HRES. NO. I,10X,4HF(I),9X,5HTR(I)  A 8320
       2/(1H0,15X,I3,9X,E11.3,5X,F7.1))                                    A 8330
  770   FORMAT (1H1)                                                        A 8340
  780   FORMAT (1H ,8A10//)                                                 A 8350
  790   FORMAT (1H ,9X,46HTAMSIV-THERMAL ANALYZER FOR MULTILAYER SYSTEMS,   A 8360
       1/,10X,21HAUG. 10, 1983 VERSION/)                                    A 8370
  800   FORMAT (1H ,9X,55HTHERMAL ANALYSIS FOR NEWTON'S LAW COOLING AT Z=0  A 8380
       1 AND C4/1H ,9X,24HSOURCES AND LEADS AT Z=0//)                       A 8390
  810   FORMAT (1H ,9X,85HTHERMAL ANALYSIS FOR NEWTON'S LAW COOLING AT Z=0  A 8400
       1 AND AN ISOTHERMAL SURFACE TA AT Z=C4/1H ,9X,24HSOURCES AND LEADS   A 8410
       2AT Z=0//)                                                          A 8420
  820   FORMAT (1H ,9X,55HTHERMAL ANALYSIS FOR NEWTON'S LAW COOLING AT Z=0  A 8430
       1 AND C4/1H ,9X,60HSOURCES AND LEADS AT Z=C1, BUT TEMPERATURE CALCU  A 8440
       2LATED AT Z=0//)                                                     A 8450
  830   FORMAT (1H ,9X,85HTHERMAL ANALYSIS FOR NEWTON'S LAW COOLING AT Z=0  A 8460
       1 AND AN ISOTHERMAL SURFACE TA AT Z=C4/1H ,9X,60HSOURCES AND LEADS   A 8470
       2AT Z=C1, BUT TEMPERATURE CALCULATED AT Z=0//)                       A 8480
  840   FORMAT (1H ,9X,55HTHERMAL ANALYSIS FOR NEWTON'S LAW COOLING AT Z=0  A 8490
       1 AND C4/1H ,9X,25HSOURCES AND LEADS AT Z=C1//)                      A 8500
  850   FORMAT (1H ,9X,85HTHERMAL ANALYSIS FOR NEWTON'S LAW COOLING AT Z=0  A 8510
       1 AND AN ISOTHERMAL SURFACE TA AT Z=C4/1H ,9X,25HSOURCES AND LEADS   A 8520
       2AT Z=C1//)                                                          A 8530
  860   FORMAT(1H0,12X,41HTEMPERATURES CALCULATED AT SOURCE CENTERS/1H ,12  A 8540
       1X,12HSOURCE NO. I,12X,23HTS(I) WITH SOURCES ONLY,9X,28HTS(I) WITH   A 8550
       2SOURCES AND RES. )                                                  A 8560
  870   FORMAT (1H0,I18,21X,F7.1)                                           A 8570
  880   FORMAT (1H+,70X,F7.1)                                               A 8580
        END                                                                A 8590-
        SUBROUTINE SST (A,B,AP,BP,X1,X2,Y1,Y2,X,Y,TERM0,AL,AM,ALM,LMAX,MMA  B   10
       1X,UNTMP)                                                            B   20
        DIMENSION AL(70), AM(70), ALM(70,70)                               B   30
C       FUNCTION DEFINITIONS                                               B   40
        DELL(V) = 2.0 * (SIN(AP * X2 * (V - 1.0)) - SIN(AP * X1 * (V        B   50
       1      - 1.0))) / (AP * (X2 - X1) * (V - 1.0))                       B   60
        DELM(V) = 2.0 * (SIN(BP * Y2 * (V - 1.0)) - SIN(BP * Y1 * (V        B   70
       1      - 1.0))) / (BP * (Y2 - Y1) * (V - 1.0))                       B   80
        SUM = 0.0                                                          B   90
        DO 50 L = 1,LMAX                                                   B  100
        DO 40 M = 1,MMAX                                                   B  110
        IF (L .EQ. 1 .AND. M .EQ. 1) GO TO 40                              B  120
        IF (L .EQ. 1) GO TO 20                                             B  130
        IF (M .EQ. 1) GO TO 10                                             B  140
        RL = L                                                             B  150
        RM = M                                                             B  160
        ANS = ALM(L,M) * DELL(RL) * DELM(RM) * COS(BP * Y * (RM            B  170
       1      - 1.0)) * COS(AP * X * (RL - 1.0))                           B  180
        GO TO 30                                                           B  190
  10    RL = L                                                             B  200
        RM = M                                                             B  210
        ANS = AL(L) * DELL(RL) * COS(AP * X * (RL - 1.0))                  B  220
        GO TO 30                                                           B  230
  20    RL = L                                                             B  240
        RM = M                                                             B  250
        ANS = AM(M) * DELM(RM) * COS(BP * Y * (RM - 1.0))                  B  260
  30    SUM = SUM + ANS                                                    B  270
  40    CONTINUE                                                           B  280
  50    CONTINUE                                                           B  290
        UNTMP = (TERM0 + SUM) / (A * B)                                    B  300
        RETURN                                                             B  310
        END                                                                B  320-
        SUBROUTINE GER (A,NC,NR,N,MC,LTEST,JC,V)                           C   10
        DIMENSION A(NR,NC),JC(1),V(2)                                      C   20
C               JC IS THE PERMUTATION VECTOR                               C   30
C               KD IS THE OPTION KEY FOR DETERMINANT EVALUATION            C   40
```

```
C                      KI IS THE OPTION KEY FOR MATRIX INVERSION         C    50
C                      L IS THE COLUMN CONTROL FOR AX=B                   C    60
C                      M IS THE COLUMN CONTOL FOR MATRIX INVERSION        C    70
C                      INITIALIZATION                                     C    80
      IW = V(1)                                                          C    90
      M = 1                                                              C   100
      S = 1.                                                             C   110
      L = N + (MC - N) * (IW / 4)                                        C   120
      KD = 2 - MOD(IW / 2,2)                                             C   130
      IF (KD .EQ. 1) V(2) = 0.                                           C   140
      KI = 2 - MOD(IW,2)                                                 C   150
      GO TO (10,30),KI                                                   C   160
C                      INITIALIZE JC FOR INVERSION                        C   170
10    DO 20 I = 1,N                                                      C   180
20    JC(I) = I                                                         C   190
C                      SEARCH FOR PIVOT ROW                              C   200
30    DO 160 I = 1,N                                                     C   210
      GO TO (50,40),KI                                                   C   220
40    M = I                                                             C   230
50    IF (I .EQ. N) GO TO 100                                            C   240
      X = - 1.                                                           C   250
      DO 60 J = I,N                                                      C   260
      IF (X .GT. ABS(A(J,I))) GO TO 60                                   C   270
      X = ABS(A(J,I))                                                    C   280
      K = J                                                              C   290
60    CONTINUE                                                          C   300
      IF (K .EQ. I) GO TO 100                                            C   310
      S = - S                                                            C   320
      V(1) = - V(1)                                                      C   330
      GO TO (70,80),KI                                                   C   340
70    MU = JC(I)                                                         C   350
      JC(I) = JC(K)                                                      C   360
      JC(K) = MU                                                         C   370
C                      INTERCHANGE ROW I AND ROW K                       C   380
80    DO 90 J = M,L                                                      C   390
      X = A(I,J)                                                         C   400
      A(I,J) = A(K,J)                                                    C   410
90    A(K,J) = X                                                         C   420
C                      TEST FOR SINGULARITY                              C   430
100   IF (ABS(A(I,I)) .GT. 0.) GO TO 110                                 C   440
C                      MATRIX IS SINGULAR                                C   450
      IF (KD .EQ. 1) V(1) = 0.                                           C   460
      JC(1) = I - 1                                                      C   470
      LTEST = 1                                                          C   480
      RETURN                                                             C   490
110   GO TO (120,130),KD                                                C   500
C                      COMPUTE THE DETERMINANT                           C   510
120   IF (A(I,I) .LT. 0.) S = - S                                        C   520
      V(2) = V(2) + ALOG(ABS(A(I,I)))                                    C   530
130   X = A(I,I)                                                         C   540
      A(I,I) = 1.                                                        C   550
C                      REDUCTION OF THE I-TH ROW                         C   560
      DO 140 J = M,L                                                     C   570
      A(I,J) = A(I,J) / X                                                C   580
140   CONTINUE                                                          C   590
C                      REDUCTION OF ALL REMAINING ROWS                   C   600
      DO 160 K = 1,N                                                     C   610
      IF (K .EQ. I) GO TO 160                                            C   620
      X = A(K,I)                                                         C   630
      A(K,I) = 0.                                                        C   640
      DO 150 J = M,L                                                     C   650
      A(K,J) = A(K,J) - X * A(I,J)                                       C   660
150   CONTINUE                                                          C   670
160   CONTINUE                                                          C   680
```

```
C               AX=B AND DET.(A) ARE NOW COMPUTED           C   690
      GO TO (170,220),KI                                    C   700
C               PERMUTATION OF THE COLUMNS FOR MATRIX INVERSION  C   710
170   DO 210 J = 1,N                                        C   720
      IF (JC(J) .EQ. J) GO TO 210                           C   730
      JJ = J + 1                                            C   740
      DO 180 I = JJ,N                                       C   750
      IF (JC(I) .EQ. J) GO TO 190                           C   760
180   CONTINUE                                              C   770
190   JC(I) = JC(J)                                         C   780
      DO 200 K = 1,N                                        C   790
      X = A(K,I)                                            C   800
      A(K,I) = A(K,J)                                       C   810
200   A(K,J) = X                                            C   820
210   CONTINUE                                              C   830
220   JC(1) = N                                             C   840
      IF (KD .EQ. 1) V(1) = S                               C   850
      RETURN                                                C   860
      END                                                   C   870-
      SUBROUTINE SST2 (A,B,AP,BP,X1,X2,Y1,Y2,X,Y,TERM0,AL,AM,ALM,LMAX,MM  D   10
     1AX,SPSUM)                                             D    20
      DIMENSION AL(70), AM(70), ALM(70,70), SPSUM(70)       D    30
C     FUNCTION DEFINITIONS                                  D    40
      DELL(V) = 2.0 * (SIN(AP * X2 * (V - 1.0)) - SIN(AP * X1 * (V  D    50
     1    - 1.0))) / (AP * (X2 - X1) * (V - 1.0))           D    60
      DELM(V) = 2.0 * (SIN(BP * Y2 * (V - 1.0)) - SIN(BP * Y1 * (V  D    70
     1    - 1.0))) / (BP * (Y2 - Y1) * (V - 1.0))           D    80
      DO 10 I = 1,LMAX                                      D    90
10    SPSUM(I) = 0.0                                        D   100
      DO 30 L = 2,LMAX                                      D   110
      RL = L                                                D   120
      ANS = AL(L) * DELL(RL) * COS(AP * X * (RL - 1.0))     D   130
      DO 20 I = L,LMAX                                      D   140
      SPSUM(I) = SPSUM(I) + ANS                             D   150
20    CONTINUE                                              D   160
30    CONTINUE                                              D   170
      DO 50 M = 2,LMAX                                      D   180
      RM = M                                                D   190
      ANS = AM(M) * DELM(RM) * COS(BP * Y * (RM - 1.0))     D   200
      DO 40 I = M,LMAX                                      D   210
      SPSUM(I) = SPSUM(I) + ANS                             D   220
40    CONTINUE                                              D   230
50    CONTINUE                                              D   240
      DO 90 L = 2,LMAX                                      D   250
      DO 90 M = 2,LMAX                                      D   260
      RL = L                                                D   270
      RM = M                                                D   280
      ANS = ALM(L,M) * DELL(RL) * DELM(RM) * COS(BP * Y * (RM  D   290
     1    - 1.0)) * COS(AP * X * (RL - 1.0))                D   300
      IF (M .GT. L .OR. M .EQ. L) GO TO 60                  D   310
      LIMIT = L                                             D   320
      GO TO 70                                              D   330
60    LIMIT = M                                             D   340
70    CONTINUE                                              D   350
      DO 80 I = LIMIT,LMAX                                  D   360
80    SPSUM(I) = SPSUM(I) + ANS                             D   370
90    CONTINUE                                              D   380
      SPSUM(1) = TERM0 / (A * B)                            D   390
      DO 100 I = 2,LMAX                                     D   400
      SPSUM(I) = (SPSUM(I) + TERM0) / (A * B)               D   410
100   CONTINUE                                              D   420
      RETURN                                                D   430
      END                                                   D   440-
      SUBROUTINE TOUT (Q1,LMAX,MMAX,SPSUM)                  E    10
```

```
        DIMENSION SPSUM(70)                                      E    20
        WRITE (3,30)                                             E    30
        DO 10 I = 1,LMAX                                         E    40
10      SPSUM(I) = Q1 * SPSUM(I)                                 E    50
        DO 20 I = 1,LMAX                                         E    60
20      WRITE (3,40) I,SPSUM(I)                                  E    70
30      FORMAT(26HSINGLE SOURCE TEMP VS LMAX)                    E    80
40      FORMAT(5(I3,5X,E10.4))                                   E    90
        RETURN                                                   E   100
        END                                                      E   110-
        SUBROUTINE LPLOT (LMAX,SPSUM)                            F    10
        DIMENSION SPSUM(70),LINE(51)                             F    20
        SMIN = SPSUM(1)                                          F    30
        SMAX = SPSUM(1)                                          F    40
        DO 10 I = 2,LMAX                                         F    50
        IF (SPSUM(I) .LT. SMIN) SMIN = SPSUM(I)                  F    60
        IF (SPSUM(I) .GT. SMAX) SMAX = SPSUM(I)                  F    70
10      CONTINUE                                                 F    80
        WRITE (4,90) SMAX,SMIN                                   F    90
        DO 20 I = 1,51                                           F   100
        LINE(I) = 1H.                                            F   110
20      CONTINUE                                                 F   120
        WRITE (4,100) LINE                                       F   130
        DO 30 I = 1,51                                           F   140
        LINE(I) = 1H                                             F   150
30      CONTINUE                                                 F   160
        LINE(1) = 1H.                                            F   170
        LTEST = 0                                                F   180
        DO 80 I = 1,LMAX                                         F   190
        LTEST = LTEST + 1                                        F   200
        J = ((SPSUM(I) - SMIN) / (SMAX - SMIN)) * 50.0 + 1       F   210
        IF (LTEST .NE. 10) GO TO 60                              F   220
        DO 40 K = 2,51                                           F   230
40      LINE(K) = 1H.                                            F   240
        LINE(J) = 1HX                                            F   250
        WRITE (4,110) I,LINE                                     F   260
        LTEST = 0                                                F   270
        DO 50 K = 2,51                                           F   280
50      LINE(K) = 1H                                             F   290
        GO TO 70                                                 F   300
60      LINE(J) = 1HX                                            F   310
        WRITE (4,120) LINE                                       F   320
        LINE(J) = 1H                                             F   330
70      LINE(1) = 1H.                                            F   340
80      CONTINUE                                                 F   350
90      FORMAT(1H1,60X,17HTMAX-TMIN VS LMAX///1H ,49X,           F   360
     1       5HTMAX=,E10.4,5X,5HTMIN=,E10.4)                     F   370
100     FORMAT(1H0,37X,1H0,2X,51A1)                              F   380
110     FORMAT(1H ,35X,I3,2X,51A1)                               F   390
120     FORMAT(1H ,40X,51A1)                                     F   400
        RETURN                                                   F   410
        END                                                      F   420-
```

Appendix VI
TNETFA Source Listing

```
        PROGRAM TNETFA (DIN,DOUT,TAPE1=DIN,TAPE2=DOUT)                  A    10
C         THE TNETFA PROGRAM AND ALL RELATED MATERIAL ARE SUPPLIED      A    11
C         AS - IS AND WITHOUT WARRANTY OR REPRESENTATION OF ANY KIND.   A    12
C         THE PROGRAM AUTHOR, HIS EMPLOYERS AND PUBLISHERS, PAST, PRESENT A  13
C         OR FUTURE MAKE NO REPRESENTATIONS RESPECTING THE PROGRAM AND  A    14
C         RELATED MATERIAL, AND EXPRESSLY DISCLAIM ANY LIABILITY FOR    A    15
C         DAMAGES WHETHER DIRECT, CONSEQUENTIAL OR OTHERWISE, ARISING   A    16
C         FROM THE USE OF THE PROGRAM OR RELATED MATERIAL, OR ANY PART  A    17
C         THEREOF.                                                      A    18
C       TNETFA - TRANSIENT NETWORK THERMAL ANALYZER -                   A    20
C       JAN. 29, 1982 VERSION                                           A    30
C       REFERENCES TO EQUATIONS, E.G. T.C.E.E., EX.XX, REFER TO:        A    31
C       GORDON N. ELLISON, 'THERMAL COMPUTATIONS IN ELECTRONIC EQUIPMENT'. A 32
C       650 NODE, 4000 CONDUCTOR VERSION                                A    40
C       50 NODE LIMIT TO MULTUPLE SURFACE RADIATION EXCHANGE            A    50
        REAL KFAC,MAXDT,KVFAC,MAXT                                      A    70
        INTEGER CURVE,TPRINT,CTYPE,ATYPE,BTYPE,BARY,UNITS,BB4           A    80
        INTEGER CTYBLD                                                  A    90
C       THE FOLLOWING DIMENSION MUST BE GREATER THAN OR EQUAL TO        A   100
C       THE TOTAL NUMBER OF GENERATOR BLOCKS                            A   110
        DIMENSION NBLD(300),NA1(300),NAS(300),NB1(300),NBS(300),CBLD(300), A 120
     1       CTYBLD(300)                                               A   130
        DIMENSION RF(50),AA1(50),BB1(50),BB2(50),BB3(50),BB4(50)        A   140
        DIMENSION ATYPE(50),BTYPE(50)                                   A   150
        INTEGER TITLE1(14),TITLE2(14),CAPCUR(650),SINCAP,STRCAP         A   160
C       NPAIRS AND CURVE MUST HAVE DIMENSIONS EQ OR GT THAN             A   170
C       NO. OF INTERPOLABLE CURVES                                      A   180
C       X AND Y MUST HAVE DIMENSIONS EQ OR GT                           A   190
C       THAN THE NO. OF DATA POINTS FOR ALL CURVES                      A   200
        DIMENSION NPAIRS(20),CURVE(20),X(100),Y(100)                    A   210
C       T,Q,NBCNT MUST HAVE DIMENSIONS EQUAL TO OR GREATER THAN         A   220
C       THE NUMBER OF NODES                                             A   230
C       NA,NB,C,CTYPE MUST HAVE DIMENSIONS GREATER THAN THE NO. OF CONDUCT A 240
        DIMENSION T(650),C(4000),NA(4000),NB(4000),Q(650),CTYPE(4000)   A   250
        DIMENSION NBCNT(650),TP(650),CAP(650),STAB(650)                 A   260
        DIMENSION NQ(20),NQPRS(20),TM(100),QT(100)                      A   270
        COMMON /BLK1/ T,C,NA,NB,Q/BLK2/NPAIRS,X,Y/BLK3/RF,CTYPE,CURVE   A   280
        COMMON /BLK4/ TZ,SIGMA,TFAC,TCON,KFAC,KVFAC,FLFAC,DFAC,HFAC     A   290
        COMMON /BLK5/ AA1,BB1,BB2,BB3,BB4,      ATYPE,BTYPE             A   300
        COMMON /BLK6/ NBCNT                                             A   310
        COMMON/BLK7/TP,CAP,STAB,CAPCUR,NQ,NQPRS,TM,QT                   A   320
C       COMMON BLKS 8-10 ARE DIMENSIONED FOR RADIATION EXCHANGE BETWEEN A   330
C       50 NODES MAXIMUM                                                A   340
C       COMMON BLK11 DIMENSIONED FOR MAXIMUM OF 50 CONSTANT             A   350
C       TEMPERAURE NODES                                                A   360
        COMMON/BLK8/IAE(50),EM(50),AR(50)                              A   370
        COMMON/BLK9/IF(2500),JF(2500),AM(50,50)                        A   380
        COMMON/BLK10/BM(50),RJM(50)/BLK11/NCTEMP(50)                   A   390
        ICSCNT = 0                                                      A   400
C       INPUT DATA                                                      A   410
C       TWO TITLE CARDS(BLANK IF NEC.) REQUIRED                         A   420
10      READ (1,780) (TITLE1(I),I=1,8)                                  A   430
        IF (EOF(1) .NE. 0) CALL EXIT                                    A   440
        READ (1,780) (TITLE2(I),I=1,8)                                  A   450
C       INITIALIZE                                                      A   460
        IRCTMX = 50                                                     A   470
        DO 20 I = 1,IRCTMX                                              A   480
        AR(I) = 0.0                                                     A   490
        EM(I) = 0.0                                                     A   500
20      CONTINUE                                                        A   510
        DO 30 I = 1,IRCTMX                                              A   520
        DO 30 J = 1,IRCTMX                                              A   530
        AM(I,J) = 0.0                                                   A   540
30      CONTINUE                                                        A   550
```

```
        IRADFLG = 0                                                        A  560
C     MODE = 0 IF SUPPRESSION OF ITERATION IS DESIRED                      A  570
C     MODE = 1 IF STEADY STATE REQUIRED                                    A  580
C     MODE = 2 IF POTENTIAL FLOW CALCULATION IS DESIRED                    A  590
C     MODE = 3 IF TRANSIENT CALC REQUIRED - FORWARD TIME DIFFERENCE        A  600
C     UNITS = 0 FOR X,L,D=FT, V=FT/SEC, Q=BTU/HR, T=DEG F,                 A  610
C     C=BTU/HR-DEG F, G=CUFT/MIN, RHOCP=BTU/CUFT-DEG. F,                   A  620
C     CAP=BTU/DEG. F, TIME=HR                                             A  630
C     UNITS = 1 FOR X,L,D=CM, V=CM/SEC, Q=WATTS, T=DEG C,                  A  640
C     C=WATTS/DEG C, G=CUCM/SEC, RHOCP=CAL/CUCM-DEG. C,                    A  650
C     CAP=JOULES/DEG. C, TIME=SEC                                          A  660
C     UNITS = 2 FOR X,L,D=IN, V=FT/MIN, Q=WATTS, T=DEG C,                  A  670
C     C=WATTS/DEG C, G=CUFT/MIN, RHOCP=CAL/CUCM-DEG. C,                    A  680
C     CAP=JOULES/DEG. C, TIME=SEC                                          A  690
        READ (1,*) MODE,UNITS,ICSE                                         A  700
C     NUMBER OF NODES(INCLUDING CONSTANT TEMP NODES),                      A  710
C     NUMBER OF CONSTANT TEMPERATURE NODES                                 A  720
C     NUMBER OF SPECIAL TEMPERATURE AND/OR SOURCE POWERS(EXCLUDING         A  730
C     CONSTANT TEMPERATURE NODES), NUMBER OF CONDUCTORS                    A  740
C     SINGLY CONSTRUCTED, NUMBER OF AUTO-GENERATOR BLOCKS, SIMPLE          A  750
C     NON-LINEAR COND. CURVES,                                             A  760
C     NATURAL CONV. COND. CARDS, FORCED CONV. COND. CARDS                  A  770
C     DATA CARDS                                                           A  780
        READ (1,*) NN,NCONST,NZS,NQCRV,NCBLC,NCS,NCRV,NNCNV,NFCNV          A  790
C     STARTING TEMPERATURES(INCLUDING CONSTANT TEMPS) AND POWER INPUTS     A  800
C     Q = VX0/DX OR VY0/DY (FLOW B.C.) FOR MODE = 2                        A  810
        READ (1,*) TSET,QSET                                               A  820
        DO 40 I = 1,NN                                                     A  830
        T(I) = TSET                                                        A  840
        Q(I) = QSET                                                        A  850
40    CONTINUE                                                             A  860
        NSUM = 0                                                           A  870
        I2 = NCONST + NZS                                                  A  880
        DO 50 I = 1,I2                                                     A  890
        NSUM = NSUM + 1                                                    A  900
        IF (NSUM .EQ. NCONST                                               A  910
     1      .OR. NSUM .LT. NCONST) READ (1,  * )NCTEMP(I),T(NCTEMP(I))     A  920
        IF (NSUM .GT. NCONST) READ (1,  * )JQ,T(JQ),Q(JQ)                  A  930
50    CONTINUE                                                             A  940
C     Q VX. TIME CURVES MUST BE ARRANGED IN ORDER OF ASCENDING             A  950
C     NODE NUMBERS                                                         A  960
C      NQ(I) = NODE NO. FOR THIS Q VS. TIME CURVE                          A  970
C     NQPRS(I) = NO. Q VS. TIME DATA PAIRS                                 A  980
        CONTINUE                                                           A  990
        IF (NQCRV .EQ. 0) GO TO 70                                         A 1000
        J1QP = 1                                                           A 1010
        DO 60 I = 1,NQCRV                                                  A 1020
        READ (1,*) NQ(I),NQPRS(I)                                          A 1030
        J2QP = J1QP + NQPRS(I) - 1                                         A 1040
        READ (1,*) (TM(J),QT(J),J=J1QP,J2QP)                              A 1050
        J1QP = J2QP + 1                                                    A 1060
60    CONTINUE                                                             A 1070
C     INITIALIZE ALL NODES WITH A SMALL CAPACITANCE                        A 1080
70    CONTINUE                                                             A 1090
        DO 80 I = 1,NN                                                     A 1100
        CAPCUR(I) = 0                                                      A 1110
        CAP(I) = 1.0E - 20                                                 A 1120
80    CONTINUE                                                             A 1130
C     CAPACITANCE INPUT                                                    A 1140
        READ (1,*) SINCAP,STRCAP                                           A 1150
        IF (SINCAP .EQ. 0) GO TO 100                                       A 1160
        DO 90 I = 1,SINCAP                                                 A 1170
        READ (1,*) J,C1,IC2                                               A 1180
        CAP(J) = C1                                                        A 1190
```

```
            CAPCUR(J) = IC2                                                  A 1200
90          CONTINUE                                                         A 1210
100         IF (STRCAP .EQ. 0) GO TO 130                                     A 1220
            DO 120 I = 1,STRCAP                                              A 1230
            READ (1,*) IACAP,IBCAP,CAP2,CAPCV2                               A 1240
            DO 110 J = IACAP,IBCAP                                           A 1250
            CAP(J) = CAP2                                                    A 1260
            CAPCUR(J) = CAPCV2                                               A 1270
110         CONTINUE                                                         A 1280
120         CONTINUE                                                         A 1290
C           NB(I) IS NEGATIVE FOR FLUID(AIR) NODE(PARTICULARLY IN NAT CONV)  A 1300
C           CTYPE(I)=-1, C(I)=(SCRIPT F(NA,NB))*AREA(NA))                    A 1310
C           CTYPE(I)=-2, C(I)=(ANGLE FACTOR(NA,NB))*AREA(NA))               A 1320
C           CTYPE(I)=ZERO IF NO INTERPOLATION ETC. IS REQUIRED,              A 1330
C           CTYPE FOR FLOW CALCULATION                                       A 1340
C             CTYPE = 0 FOR X DIRECTION, C(I) = 1/DX**2                      A 1350
C             CTYPE = 1 FOR Y DIRECTION, C(I) = 1/DY**2                      A 1360
C           C(I)=CONDUCTANCE                                                 A 1370
C           VALUE OF CTYPE(I) REFERENCES CURVE(I)                           A 1380
C           CTYPE(I) GT ZERO AND LT 101 FOR T-DEPENDENT CONDUCTANCE AND OR   A 1390
C           CAPCITANCE                                                       A 1400
C           C(I)=COND CROSS SECTIONAL AREA/PATH LENGTH                       A 1410
C           CTYPE(I) MUST EQUAL CURVE(I) BELOW FOR NCRV DATA                 A 1420
C           CTYPE(I) MUST BE GT 100 AND LT 201 FOR NAT CONV                  A 1430
C           VALUE OF CTYPE(I) - 100 REFERENCES ARGUMENT OF AA               A 1440
C           CTYPE(I) MUST BE GT 200 FOR FOR FORCED CONV,                     A 1450
C           C(I)=COND CROSS-SEC. AREA                                        A 1460
C           VALUE OF CTYPE(I) - 200 REFERENCES ARGUMENT OF BB               A 1470
C           WHEN BTYPE = 5, 6    SUBSTRATE NODES MUST BE SEQUENTIALLY        A 1480
C           NUMBERED FROM LEADING EDGE NODE                                  A 1490
C           IF MORE THAN 1 SUBSTRATE OF THIS TYPE, THE ADDITIONAL            A 1500
C           SUBSTRATE NODE NOS. MUST FOLLOW WITHOUT BREAK AND IN SEQUENCE    A 1510
C           CTYPE(I) MUST BE GT 300 AND LT 401 FOR FLUID CONDUCTORS          A 1520
C           C(I)=MASS FLOW RATE G                                            A 1530
C           CTYPE FOR CFM, PRESSURE CALCULATION ---- C(I)=R                  A 1540
C           FOR P=R*G**2 OR R*G                                              A 1550
C             CTYPE = 401 FOR LAMINAR FLOW                                   A 1560
C             CTYPE = 402 FOR TURBULENT FLOW                                 A 1570
130         JSTOR1 = 0                                                       A 1580
            JSTOR2 = 0                                                       A 1590
            IF (NCBLC .EQ. 0) GO TO 180                                      A 1600
            DO 140 N = 1,NCBLC                                               A 1610
            READ (1,*) NBLD(N),NA1(N),NAS(N),NB1(N),NBS(N),CBLD(N),CTYBLD(N) A 1620
140         CONTINUE                                                         A 1630
            J = 0                                                            A 1640
C           AUTOMATIC CONDUCTOR GENERATION                                   A 1650
            DO 170 N = 1,NCBLC                                               A 1660
C           CONDUCTOR GENERATION FOR N-TH BLOCK                              A 1670
            ISUMNA = 0                                                       A 1680
            ISUMNB = 0                                                       A 1690
            NBD = NBLD(N)                                                    A 1700
            DO 160 L = 1,NBD                                                 A 1710
            J = J + 1                                                        A 1720
            NA(J) = NA1(N) + ISUMNA                                          A 1730
            NB(J) = NB1(N) + ISUMNB                                          A 1740
            C(J) = CBLD(N)                                                   A 1750
            CTYPE(J) = CTYBLD(N)                                             A 1760
C           TEMPORARY STORAGE AND ABSOLUTE VALUE OF NA,NB                    A 1770
            NASTOR = IABS(NA(J))                                             A 1780
            NBSTOR = IABS(NB(J))                                             A 1790
C           DETERMINATION OF RETURN CONDUCTOR                                A 1800
C           DO NOT REPEAT DIAGONAL ELEMENT FOR MULTI-SURF RAD. EXCH.         A 1810
            IF (IABS(NA(J)) .EQ. IABS(NB(J)) .AND. CTYPE(J) .E               A 1820
          1    Q. - 2) GO TO 150                                            A 1830
```

```
         J = J + 1                                                          A 1840
         NB(J) = NASTOR                                                     A 1850
         NA(J) = NBSTOR                                                     A 1860
         C(J) = CBLD(N)                                                     A 1870
         CTYPE(J) = CTYBLD(N)                                               A 1880
         IF (CTYBLD(N) .LT. 301 .OR. CTYBLD(N) .GT. 400) GO TO 150          A 1890
         C(J) = 0.0                                                         A 1900
150      ISUMNA = ISUMNA + NAS(N)                                          A 1910
         ISUMNB = ISUMNB + NBS(N)                                          A 1920
C         STORE TOTAL NO. OF CONDUCTORS       GENERATED UP TO HERE          A 1930
         JSTOR1 = J                                                         A 1940
160      CONTINUE                                                          A 1950
170      CONTINUE                                                          A 1960
180      IF (NCS .EQ. 0) GO TO 210                                         A 1970
         J = JSTOR1 + 1                                                     A 1980
         DO 200 I = 1,NCS                                                   A 1990
         JSTOR2 = JSTOR2 + 1                                                A 2000
         READ (1,*) NA(J),NB(J),C(J),CTYPE(J)                              A 2010
C         DO NOT REPEAT DIAG ELEMENT IF                                     A 2020
         IF (IABS(NA(J)) .EQ. IABS(NB(J)) .AND. CTYPE(J) .E                 A 2030
1            Q. - 2) GO TO 190                                             A 2040
         JSTOR2 = JSTOR2 + 1                                                A 2050
         NA(J + 1) = IABS(NB(J))                                           A 2060
         NB(J + 1) = IABS(NA(J))                                           A 2070
         C(J + 1) = C(J)                                                    A 2080
         IF (CTYPE(J) .GT. 300 .AND. CTYPE(J) .LT. 401) C(J + 1) = 0.0      A 2090
         CTYPE(J + 1) = CTYPE(J)                                           A 2100
         J = J + 2                                                          A 2110
         GO TO 200                                                         A 2120
190      J = J + 1                                                          A 2130
200      CONTINUE                                                          A 2140
C         CONDUCTOR ORDERING SCHEME - ARRANGES NA S IN GROUPS               A 2150
C         OF ASCENDING VALUE                                                A 2160
210      JJ2 = JSTOR1 + JSTOR2 - 1                                         A 2170
         I2 = JSTOR1 + JSTOR2                                               A 2180
         NCND = I2                                                         A 2190
         DO 220 J = 1,JJ2                                                   A 2200
         IP1 = J + 1                                                        A 2210
         DO 220 I = IP1,I2                                                  A 2220
         IF (IABS(NA(J)) .LT. IABS(NA(I))) GO TO 220                       A 2230
         ITEMPA = NA(J)                                                     A 2240
         ITEMPB = NB(J)                                                     A 2250
         TEMPC = C(J)                                                       A 2260
         ITECTP = CTYPE(J)                                                  A 2270
         NA(J) = NA(I)                                                      A 2280
         NB(J) = NB(I)                                                      A 2290
         C(J) = C(I)                                                        A 2300
         CTYPE(J) = CTYPE(I)                                               A 2310
         NA(I) = ITEMPA                                                     A 2320
         NB(I) = ITEMPB                                                     A 2330
         C(I) = TEMPC                                                       A 2340
         CTYPE(I) = ITECTP                                                  A 2350
220      CONTINUE                                                          A 2360
C         ASSIGN NEGATIVE SIGN TO NODE NUMBERS NA IF CONSTANT               A 2370
C         TEMPERATURE NODE                                                  A 2380
         CALL NEGNODE (NCONST,NCND)                                         A 2390
C         SCAN TO CHECK FOR MULTI. SURF. RADIATION EXCHANGE                 A 2400
         DO 240 I = 1,NCND                                                  A 2420
         IF (CTYPE(I) .NE. - 2) GO TO 240                                  A 2430
         IRADFLG = 1                                                        A 2440
         GO TO 250                                                         A 2450
240      CONTINUE                                                          A 2460
230      IF (IRADFLG .EQ. 0) GO TO 300                                     A 2470
C         NODE NO., EMISSIVITY, AREA INPUT FOR MULT. SURF. RAD.             A 2480
```

```
250       READ (1, *)ISINA,ISTRA                                          A 2490
          IF (ISINA .EQ. 0) GO TO 270                                     A 2500
C         SINGLY INPUT VALUES OF AREA(ARR), EMISSIVITY(EMM) FOR NODE J     A 2510
          DO 260 I = 1,ISINA                                              A 2520
          READ (1, *)J,ARR,EMM                                            A 2530
          IAE(I) = J                                                      A 2540
          AR(I) = ARR                                                     A 2550
          EM(I) = EMM                                                     A 2560
260       CONTINUE                                                        A 2570
270       IF (ISTRA .EQ. 0) GO TO 300                                     A 2580
          J1 = ISINA + 1                                                  A 2590
C         STRING INPUT OF AREA, EMISSIVITY FOR NODES IAA TO IBA SEQUENTIALLY A 2600
          DO 290 I = 1,ISTRA                                              A 2610
          READ (1, *)IAA,IBA,ARR,EMM                                      A 2620
          J2 = J1 + (IBA - IAA)                                           A 2630
          IAE1 = IAA                                                      A 2640
          DO 280 J = J1,J2                                                A 2650
          IAE(J) = IAE1                                                   A 2660
          AR(J) = ARR                                                     A 2670
          EM(J) = EMM                                                     A 2680
          IAE1 = IAE1 + 1                                                 A 2690
280       CONTINUE                                                        A 2700
          J1 = J2 + 1                                                     A 2710
290       CONTINUE                                                        A 2720
C         CURVE DATA - NPAIRS(I)=NO. OF XY PAIRS FOR ITH CURVE            A 2730
300       IF (NCRV .EQ. 0) GO TO 320                                      A 2740
          J1NP = 1                                                        A 2750
          DO 310 I = 1,NCRV                                               A 2760
          READ (1,*) CURVE(I),NPAIRS(I)                                   A 2770
          J2NP = J1NP + NPAIRS(I) - 1                                     A 2780
          READ (1,*) (X(J),Y(J),J=J1NP,J2NP)                             A 2790
          J1NP = J2NP + 1                                                 A 2800
310       CONTINUE                                                        A 2810
320       IF (NNCNV .EQ. 0) GO TO 340                                     A 2820
C         CTYPE(J) - 100 = ARG I IN AA(I)                                 A 2830
C         ATYPE DENOTES SPECIFIC NATURAL CONVECTION MODE                  A 2840
C         AA1=SIGNIFICANT DIMENSION                                       A 2850
          DO 330 I = 1,NNCNV                                              A 2860
          READ (1,*) ATYPE(I),AA1(I)                                      A 2870
330       CONTINUE                                                        A 2880
C         CTYPE(J) - 200 = ARG I IN BB(I)                                 A 2890
C         BTYPE DENOTES SPECIFIC FORCED CONVECTION MODE                   A 2900
C         DX = CONSTANT NODE LENGTH                                       A 2910
C         BB1 = V, BB2 = D, BB3 = L,DX, BB4 = FIRST F.P. NODE             A 2920
340       IF (NFCNV .EQ. 0) GO TO 360                                     A 2930
          DO 350 I = 1,NFCNV                                              A 2940
          READ (1,*) BTYPE(I),BB1(I),BB2(I),BB3(I),BB4(I)                A 2950
350       CONTINUE                                                        A 2960
360       IF (MODE .NE. 2) GO TO 370                                      A 2970
          READ (1,*) V0                                                   A 2980
C         NLOOP=NO. OF STEADY STATE ITERATIONS                            A 2990
C         BETA=STEADY STATE OVER RELAXATION CONSTANT                      A 3000
C         ALDT=STEADY STATE MAX TEMP CHANGE/ITERATION TERMINATE CRITERIA  A 3010
C         LOOPEN=0---------SUPPRESS NODE DETAIL                           A 3020
C         LOOPEN=1---------PRINT NODE DETAIL                              A 3030
C         DELT=TRANSIENT TIME STEP                                        A 3040
C         MAXT=TRANSIENT MAX TIME                                         A 3050
C         TPRINT=NO. OF STEADY STATE ITERATIONS OR TRANSIENT TIME STEPS   A 3060
C             BETWEEN TEMP PRINTOUTS                                      A 3070
370       READ (1,*) NLOOP,BETA,ALDT,LOOPEN                              A 3080
          READ (1,*) DELT,MAXT                                            A 3090
          READ (1,*) TPRINT,NPRINT                                        A 3100
          WRITE (2,760)                                                   A 3110
          WRITE (2,770)                                                   A 3120
```

```
      WRITE (2,800) (TITLE1(I),I=1,8)                         A 3130
      WRITE (2,800) (TITLE2(I),I=1,8)                         A 3140
      IF (UNITS .EQ. 0) WRITE (2,810)                         A 3150
      IF (UNITS .EQ. 1) WRITE (2,820)                         A 3160
      IF (UNITS .EQ. 2) WRITE (2,830)                         A 3170
      WRITE (2,790) NN,NCND                                   A 3180
      WRITE (2,840) NLOOP,TPRINT,NPRINT,LOOPEN,ALDT,BETA      A 3190
      IF (NCRV .EQ. 0) GO TO 390                              A 3200
      WRITE (2,850)                                           A 3210
      J1NP = 1                                                A 3220
      DO 380 I = 1,NCRV                                       A 3230
      WRITE (2,880) CURVE(I),NPAIRS(I)                        A 3240
      J2NP = J1NP + NPAIRS(I) - 1                             A 3250
      WRITE (2,890) (X(J),Y(J),J=J1NP,J2NP)                   A 3260
      J1NP = J2NP + 1                                         A 3270
380   CONTINUE                                                A 3280
390   IF (NQCRV .EQ. 0) GO TO 410                             A 3290
      WRITE (2,860)                                           A 3300
      J1QP = 1                                                A 3310
      DO 400 I = 1,NQCRV                                      A 3320
      WRITE (2,870) NQ(I),NQPRS(I)                            A 3330
      J2QP = J1QP + NQPRS(I) - 1                              A 3340
      WRITE (2,890) (TM(J),QT(J),J=J1QP,J2QP)                 A 3350
      J1QP = J2QP + 1                                         A 3360
400   CONTINUE                                                A 3370
410   IF (NFCNV .EQ. 0) GO TO 490                             A 3380
      WRITE (2,900)                                           A 3390
      DO 480 I = 1,NFCNV                                      A 3400
      BARY = BTYPE(I)                                         A 3410
      GO TO (420,430,440,460,450,470),BARY                   A 3420
420   WRITE (2,910) I,BB1(I),BB2(I),BB3(I)                   A 3430
      GO TO 480                                               A 3440
430   WRITE (2,920) I,BB1(I),BB2(I),BB3(I)                   A 3450
      GO TO 480                                               A 3460
440   WRITE (2,930) I,BB1(I),BB3(I)                          A 3470
      GO TO 480                                               A 3480
450   WRITE (2,940) I,BB1(I),BB3(I),BB4(I)                   A 3490
      GO TO 480                                               A 3500
460   WRITE (2,950) I,BB1(I),BB3(I)                          A 3510
      GO TO 480                                               A 3520
470   WRITE (2,960) I,BB1(I),BB3(I),BB4(I)                   A 3530
480   CONTINUE                                                A 3540
490   IF (NNCNV .EQ. 0) GO TO 570                             A 3550
      WRITE (2,690)                                           A 3560
      DO 560 I = 1,NNCNV                                      A 3570
      BARY = ATYPE(I)                                         A 3580
      GO TO (500,510,520,530,540,550,551,552),BARY           A 3590
500   WRITE (2,700) I,AA1(I)                                  A 3600
      GO TO 560                                               A 3610
510   WRITE (2,710) I,AA1(I)                                  A 3620
      GO TO 560                                               A 3630
520   WRITE (2,720) I,AA1(I)                                  A 3640
      GO TO 560                                               A 3650
530   WRITE (2,730) I,AA1(I)                                  A 3660
      GO TO 560                                               A 3670
540   WRITE (2,740) I,AA1(I)                                  A 3680
      GO TO 560                                               A 3690
550   WRITE (2,750) I,AA1(I)                                  A 3700
      GO TO 560                                               A 3701
551   WRITE(2,751) I, AA1(I)                                  A 3702
      GO TO 560                                               A 3703
552   WRITE(2,752) I, AA1(I)                                  A 3704
560   CONTINUE                                                A 3710
570   IF (IRADFLG .EQ. 0) GO TO 590                           A 3720
```

```
C         COUNT NO. OF NON-ZERO AR TO GET NO. OF MULTI. SURFS.        A 3730
C         FOR RADIATION EXCHANGE                                      A 3740
          IRCNT = 0                                                   A 3750
          DO 580 I = 1, IRCTMX                                        A 3760
          IF (AR(I) .NE. 0.0) IRCNT = IRCNT + 1                       A 3770
580       CONTINUE                                                    A 3780
590       IUNITS = UNITS + 1                                          A 3790
          GO TO (600,610,620), IUNITS                                 A 3800
600       TZ = 460.0                                                  A 3810
          SIGMA = 0.1714                                              A 3820
          TFAC = 1.0                                                  A 3830
          TCON = 0.0                                                  A 3840
          KFAC = 1.0                                                  A 3850
          KVFAC = 1.0                                                 A 3860
          FLFAC = 60.0                                                A 3870
          DFAC = 1.0                                                  A 3880
          HFAC = 1.0                                                  A 3890
          GO TO 630                                                   A 3900
610       TZ = 273.0                                                  A 3910
          SIGMA = 5.669E - 04                                         A 3920
          TFAC = 1.8                                                  A 3930
          TCON = 32.0                                                 A 3940
          KFAC = 1.0 / 57.79                                          A 3950
          KVFAC = 928.0308                                            A 3960
          FLFAC = 4.184 / 62.43                                       A 3970
          DFAC = 1.0 / (2.54 * 12.0)                                  A 3980
          HFAC = 1.0 / 1761.0                                         A 3990
          GO TO 630                                                   A 4000
620       TZ = 273.0                                                  A 4010
          SIGMA = 3.657E - 3                                          A 4020
          TFAC = 1.8                                                  A 4030
          TCON = 32.0                                                 A 4040
          KFAC = 2.54 / 57.79                                         A 4050
          KVFAC = 12.0 * 60.0                                         A 4060
          FLFAC = ((12.0 * 2.54) ** 3) * 4.184 / (62.43 * 60.0)       A 4070
          DFAC = 1.0 / 12.0                                           A 4080
          HFAC = (2.54 ** 2) / 1761.0                                 A 4090
630       IF(IRADFLG.EQ.0) GO TO 635                                  A 4095
          CALL NSCAN (NCND, IRCNT)                                    A 4100
          CALL AADJ (IRCNT)                                           A 4110
          CALL INVAM (IRCNT)                                          A 4120
          CALL SETBM (IRCNT)                                          A 4130
          CALL RADIOS (IRCNT)                                         A 4140
635       IF (MODE .NE. 0) GO TO 640                                  A 4150
          CALL TEMPRT (NN, LOOPCT, MAXDT, MODE)                       A 4160
          CALL NDET (IRADFLG, IRCNT, NCND, LOOPCT, MAXDT, MODE)       A 4170
          GO TO 660                                                   A 4180
C         COUNT NO. OF NB FOR EACH NA AND STORE AT NBCNT( )           A 4190
640       NNCND = NCND + 1                                            A 4200
          NA(NNCND) = 0                                               A 4210
          CALL COUNT (NN, NNCND)                                      A 4220
          GO TO (650,650,670), MODE                                   A 4230
650       CALL STSTA (NLOOP, LOOPCT, NN, ALDT, NCND, TPRINT, LOOPEN, NPRINT, MODE, A 4240
        1 V0, ICSE, BETA, MAXDT, IRADFLG, IRCNT)                      A 4250
          IF (ICSE .EQ. 0) GO TO 660                                  A 4260
          ICSCNT = ICSCNT + 1                                         A 4270
          IF (ICSCNT.EQ.ICSE.OR.ICSCNT.LT.ICSE) GO TO 230             A 4280
660       WRITE (2,760)                                               A 4290
          GO TO 680                                                   A 4300
670       CALL TRANF (NN, NQCRV, DELT, MAXT, NPRINT, TPRINT, NCND, IRADFLG, IRCNT) A 4310
680       CONTINUE                                                    A 4320
          GO TO 10                                                    A 4330
690       FORMAT (1H0,4X,28HNATURAL CONVECTION PARAMETER/)            A 4340
700       FORMAT (1H ,4X,I3,4X,31HVERTICAL FLAT PLATE OR CYLINDER,62X,2HP=, A 4350
```

```
       1         E10.4)                                                        A 4360
710    FORMAT (1H ,4X,I3,4X,83HHORIZONTAL FLAT PLATE OR CYLINDER, HEATED       A 4370
      1SIDE FACING UP OR COOLED SIDE FACING DOWN,10X,2HP=,E10.4)               A 4380
720    FORMAT (1H ,4X,I3,4X,83HHORIZONTAL FLAT PLATE OR CYLINDER, HEATED       A 4390
      1SIDE FACING DOWN OR COOLED SIDE FACING UP,10X,2HP=,E10.4)               A 4400
730    FORMAT(1H ,4X,I3,4X,26HHORIZONTAL PARALLEL PLATES,                      A 4410
      110X,2HB=,E10.4)                                                         A 4420
740    FORMAT (1H ,4X,I3,4X,24HVERTICAL PARALLEL PLATES,                       A 4430
      110X,2HB=,E10.4)                                                         A 4440
750    FORMAT(1H ,4X,I3,4X,23HSMALL SURFACE, VERTICAL,  10X,2HP=,E10.4)        A 4450
751    FORMAT(1H ,4X,I3,4X,74HSMALL HORIZONTAL SURFACE, HEATED SIDE FACIN      A 4471
      1G UP OR COOLED SIDE FACING DOWN,10X,2HP=,E10.4)                         A 4472
752    FORMAT(1H ,4X,I3,4X,74HSMALL HORIZONTAL SURFACE, HEATED SIDE FACIN      A 4473
      1G DOWN OR COOLED SIDE FACING UP,10X,2HP=,E10.4)                         A 4474
760    FORMAT (1H1)                                                            A 4470
770    FORMAT (1H ,48X,24HNETWORK THERMAL ANALYSIS//)                         A 4480
780    FORMAT (8A10)                                                           A 4490
790    FORMAT (1H0,4X,16HNUMBER OF NODES=,I4,5X,21HNUMBER OF CONDUCTORS=,      A 4500
      1         I4)                                                            A 4510
800    FORMAT (1H ,8A10)                                                       A 4520
810    FORMAT(1H0,4X,7HUNITS=0)                                                A 4530
820    FORMAT(1H0,4X,7HUNITS=1)                                                A 4540
830    FORMAT(1H0,4X,7HUNITS=2)                                                A 4550
840    FORMAT (1H0,4X,6HNLOOP=,I5,5X,7HTPRINT=,I5,5X,7HNPRINT=,I5,5X,          A 4560
      1    7HLOOPEN=,I5,5X,                                                    A 4570
      2         5HALDT=,E10.4,5X,5HBETA=,F5.2)                                 A 4580
850    FORMAT (1H0,4X,10HARRAY DATA)                                           A 4590
860    FORMAT(1H0,4X,17HTIME, POWER ARRAY)                                     A 4600
870    FORMAT(1H0,4X,4HNODE,I4,5X,I3,2X,9HT-Q PAIRS/)                          A 4610
880    FORMAT (1H0,4X,5HARRAY,I3,5X,I3,2X,9HX-Y PAIRS/)                        A 4620
890    FORMAT (1H ,4X,E10.4,1H,,2X,E10.4,4X,E10.4,1H,,2X,E10.4,4X,             A 4630
      1E10.4,1H,,2X,E10.4,4X,E10.4,1H,,2X,E10.4)                               A 4640
900    FORMAT (1H0,4X,                                                         A 4650
      1    55HDUCT AND/OR FLAT PLATE PARAMETERS FOR FORCED CONVECTION/         A 4660
      2    )                                                                   A 4670
910    FORMAT (1H ,4X,I3,4X,19HLAMINAR FLOW - DUCT,10X,4HVEL=,                 A 4680
      1    E10.4,5X,2HD=,E10.4,5X,2HL=,E10.4)                                  A 4690
920    FORMAT (1H ,4X,I3,4X,21HTURBULENT FLOW - DUCT,10X,4HVEL=,               A 4700
      1    E10.4,5X,2HD=,E10.4,5X,2HL=,E10.4)                                  A 4710
930    FORMAT (1H ,4X,I3,4X,36HLAMINAR FLOW OVER FLAT PLATE - AVE H,12X,       A 4720
      1    4HVEL=,E10.4,5X,2HL=,E10.4)                                         A 4730
940    FORMAT (1H ,4X,I3,4X,38HLAMINAR FLOW OVER FLAT PLATE - LOCAL H,         A 4740
      1    10X,4HVEL=,E10.4,5X,3HDX=,E10.4,5X,4HSTN=,I4)                       A 4750
950    FORMAT (1H ,4X,I3,4X,38HTURBULENT FLOW OVER FLAT PLATE - AVE H,         A 4760
      1    10X,4HVEL=,E10.4,5X,2HL=,E10.4)                                     A 4770
960    FORMAT (1H ,4X,I3,4X,40HTURBULENT FLOW OVER FLAT PLATE - LOCAL H,       A 4780
      1    10X,4HVEL=,E10.4,5X,3HDX=,E10.4,5X,4HSTN=,I4)                       A 4790
       END                                                                    A 4800-
       SUBROUTINE STSTA (NLOOP,LOOPCT,NN,ALDT,NCND,TPRINT,LOOPEN,NPRINT,M  B     10
      1ODE,V0,ICSE,BETA,MAXDT,IRADFLG,IRCNT)                                B     20
C      STEADY-STATE SOLUTION SCHEME BASED ON T.C.E.E., E1.16.              B     21
       REAL KFAC,MAXDT,KVFAC                                               B     30
       INTEGER CURVE,TPRINT,CTYPE,ATYPE,BTYPE,BARY,UNITS,BB4,CTYBLD        B     40
       DIMENSION NBLD(300),NA1(300),NAS(300),NB1(300),CBLD(300),          B     50
      1         CTYBLD(300)                                                B     60
       DIMENSION RF(50),AA1(50),BB1(50),BB2(50),BB3(50),BB4(50)           B     70
       DIMENSION ATYPE(50),BTYPE(50)                                      B     80
       INTEGER TITLE1(14),TITLE2(14),CAPCUR(650),SINCAP,STRCAP            B     90
       DIMENSION T(650),C(4000),NA(4000),NB(4000),Q(650),CTYPE(4000)      B    100
       DIMENSION NBCNT(650),TP(650),CAP(650),STAB(650),NQ(20),NQPRS(20)   B    110
      1,TM(100)          ,QT(100)                                         B    120
       DIMENSION NPAIRS(20),CURVE(20),X(100),Y(100)                       B    130
       COMMON /BLK1/ T,C,NA,NB,Q/BLK2/NPAIRS,X,Y/BLK3/RF,CTYPE,CURVE      B    140
       COMMON /BLK4/ TZ,SIGMA,TFAC,TCON,KFAC,KVFAC,FLFAC,DFAC,HFAC        B    150
```

```
        COMMON /BLK5/ AA1,BB1,BB2,BB3,BB4,ATYPE,BTYPE              B  160
        COMMON /BLK6/ NBCNT                                        B  170
        COMMON /BLK7/TP,CAP,STAB,CAPCUR,NQ,NQPRS,TM,QT             B  180
C       INITIALIZE                                                B  190
        LOOPCT = 0                                                B  200
        NTPRIN = TPRINT                                           B  210
        NEPRIN = LOOPEN                                           B  220
        CALL TEMPRT (NN,LOOPCT,MAXDT,1)                           B  230
C       START ITERATION LOOPING                                   B  240
        DO 120 L = 1,NLOOP                                        B  250
C       CALCULATE BM MATRIX AND RADIOSITY BEFORE EACH SET OF      B  260
C       ITERATIONS IF MULTI-SURFACE RADIATION EXCHANGE IS REQUIRED B 270
        IF (IRADFLG .EQ. 1) CALL SETBM (IRCNT)                    B  280
        IF (IRADFLG .EQ. 1) CALL RADIOS (IRCNT)                   B  290
        MAXDT = 0.0                                               B  300
        K1 = 1                                                    B  310
        LOOPCT = LOOPCT + 1                                       B  320
C       START LOOP TO DETERMINE T(I)                              B  330
        DO 90 I = 1,NN                                            B  340
C       STORE T(I) FOR COMPARISON TEST LATER                      B  350
        TOLD = T(I)                                               B  360
C       IS THIS A CONSTANT TEMP OR NN+1 NODE. IF ANSWER IS YES,   B  370
C       DO NOT ITERATE THIS NODE                                  B  380
        IF (NA(K1) .EQ. 0) GO TO 100                              B  390
C       INITIALIZATION FOR EACH ITERATION OF T(I)                 B  400
        RNUM = 0.0                                                B  410
        DENOM = 0.0                                               B  420
        IF (NA(K1) .LT. 0) GO TO 10                               B  430
        GO TO 20                                                  B  440
10      K1 = K1 + NBCNT(I)                                        B  450
        GO TO 90                                                  B  460
C       CALC OF T(I)                                              B  470
20      J2 = NBCNT(I) + K1 - 1                                    B  480
        DO 80 J = K1,J2                                           B  490
        N = IABS(NB(J))                                           B  500
        IF (CTYPE(J)) 30,70,40                                    B  510
30      IF (CTYPE(J) .EQ. - 2) CALL REX (I,N,IRCNT,C(J),CC)       B  520
        IF (CTYPE(J) .EQ. - 1) CALL RCOND (I,N,C(J),CC)           B  530
        GO TO 60                                                  B  540
40      IF (CTYPE(J) .GT. 100) GO TO 50                           B  550
        LIM1 = CTYPE(J)                                           B  560
        CALL COND (CURVE(LIM1),C(J),T(I),T(N),CC)                 B  570
        GO TO 60                                                  B  580
50      CALL CONV (J,CC,RE,RKINV,RLX)                             B  590
60      RNUM = RNUM + CC * T(N)                                   B  600
        DENOM = DENOM + CC                                        B  610
        GO TO 80                                                  B  620
70      RNUM = RNUM + C(J) * T(N)                                 B  630
        DENOM = DENOM + C(J)                                      B  640
80      CONTINUE                                                  B  650
        K1 = J2 + 1                                               B  660
        RNUM = RNUM + Q(I)                                        B  670
        T(I) = RNUM / DENOM                                       B  680
C       STORE NEW VALUE OF MAX TEMP CHANGE(MAXDT) IF NECESSARY    B  690
        DT = T(I) - TOLD                                          B  700
        T(I) = TOLD + BETA * DT                                   B  710
        DT = T(I) - TOLD                                          B  720
        DT = ABS(DT)                                              B  730
        IF (MAXDT .GT. DT) GO TO 90                               B  740
        MAXDT = DT                                                B  750
90      CONTINUE                                                  B  760
C       TEST FOR PROBLEM TERMINATION BASED ON TEMP CHANGE         B  770
100     IF (MAXDT .LT. ALDT) GO TO 140                            B  780
C       TEST FOR PRINT CALLS FOLLOW                               B  790
```

```
        IF (LOOPCT .NE. NTPRIN) GO TO 110                            B  800
        NTPRIN = NTPRIN + TPRINT                                     B  810
        CALL TEMPRT (NN,LOOPCT,MAXDT,1)                              B  820
110     IF (LOOPCT .NE. NEPRIN) GO TO 120                            B  830
        NEPRIN = NEPRIN + LOOPEN                                     B  840
        CALL BAL (NCND,IRCNT)                                        B  850
120     CONTINUE                                                     B  860
        IF (IRADFLG .EQ. 0) GO TO 130                                B  870
        CALL SETBM (IRCNT)                                           B  880
        CALL RADIOS (IRCNT)                                          B  890
130     IF (NPRINT.NE.0) CALL NDET(IRADFLG,IRCNT,NCND,LOOPCT,MAXDT,1) B  900
        GO TO 160                                                    B  920
C       FINAL PRINT CALL BEFORE PROBLEM TERMINATION                  B  930
140     CALL TEMPRT (NN,LOOPCT,MAXDT,1)                              B  940
C       PRINT CONDUCTORS AT PROB TERMINATION IF REQUESTED            B  950
        IF (NPRINT.NE.0) CALL NDET(IRADFLG,IRCNT,NCND,LOOPCT,MAXDT,1) B  970
        GO TO 160                                                    B  980
150     IF (LOOPEN.NE.0) CALL BAL(NCND,IRCNT)                        B  990
160     IF (MODE.EQ.2) CALL FLOW(NCND,V0)                            B 1010
        WRITE (2,180)                                                B 1030
C                                                                    B 1040
180     FORMAT(1H1)                                                  B 1050
        RETURN                                                       B 1060
        END                                                          B 1070-
        SUBROUTINE TRANF (NN,NQCRV,DELTM,RTM,NPRINT,TPRINT,NCND,IRADFLG,IR C   10
       1CNT)                                                         C   20
C       TIME-DEPENDENT SOLUTION SCHEME BASED ON T.C.E.E., E1.18.     C   21
        REAL KFAC,MAXDT,KVFAC                                        C   30
        INTEGER CURVE,TPRINT,CTYPE,ATYPE,BTYPE,BARY,UNITS,BB4,CTYBLD C   40
        DIMENSION NBLD(300),NA1(300),NAS(300),NB1(300),CBLD(300),    C   50
       1          CTYBLD(300)                                        C   60
        DIMENSION RF(50),AA1(50),BB1(50),BB2(50),BB3(50),BB4(50)     C   70
        DIMENSION ATYPE(50),BTYPE(50)                                C   80
        INTEGER TITLE1(14),TITLE2(14),CAPCUR(650),SINCAP,STRCAP      C   90
        DIMENSION NPAIRS(20),CURVE(20),X(100),Y(100)                 C  100
        DIMENSION T(650),C(4000),NA(4000),NB(4000),Q(650),CTYPE(4000) C  110
        DIMENSION NBCNT(650),TP(650),CAP(650),STAB(650),NQ(20),NQPRS(20), C  120
       1TM(100),QT(100)                                             C  130
        COMMON /BLK1/ T,C,NA,NB,Q/BLK2/NPAIRS,X,Y/BLK3/RF,CTYPE,CURVE C  140
        COMMON /BLK4/ TZ,SIGMA,TFAC,TCON,KFAC,KVFAC,FLFAC,DFAC,HFAC  C  150
        COMMON /BLK5/ AA1,BB1,BB2,BB3,BB4,ATYPE,BTYPE                C  160
        COMMON /BLK6/ NBCNT                                          C  170
        COMMON /BLK7/TP,CAP,STAB,CAPCUR,NQ,NQPRS,TM,QT               C  180
        NTPRIN = TPRINT                                              C  190
        TIME = 0.0                                                   C  200
        ISTEP = 0                                                    C  210
        STASV = 0.0                                                  C  220
C       PRINT TEMPS AT ZERO TIME                                     C  230
        CALL TEMPRT (NN,0,0.0,3)                                     C  240
10      J2 = 0                                                       C  250
C       START NEW TIME                                              C  260
        TIME = TIME + DELTM                                          C  270
        IF (TIME .GT. RTM) GO TO 140                                 C  280
        ISTEP = ISTEP + 1                                           C  290
        M1 = 1                                                       C  300
        K1 = 1                                                       C  310
        DO 110 I = 1,NN                                              C  320
        IF (NA(K1) .LT. 0) GO TO 100                                 C  330
        J2 = K1 - 1 + NBCNT(I)                                       C  340
C       CALC SUM OF CONDUCTANCES, SUM OF TEMP*COND                   C  350
        CSUM = 0.0                                                   C  360
        CTSUM = 0.0                                                  C  370
        DO 70 J = K1,J2                                              C  380
        N = IABS(NB(J))                                              C  390
```

```
         IF (CTYPE(J)) 20,30,40                                    C  400
20       IF (IRADFLG .EQ. 1) CALL REX (I,N,IRCNT,C(J),CC)          C  410
         IF (IRADFLG .EQ. 0) CALL RCOND (I,N,C(J),CC)              C  420
         GO TO 60                                                  C  430
30       CC = C(J)                                                 C  440
         GO TO 60                                                  C  450
40       IF (CTYPE(J) .GT. 100) GO TO 50                           C  460
         LIM1 = CTYPE(J)                                           C  470
         CALL COND (CURVE(LIM1),C(J),T(I),T(N),CC)                 C  480
         GO TO 60                                                  C  490
50       CALL CONV (J,CC,RE,RKINV,RLX)                             C  500
60       CSUM = CSUM + CC                                          C  510
         CTSUM = CTSUM + CC * T(N)                                 C  520
70       CONTINUE                                                  C  530
C        GET CAPCITANCE AND POWER DISSIP. FOR NODE I              C  540
         IF (NQCRV .EQ. 0) GO TO 80                                C  550
         CALL QP (I,TIME,NQCRV,QPI)                                C  560
         GO TO 90                                                  C  570
80       QPI = Q(I)                                                C  580
90       CALL CAPC (I,RDVCP)                                       C  590
         DTDCA = DELTM / RDVCP                                     C  600
C        CALC STABILITY CONSTANT                                   C  610
         STAB(I) = DTDCA * CSUM                                    C  620
C        CALC TEMP AT TIME STEP                                    C  630
         TP(I) = T(I) * (1.0 - STAB(I)) + DTDCA * CTSUM + DTDCA * QPI   C  640
         IF (STAB(I) .GT. STASV) STASV = STAB(I)                   C  650
100      K1 = K1 + NBCNT(I)                                        C  660
110      CONTINUE    .                                             C  670
         K = 1                                                     C  680
         DO 130 J = 1,NN                                           C  690
         IF (NA(K) .LT. 0) GO TO 120                               C  700
         T(J) = TP(J)                                              C  710
120      K = K + NBCNT(J)                                          C  720
130      CONTINUE                                                  C  730
         IF (IRADFLG .EQ. 1) CALL SETBM (IRCNT)                    C  740
         IF (IRADFLG .EQ. 1) CALL RADIOS (IRCNT)                   C  750
         IF (ISTEP .NE. NTPRIN) GO TO 10                           C  760
         CALL TEMPRT (NN,ISTEP,TIME,3)                             C  770
         NTPRIN = NTPRIN + TPRINT                                  C  780
         GO TO 10                                                  C  790
140      IF (NPRINT .EQ. 0) GO TO 150                              C  800
C        RESET TO TIME AT LAST STEP                                C  810
         TIME = TIME - DELTM                                       C  820
         CALL NDET (IRADFLG,IRCNT,NCND,ISTEP,TIME,3)               C  830
150      CONTINUE                                                  C  840
         RETURN                                                    C  850
         END                                                       C  860-
         SUBROUTINE CAPC (II,RDVCP)                                D   10
         REAL KFAC,MAXDT,KVFAC                                     D   20
         INTEGER CURVE,TPRINT,CTYPE,ATYPE,BTYPE,BARY,UNITS,BB4,CTYBLD   D   30
         DIMENSION NBLD(300),NA1(300),NAS(300),NB1(300),CBLD(300),     D   40
        1          CTYBLD(300)                                    D   50
         DIMENSION RF(50),AA1(50),BB1(50),BB2(50),BB3(50),BB4(50) D   60
         DIMENSION ATYPE(50),BTYPE(50)                            D   70
         INTEGER TITLE1(14),TITLE2(14),CAPCUR(650),SINCAP,STRCAP  D   80
         DIMENSION T(650),C(4000),NA(4000),NB(4000),Q(650),CTYPE(4000)  D   90
         DIMENSION NBCNT(650),TP(650),CAP(650),STAB(650),NQ(20),NQPRS(20),  D  100
        1TM(100),QT(100)                                          D  110
         COMMON /BLK1/ T,C,NA,NB,Q/BLK2/NPAIRS,X,Y/BLK3/RF,CTYPE,CURVE   D  120
         COMMON /BLK4/ TZ,SIGMA,TFAC,TCON,KFAC,KVFAC,FLFAC,DFAC,HFAC     D  130
         COMMON /BLK5/ AA1,BB1,BB2,BB3,BB4,ATYPE,BTYPE            D  140
         COMMON /BLK6/ NBCNT                                      D  150
         COMMON /BLK7/TP,CAP,STAB,CAPCUR,NQ,NQPRS,TM,QT           D  160
         DIMENSION NPAIRS(20),CURVE(20),X(100),Y(100)            D  170
```

```
        J = CAPCUR(II)                                                D   180
        IF (J .NE. 0) CALL COND (CURVE(J),CAP(II),T(II),T(II),RDVCP)   D   190
        IF (J .EQ. 0) RDVCP = CAP(II)                                 D   200
C     CAPACITANCE IN K ARRAY CURVES USING TEMP OF NODE II             D   210
        RETURN                                                        D   220
        END                                                           D   230-
        SUBROUTINE QP (I,TIME,NQCRV,QPI)                              E    10
        REAL KFAC,MAXDT,KVFAC                                         E    20
        INTEGER CURVE,TPRINT,CTYPE,ATYPE,BTYPE,BARY,UNITS,CTYBLD      E    30
        DIMENSION NBLD(300),NA1(300),NAS(300),NB1(300),CBLD(300),     E    40
       1          CTYBLD(300)                                         E    50
        DIMENSION        AA1(50),BB1(50),BB2(50),BB3(50),BB4(50)      E    60
        DIMENSION ATYPE(50),BTYPE(50)                                 E    70
        INTEGER TITLE1(14),TITLE2(14),CAPCUR(650),SINCAP,STRCAP       E    80
        DIMENSION T(650),C(4000),NA(4000),NB(4000),Q(650)            E    90
        DIMENSION NBCNT(650),TP(650),CAP(650),STAB(650),NQ(20),       E   100
       1NQPRS(20),TM(100),QT(100)                                     E   110
        DIMENSION NPAIRS(20),CURVE(20),X(100),Y(100)                  E   115
        COMMON /BLK1/ T,C,NA,NB,Q/BLK2/NPAIRS,X,Y/BLK3/RF,CTYPE,CURVE E   120
        COMMON /BLK4/ TZ,SIGMA,TFAC,TCON,KFAC,KVFAC,FLFAC,DFAC,HFAC   E   130
        COMMON /BLK5/ AA1,BB1,BB2,BB3,BB4,ATYPE,BTYPE                 E   140
        COMMON/BLK6/NBCNT                                             E   150
        COMMON /BLK7/TP,CAP,STAB,CAPCUR,NQ,NQPRS,TM,QT                E   160
C     THIS SUBROUTINE RETURNS A HEAT DISSIPATION QPI FOR NODE I       E   170
C     AT THIS POINT, TIME DEPENDENT Q REQUIRED                       E   180
C     M1 IS CURVE NUMBER                                             E   190
C     DETERMINE FIRST/LAST VALUES NP1/NP2 OF I FOR TM(I),QT(I)       E   200
        M1 = 0                                                        E   210
        NP2 = 0                                                       E   220
C     FIND Q CURVE NO. FOR NODE I                                    E   230
        DO 10 J = 1,NQCRV                                             E   240
        IF (NQ(J) .EQ. I) M1 = J                                      E   250
10      CONTINUE                                                      E   260
C     IF NO CURVE FOUND, M1 IS STILL 0                               E   270
        IF (M1 .EQ. 0) GO TO 90                                       E   280
        DO 20 L = 1,M1                                                E   290
20      NP2 = NP2 + NQPRS(L)                                          E   300
        NP1 = NP2 - NQPRS(M1) + 1                                     E   310
        IF (TIME .GT. TM(NP1)) GO TO 30                               E   320
        QPI = QT(NP1)                                                 E   330
        GO TO 100                                                     E   340
30      NP1 = NP1 + 1                                                 E   350
        DO 60 J = NP1,NP2                                             E   360
        DT = TIME - TM(J)                                             E   370
        IF (DT) 80,40,50                                              E   380
40      QPI = QT(J)                                                   E   390
        GO TO 100                                                     E   400
50      IF (J .EQ. NP2) GO TO 70                                      E   410
60      CONTINUE                                                      E   420
        GO TO 100                                                     E   430
70      QPI = QT(NP2)                                                 E   440
        GO TO 100                                                     E   450
80      QPI = QT(J - 1) + ((QT(J) - QT(J - 1)) / (TM(J) - TM(J        E   460
       1     - 1))) *.(TIME - TM(J - 1))                              E   470
        GO TO 100                                                     E   480
90      QPI = 0.0                                                     E   490
100     CONTINUE                                                      E   500
        RETURN                                                        E   510
        END                                                           E   520-
        SUBROUTINE NEGNODE (NCONST,NCND)                              F    10
C     ASSIGN NEGATIVE TO NODE NA FOR CONSTANT TEMPERATURE NODES       F    20
        COMMON/BLK1/T(650),C(4000),NA(4000),NB(4000),Q(650)          F    30
        COMMON/BLK11/NCTEMP(50)                                       F    40
        DO 10 I = 1,NCONST                                            F    50
```

```
      ICK = NCTEMP(I)                                              F   60
      ICK = IABS(ICK)                                              F   70
      DO 10 J = 1,NCND                                             F   80
      N = IABS(NA(J))                                              F   90
      IF (N .EQ. ICK) NA(J) = - N                                 F  100
   10 CONTINUE                                                     F  110
      RETURN                                                       F  120
      END                                                          F  130-
      SUBROUTINE COUNT (ND,NC)                                     G   10
C     THIS SUBROUTINE COUNTS THE NO. OF NB FOR EACH NA             G   20
      COMMON /BLK1/ T,C,NA,NB,Q/BLK6/NBCNT                         G   30
      DIMENSION T(650),C(4000),NA(4000),NB(4000),Q(650),NBCNT(650) G   40
      J1 = 1                                                       G   50
      DO 40 I = 1,ND                                               G   60
      NCNT = 0                                                     G   70
      DO 20 J = J1,NC                                              G   80
      IF (NA(J) .EQ. NA(J1)) GO TO 10                             G   90
      GO TO 30                                                     G  100
   10 NCNT = NCNT + 1                                              G  110
   20 CONTINUE                                                     G  120
   30 NBCNT(I) = NCNT                                              G  130
      J1 = J1 + NCNT                                               G  140
   40 CONTINUE                                                     G  150
      RETURN                                                       G  160
      END                                                          G  170-
      SUBROUTINE TEMPRT (NN,LCT,RMXDT,MM)                          H   10
C     THIS SUBROUTINE PRINTS TEMPERATURES                         H   20
      COMMON/BLK1/T,C,NA,NB,Q                                      H   30
      DIMENSION T(650),C(4000),NA(4000),NB(4000),Q(650)           H   40
      IF (NN .LT. 6) GO TO 10                                     H   50
      GO TO 20                                                     H   60
   10 I1 = 1                                                       H   70
      IF (MM .EQ. 3) WRITE (2,100) LCT,RMXDT                       H   80
      IF (MM .NE. 3) WRITE (2,90) LCT,RMXDT                        H   90
      WRITE (2,110)                                                H  100
      GO TO (30,40,50,60,70),NN                                    H  110
   20 IREM = MOD(NN,6)                                             H  120
      IT = NN - IREM                                               H  130
      I1 = IT + 1                                                  H  140
      IF (MM .EQ. 3) WRITE (2,100) LCT,RMXDT                       H  150
      IF (MM .NE. 3) WRITE (2,90) LCT,RMXDT                        H  160
      WRITE (2,110)                                                H  170
      WRITE (2,180) (I,T(I),I=1,IT)                                H  180
      IF (IREM .EQ. 0) GO TO 80                                    H  190
      GO TO (30,40,50,60,70),IREM                                  H  200
   30 WRITE (2,130) (I,T(I),I=I1,NN)                               H  210
      GO TO 80                                                     H  220
   40 WRITE (2,140) (I,T(I),I=I1,NN)                               H  230
      GO TO 80                                                     H  240
   50 WRITE (2,150) (I,T(I),I=I1,NN)                               H  250
      GO TO 80                                                     H  260
   60 WRITE (2,160) (I,T(I),I=I1,NN)                               H  270
      GO TO 80                                                     H  280
   70 WRITE (2,170) (I,T(I),I=I1,NN)                               H  290
   80 WRITE (2,120)                                                H  300
   90 FORMAT(1H0,40X,7HLOOPCT=,I5,10X,6HMAXDT=,E11.4/)            H  310
  100 FORMAT(1H0,40X,7HLOOPCT=,I5,10X,5HTIME=,E11.4/)             H  320
  110 FORMAT(1H ,54X,12HTEMPERATURES/)                           H  330
  120 FORMAT(1H0)                                                  H  340
  130 FORMAT(1H ,4X,2HT(,I3,2H)=,E10.4)                           H  350
  140 FORMAT(1H ,4X,2HT(,I3,2H)=,E10.4,4X,2HT(,I3,2H)=,E10.4)     H  360
  150 FORMAT(1H ,4X,2HT(,I3,2H)=,E10.4,4X,2HT(,I3,2H)=,E10.4,     H  370
     1            4X,2HT(,I3,2H)=,E10.4)                          H  380
  160 FORMAT(1H ,4X,2HT(,I3,2H)=,E10.4,4X,2HT(,I3,2H)=,E10.4,     H  390
```

```
      1                 4X,2HT(,I3,2H)=,E10.4,4X,2HT(,I3,2H)=,E10.4)        H   400
170   FORMAT(1H ,4X,2HT(,I3,2H)=,E10.4,4X,2HT(,I3,2H)=,E10.4,             H   410
      1                 4X,2HT(,I3,2H)=,E10.4,4X,2HT(,I3,2H)=,E10.4,       H   420
      2                 4X,2HT(,I3,2H)=,E10.4)                            H   430
180   FORMAT(1H ,4X,2HT(,I3,2H)=,E10.4,4X,2HT(,I3,2H)=,E10.4,             H   440
      1                 4X,2HT(,I3,2H)=,E10.4,4X,2HT(,I3,2H)=,E10.4,       H   450
      2                 4X,2HT(,I3,2H)=,E10.4,4X,2HT(,I3,2H)=,E10.4)       H   460
      RETURN                                                             H   470
      END                                                                H   480-
      SUBROUTINE COND (L2,C,A,B,CON)                                     I    10
C     THIS SUBROUTINE RETURNS A CONDUCTANCE USING K DETERMINED BY LINEAR I    20
C     INTERPOLATION (BY A CALL TO LNINTR).                               I    30
C     THE NUMBER ORIGINALLY STORED AT C(I) (C,A,OR A/L) IS NOT DISTURBED I    40
      COMMON /BLK2/ NPAIRS,X,Y                                           I    50
      DIMENSION NPAIRS(20),X(100),Y(100)                                I    60
C     DETERMINE FIRST VALUE NP1 AND LAST VALUE NP2 OF I FOR X(I),Y(I)    I    70
      NP2 = 0                                                            I    80
      DO 10 L = 1,L2                                                     I    90
10    NP2 = NP2 + NPAIRS(L)                                              I   100
      NP1 = NP2 - NPAIRS(L2) + 1                                         I   110
C     DETERMINE AVE TEMP OF NODES A AND B                               I   120
      XI = 0.5 * (A + B)                                                 I   130
      CALL LNINTR (NP1,NP2,XI,YI)                                        I   140
      CON = C * YI                                                       I   150
      RETURN                                                             I   160
      END                                                                I   170-
      SUBROUTINE LNINTR (J1,J2,XI,YI)                                   J    10
C     THIS SUBROUTINE PERFORMS A LINEAR INTERPOLATION AT XI TO RETURN YI J    20
C     FIRST X,Y PAIR IS X(J1),Y(J1)                                     J    30
C     LAST  X,Y PAIR IS X(J2),Y(J2)                                     J    40
C     IF  XI LESS THAN OR EQUALS X(J1), SET YI=Y(J1)                     J    50
C     IF  XI GREATER THAN OR EQUALS X(J2), SET YI=Y(J2)                  J    60
      COMMON /BLK2/ NPAIRS,X,Y                                           J    70
      DIMENSION NPAIRS(20),X(100),Y(100)                                J    80
      IF (XI .GT. X(J1)) GO TO 10                                        J    90
      YI = Y(J1)                                                         J   100
      GO TO 70                                                           J   110
10    J1 = J1 + 1                                                        J   120
      DO 40 J = J1,J2                                                    J   130
      DX = XI - X(J)                                                     J   140
      IF (DX) 60,20,30                                                   J   150
20    YI = Y(J)                                                          J   160
      GO TO 70                                                           J   170
30    IF (J .EQ. J2) GO TO 50                                            J   180
      GO TO 40                                                           J   190
40    CONTINUE                                                           J   200
      GO TO 70                                                           J   210
50    YI = Y(J2)                                                         J   220
      GO TO 70                                                           J   230
60    YI = Y(J - 1) + ((Y(J) - Y(J - 1)) / (X(J) - X(J - 1))) * (XI     J   240
      1    - X(J - 1))                                                   J   250
70    RETURN                                                             J   260
      END                                                                J   270-
      SUBROUTINE RCOND (II,NN,A,TRUEC)                                  K    10
C     THIS SUBROUTINE CALCULATES THE RADIATION CONDUCTOR FROM T(I) TO T( K    20
      REAL KFAC,KVFAC,FLFAC                                              K    30
      COMMON /BLK1/ T,C,NA,NB,Q/BLK3/RF,CTYPE,CURVE                      K    40
      COMMON /BLK4/ TZ,SIGMA,TFAC,TCON,KFAC,KVFAC,FLFAC,DFAC,HFAC        K    50
      DIMENSION T(650),C(4000),NA(4000),NB(4000),Q(650),RF(50)          K    60
      DIMENSION CTYPE(4000),CURVE(20)                                    K    70
      T1 = T(II) + TZ                                                    K    80
      T2 = T(NN) + TZ                                                    K    90
      T3 = T1 / 100.0                                                    K   100
      T4 = T2 / 100.0                                                    K   110
```

```
      TRUEC = SIGMA * A * (T3 ** 3 + (T3 ** 2) * T4 + T3 * (T4 ** 2)    K   120
     1      + T4 ** 3) / 100.0                                          K   130
      RETURN                                                            K   140
      END                                                              K   150-
      SUBROUTINE CONV (JJ,CNDD,RE,KINV,L)                              L    10
C     THIS SUBROUTINE CALCULATES CONVECTIVE CONDUCTORS                 L    20
C     NB(JJ) IS FLUID NODE                                            L    30
      DIMENSION T(650),C(4000),NA(4000),NB(4000),Q(650)               L    40
      DIMENSION RF(50),CTYPE(4000),CURVE(20)                          L    50
      DIMENSION AA1(50),BB1(50),BB2(50),BB3(50),BB4(50)               L    60
      DIMENSION ATYPE(50),BTYPE(50)                                   L    70
      INTEGER CTYPE,CURVE,ATYPE,BTYPE,BBTEST,AATEST,BB4               L    80
      REAL L,KINV,KA                                                  L    90
      COMMON /BLK1/ T,C,NA,NB,Q/BLK4/TZ,SIGMA,TFAC,TCON,KFAC,KVFAC,   L   100
     1      FLFAC,DFAC,HFAC/BLK3/RF,CTYPE,CURVE/BLK5/AA1,BB1,BB2,      L   110
     2      BB3,BB4,ATYPE,BTYPE                                        L   120
C     NEG(SIGN) NB(JJ) MAINTAINS NB AS AIR NODE                       L   130
      IF (NB(JJ)) 10,10,20                                            L   140
10    N1 = IABS(NA(JJ))                                               L   150
      N2 = IABS(NB(JJ))                                               L   160
      GO TO 30                                                        L   170
20    N1 = IABS(NB(JJ))                                               L   180
      N2 = IABS(NA(JJ))                                               L   190
30    TAVE = 0.5 * (T(N1) + T(N2))                                    L   200
      IF (CTYPE(JJ) .LT. 201) GO TO 170                               L   210
      IF (CTYPE(JJ) .GT. 300) GO TO 140                               L   220
      BBTEST = CTYPE(JJ) - 200                                        L   230
      VEL = BB1(BBTEST)                                               L   240
      D = BB2(BBTEST)                                                 L   250
      NBT = BTYPE(BBTEST)                                             L   260
      GO TO (40,50,60,100,60,100),NBT                                 L   270
C     LAMINAR FLOW - DUCT                                             L   280
40    L = BB3(BBTEST)                                                 L   290
      CALL KAIR (T(N2),KA)                                            L   300
      CALL PRAIR (T(N2),PRDT)                                         L   310
      CALL KV (T(N2),KINV)                                            L   320
      RE = VEL * D / KINV                                             L   330
      CALL LAMSD (PRDT,RE,D,L,KA,T(N2),T(N1),H)                       L   340
      GO TO 180                                                       L   350
C     TURBULENT FLOW IN A DUCT                                        L   360
50    CALL KAIR (T(N2),KA)                                            L   370
      CALL KV (T(N2),KINV)                                            L   380
      CALL PRAIR (T(N2),PRDT)                                         L   390
C     H FOR FULLY DEVELOPED FLOW                                      L   400
      L = BB3(BBTEST)                                                 L   410
      V = BB1(BBTEST)                                                 L   420
      D = BB2(BBTEST)                                                 L   430
      RE = V * D / KINV                                               L   440
      CALL TURBD (KA,KINV,VEL,T(N2),T(N1),D,HD)                       L   450
C     SHORT DUCT CORRECTION                                           L   460
C     T.C.E.E., E2.39                                                 L   461
      IF (L / D .GT.20.) H = HD * (1.0 + 6.0 * D / L)                 L   470
C     T.C.E.E., E2.38                                                 L   471
      IF (L / D .EQ.20.                                               L   480
     1    .OR. L / D .LT.20.) H = HD * (1.0 +1.68 * ((D / L) ** 0.58))  L   490
      GO TO 180                                                       L   500
C     LAMINAR FLOW OVER FLAT PLATE                                    L   510
60    CALL KAIR (TAVE,KA)                                             L   520
      CALL KV (TAVE,KINV)                                             L   530
      CALL PRAIR (TAVE,PRDT)                                          L   540
      IF (NBT .EQ. 5) GO TO 70                                        L   550
C     H AVERAGED OVER L. T.C.E.E., E2.20, E2.21, E2.22                L   560
      L = BB3(BBTEST)                                                 L   570
      RE = VEL * L / KINV                                             L   580
```

```
           H = 0.664 * KA * (SQRT(RE)) * (PRDT ** 0.33) / L           L   590
           GO TO 180                                                  L   600
C          H LOCAL. T.C.E.E., E2.20                                   L   610
C          NST IS FIRST UPSTREAM NODE NO. OF PLATE CONTAINING NODE NO. NA   L   620
70         NST = BB4(BBTEST)                                          L   630
           DX = BB3(BBTEST)                                           L   640
C          RESET OF N1,N2                                             L   650
           IF (NA(JJ) .LT. 0) GO TO 80                                L   660
           N1 = IABS(NA(JJ))                                          L   670
           N2 = IABS(NB(JJ))                                          L   680
           GO TO 90                                                   L   690
80         N1 = IABS(NB(JJ))                                          L   700
           N2 = IABS(NA(JJ))                                          L   710
90         RNST = NST                                                 L   720
           RN1 = N1                                                   L   730
           X = (RN1 - RNST + 0.5) * DX                                L   740
           L = X                                                      L   750
           RE = VEL * X / KINV                                        L   760
           H = 0.332 * KA * (SQRT(RE)) * (PRDT ** 0.33) / X           L   770
           GO TO 180                                                  L   780
C          TURBULENT FLOW OVER A FLAT PLATE                           L   790
100        CALL KAIR (TAVE,KA)                                        L   800
           CALL KV (TAVE,KINV)                                        L   810
           CALL PRAIR (TAVE,PRDT)                                     L   820
           IF (NBT .EQ. 6) GO TO 110                                  L   830
           L = BB3(BBTEST)                                            L   840
C          H AVERAGED OVER L. T.C.E.E., TABLE 10-5                    L   850
           RE = VEL * L / KINV                                        L   860
           H = 0.036 * KA * (PRDT ** 0.33) * (RE ** 0.8) / L          L   870
           GO TO 180                                                  L   880
C          H LOCAL                                                    L   899
110        NST = BB4(BBTEST)                                          L   900
           DX = BB3(BBTEST)                                           L   910
C          RESET OF N1,N2                                             L   920
           IF (NA(JJ) .LT. 0) GO TO 120                               L   930
           N1 = IABS(NA(JJ))                                          L   940
           N2 = IABS(NB(JJ))                                          L   950
           GO TO 130                                                  L   960
120        N1 = IABS(NB(JJ))                                          L   970
           N2 = IABS(NA(JJ))                                          L   980
130        RNST = NST                                                 L   990
           RN1 = N1                                                   L  1000
           X = (RN1 - RNST + 0.5) * DX                                L  1010
           L = X                                                      L  1020
C          H LOCAL. T.C.E.E., TABLE 10-5                              L  1021
           RE = VEL * X / KINV                                        L  1030
           H = 0.0288 * KA * (PRDT ** 0.3333) * (RE ** 0.8) / X       L  1040
           GO TO 180                                                  L  1050
C          NST IS FIRST UPSTREAM NODE NO. OF PLATE CONTAINING NO. NA  L  1060
C          FLUID NODES                                                L  1070
140        IF (CTYPE(JJ) .GT. 400) CALL PRESS (JJ,CNDD)               L  1080
           IF (CTYPE(JJ) .GT. 400) GO TO 190                          L  1090
           IF (CTYPE(JJ) .GT. 310) GO TO 150                          L  1100
           CALL RHOCPA (TAVE,RHOCP)                                   L  1110
C          FLUID OTHER THAN AIR - RHOCP IN ARRAY -                    L  1120
C                UNITS=0 - BTU/CUFT DEG F                             L  1130
C                UNITS=1 - CAL/CUCM DEG C                             L  1140
C                UNITS=2 - CAL/CUCM DEG C                             L  1150
C                        - G IN COND. INPUT -                         L  1160
C                UNITS=0 - CUFT/MIN                                   L  1170
C                UNITS=1 - CUCM/SEC                                   L  1180
C                UNITS=2 - CUFT/MIN                                   L  1190
           GO TO 160                                                  L  1200
150        LIM1 = CTYPE(JJ)                                           L  1210
```

```
        LIM1 = LIM1 - 310                                               L 1220
        CALL COND (CURVE(LIM1),1.0,T(N1),T(N2),RHOCP)                    L 1230
C       REMOVE 62.43 FROM FLFAC FOR UNITS=1,2                           L 1240
C       EXACT NO. TO COMPARE WITH FLFAC IS 60.0, USE 50.0              L 1250
        COR = 1.0                                                       L 1260
        IF (FLFAC .LT. 50.0) COR = 62.43                                L 1270
        RHOCP = RHOCP * COR                                             L 1280
C       C(I) IS MASS FLOW RATE CUFT/MIN OR CUCM/SEC                    L 1290
160     H = RHOCP * FLFAC                                               L 1300
        GO TO 180                                                       L 1310
C       NATURAL CONVECTION                                              L 1320
170     AATEST = CTYPE(JJ) - 100                                        L 1330
        L = AA1(AATEST)                                                 L 1340
        NAT = ATYPE(AATEST)                                             L 1350
        CALL FREECV (TAVE,N1,N2,L,NAT,H)                                L 1360
180     CNDD = H * C(JJ)                                                L 1370
190     CONTINUE                                                        L 1380
        RETURN                                                          L 1390
        END                                                             L 1400-
        SUBROUTINE SMDEV (NAT,T1,T2,P,H)                                M   10
C       THIS SUBROUTINE CALCULATES THE NATURAL CONVECTION H FOR         M   20
C       ANY ORIENTATION OF A SMALL DEVICE                              M   30
C       THE FORMULA IS BASED ON EXPERIMENT. T.C.E.E., E2.14            M   40
        REAL KFAC,KVFAC                                                 M   50
        COMMON/BLK4/TZ,SIGMA,TFAC,TCON,KFAC,KVFAC,FLFAC,DFAC,HFAC       M   60
C       H UNITS ARE BTU/(HR-FTSQ-DEG F) FOR TFAC,LFAC,HFAC ALL 1.0     M   70
        DT = (T1 - T2) * TFAC/1.8                                       M   80
        IF (DT .EQ. 0.0) DT = 0.0001 * T1                              M   90
        DT = ABS(DT)                                                    M  100
        PP = P * DFAC * 12.0                                            M  110
        IF(NAT.EQ.6) H=((DT/PP)**0.35)*0.0022*1761.0/(2.54**2)          M  120
        IF(NAT.EQ.7) H=((DT/PP)**0.33)*0.0018*1761.0/(2.54**2)          M  121
        IF(NAT.EQ.8) H=((DT/PP)**0.33)*0.0009*1761.0/(2.54**2)          M  122
        H = H * HFAC                                                    M  130
        RETURN                                                          M  140
        END                                                             M  150-
        SUBROUTINE LAMSD (PR,RED,D,L,K,TB,TS,H)                         N   10
C       THIS SUBROUTINE CALCULATES THE HEAT TRANSFER COEFFICIENT FOR AIR N  20
C       FOR LAMINAR FLOW IN A DUCT USING T.C.E.E., E2.34               N   30
C       AND CORRECTION OF T.C.E.E. REF.7, EQN. 8-29.                   N   40
C       EVALUATE PHYSICAL PROP. AT BULK MEAN TEMP                      N   50
C       REFERENCE TEMP - BULK MEAN TEMP                                N   60
        REAL K,L,KFAC,KVFAC                                            N   70
        COMMON /BLK4/ TZ,SIGMA,TFAC,TCON,KFAC,KVFAC,FLFAC,DFAC,HFAC     N   80
        T1 = TS * TFAC + TCON                                           N   90
        T2 = TB * TFAC + TCON                                           N  100
        TC = (T2 + 460.0) / (T1 + 460.0)                               N  110
        IF (T1 - T2) 10,10,20                                          N  120
C       GAS COOLING IN A TUBE                                          N  130
10      TCEXP = 0.08                                                   N  140
        GO TO 30                                                       N  150
C       GAS HEATING IN A TUBE                                          N  160
20      TCEXP = 0.25                                                   N  170
30      TC = TC ** TCEXP                                               N  180
        X = (RED * PR * D / L)                                         N  190
        Y= 3.66 + (0.104*X)/(1.0 + 0.016*X**0.8)                       N  200
50      H = Y * K * TC / D                                             N  270
        RETURN                                                         N  280
        END                                                            N  290-
        SUBROUTINE TURBD (KA,KINV,VEL,TB,TS,DIA,H)                     O   10
C       THIS SUBROUTINE CALCULATES THE HEAT TRANSFER COEFFICIENT       O   20
C       FOR FULLY DEVELOPED TURBULENT FLOW IN A DUCT. T.C.E.E., E2.37. O   30
        COMMON /BLK4/ TZ,SIGMA,TFAC,TCON,KFAC,KVFAC,FLFAC,DFAC,HFAC     O   40
        REAL KA,KINV,KFAC,KVFAC                                        O   50
```

```
        T1 = TS • TFAC + TCON                                            0    60
        T2 = TB • TFAC + TCON                                            0    70
        TC = (T2 + 460.0) / (T1 + 460.0)                                0    80
        IF (T1 - T2) 10,10,20                                           0    90
C       GAS  COOLING IN A TUBE                                          0   100
10      TCEXP = 0.15                                                    0   110
        GO TO 30                                                        0   120
C       GAS HEATING IN A TUBE                                           0   130
20      TCEXP = 0.575                                                   0   140
30      TC = TC ** TCEXP                                                0   150
        D = DIA                                                         0   160
        RE = VEL • D / KINV                                             0   170
        H = TC • 0.023 • (RE ** 0.8) • KA / D                          0   180
        RETURN                                                          0   190
        END                                                             0   200-
        SUBROUTINE KAIR (X,Y)                                           P    10
C       THIS SUBROUTINE CALCULATES THE THERMAL CONDUCTIVITY OF AIR      P    20
C       BETWEEN 0 DEG F AND 1000 DEG F                                  P    30
        REAL KFAC,KVFAC                                                 P    40
        COMMON /BLK4/ TZ,SIGMA,TFAC,TCON,KFAC,KVFAC,FLFAC,DFAC,HFAC      P    50
C       X-DEG F,TFAC=1,TCON=0                                           P    60
C       X-DEGC,TFAC=9/5,TCON=32                                         P    70
        T = TFAC • X + TCON                                             P    80
        Y = 1.332594267E - 02 + (1.981727295E - 09 • T + 2.061105698E   P    90
     1       - 05) • T                                                  P   100
        Y = Y • KFAC                                                    P   110
        RETURN                                                          P   120
        END                                                             P   130-
        SUBROUTINE KV (X,Y)                                             Q    10
C       THIS SUBROUTINE CALCULATES THE KINEMATIC VISCOSITY OF AIR       Q    20
C       FOR TEMPERATURES BETWEEN 0 DEG F AND 1000 DEG F                 Q    30
        REAL KFAC,KVFAC                                                 Q    40
        COMMON /BLK4/ TZ,SIGMA,TFAC,TCON,KFAC,KVFAC,FLFAC,DFAC,HFAC      Q    50
C       X-DEG F,TFAC=1,TCON=0                                           Q    60
C       X-DEG C,TFAC=9/5,TCON=32                                        Q    70
        T = TFAC • X + TCON                                             Q    80
        Y = 0.1271812987 + (2.717815575E - 07 • T + 5.195566004E - 04) • T  Q    90
        EX = 1.0E - 03                                                  Q   100
        Y = Y • EX                                                      Q   110
        Y = KVFAC • Y                                                   Q   120
        RETURN                                                          Q   130
        END                                                             Q   140-
        SUBROUTINE PRAIR (X,Y)                                          R    10
C       THIS SUBROUTINE CALCULATES THE PRANDTL NUMBER Y                 R    20
C       OF AIR FOR TEMPERATURES FROM 0 DEG F TO 1000 DEG F              R    30
        COMMON /BLK4/ TZ,SIGMA,TFAC,TCON,KFAC,KVFAC,FLFAC,DFAC,HFAC      R    40
C       X-DEG F ,TFAC=1,TCON=0                                          R    50
C       X-DEG C ,TFAC=9/5,TCON=32                                       R    60
        T = TFAC • X + TCON                                             R    70
        Y = 0.728316847 + ((0.1352944828E - 10 • T - 7.337277187E       R    80
     1       - 08) • T - 7.312221844E - 05) • T                         R    90
        RETURN                                                          R   100
        END                                                             R   110-
        SUBROUTINE GRAS (L,TAVE,TS,TB,GR)                               S    10
C       THIS SUBROUTINE CALCULATES THE GRASHOF NO. FOR AIR              S    20
        COMMON /BLK4/ TZ,SIGMA,TFAC,TCON,KFAC,KVFAC,FLFAC,DFAC,HFAC      S    30
        REAL L                                                          S    40
        T = TAVE • TFAC + TCON                                          S    50
        Y1 = 6.621367986 + (( - 9.699265237E - 10 • T + 2.736451029E    S    60
     1       - 06) • T - 3.968288304E - 03) • T                         S    70
        Y = 10.0 ** Y1                                                  S    80
        X = L • DFAC                                                    S    90
        TDIF = TS - TB                                                  S   100
        TDIF = ABS(TDIF)                                                S   110
```

```
      IF (TDIF .EQ. 0.0) TDIF = TDIF + 0.001 * TS          S  120
      GR = Y * (X ** 3) * (TDIF) * TFAC                     S  130
      RETURN                                                S  140
      END                                                   S  150-
      SUBROUTINE RHOCPA (X,Y)                               T   10
C     THIS SUBROUTINE CALCULATES THE PRODUCT OF DENSITY AND T   20
C     SPECIFIC HEAT(CONST PRESS) FOR AIR                    T   30
C     FOR T DEG F, TCON=0.0, TFAC=1.0, Y=BTU/CUFT-DEG F     T   40
      COMMON /BLK4/ TZ,SIGMA,TFAC,TCON,KFAC,KVFAC,FLFAC,DFAC,HFAC  T  50
      REAL KFAC,KVFAC                                       T   60
      T = TFAC * X + TCON                                   T   70
      Y = 2.048207889E - 02 + (( - 1.69669411E - 11 * T     T   80
     1   + 4.015439168E - 08) * T - 3.667583513E - 05) * T  T   90
      RETURN                                                T  100
      END                                                   T  110-
      SUBROUTINE FREECV (TAVE,N1,N2,L,NAT,HTCOEF)           U   10
C     THIS SUBROUTINE CALCULATES NATURAL CONVECTION         U   20
C     HEAT TRANSFER COEFFICIENTS                            U   30
      REAL L,KA                                             U   40
      COMMON /BLK1/ T,C,NA,NB,Q                             U   50
      DIMENSION T(650),C(4000),NA(4000),NB(4000),Q(650)     U   60
      CALL PRAIR (TAVE,PR)                                  U   70
      CALL GRAS (L,TAVE,T(N1),T(N2),GR)                     U   80
      CALL KAIR (TAVE,KA)                                   U   90
      PROD = PR * GR                                        U  100
      GO TO (10,30,50,60,110,150,150,150),NAT               U  110
C     VERTICAL FLAT PLATE OR CYLINDER                       U  120
10    IF (PROD .GT. 1.0E09) GO TO 20                        U  130
C     T.C.E.E., E2.7, TABLE 2-2 .                           U  131
      HTCOEF = (KA / L) * 0.590 * ((PROD) ** 0.25)          U  140
      GO TO 170                                             U  150
20    HTCOEF = (KA / L) * 0.130 * ((PROD) ** 0.33)          U  160
      GO TO 170                                             U  170
C     HORIZONTAL FLAT PLATE OR CYLINDER,                    U  180
C     HEATED SIDE FACING UP, COOLED SIDE FACING DOWN        U  190
30    IF (GR .GT. 8.0E06) GO TO 40                          U  200
      HTCOEF = (KA / L) * 0.54 * ((PROD) ** 0.25)           U  210
      GO TO 170                                             U  220
40    HTCOEF = (KA / L) * 0.15 * ((PROD) ** 0.33)           U  230
      GO TO 170                                             U  240
C     HORIZONTAL FLAT PLATE OR CYLINDER                     U  250
C     HEATED SIDE FACING DOWN, COOLED SIDE FACING UP        U  260
50    HTCOEF = (KA / L) * 0.27 * ((PROD) ** 0.25)           U  270
      GO TO 170                                             U  280
C     PARRALLEL HORIZONTAL PLATES. SEE T.C.E.E., TABLE 10-4. U  290
C     GR MODIFIED BY FACTOR OF 2.0 TO APPROX. ACCOUNT FOR   U  300
C     NODE BETWEEN PLATES SPLITTING DT                      U  310
60    GR = GR * 2.0                                         U  320
      IF (GR .GT. 1.58E3) GO TO 70                          U  330
      PROD = 1.0                                            U  340
      GO TO 100                                             U  350
70    IF (GR .GT. 5.62E4) GO TO 80                          U  360
      PROD = 0.0731 * GR ** 0.355                           U  370
      GO TO 100                                             U  380
80    IF (GR .GT. 2.0E5) GO TO 90                           U  390
      PROD = 3.63                                           U  400
      GO TO 100                                             U  410
90    PROD = 0.0426 * GR ** 0.37                            U  420
C     HTCOEF ALSO MODIFIED BY 2.0                           U  430
100   HTCOEF = 2.0 * (KA / L) * PROD                        U  440
      GO TO 170                                             U  450
C     PARALLEL VERTICAL PLATES.                             U  460
110   GR = GR * 2.0                                         U  470
      IF (GR .GT. 6.03) GO TO 120                           U  480
```

```
        PROD = 1.0                                                    U   490
        GO TO 160                                                     U   500
120     IF (GR .GT. 5.62E4) GO TO 130                                 U   510
        PROD = 0.0305 * GR ** 0.402                                   U   520
        GO TO 160                                                     U   530
130     IF (GR .GT. 2.055) GO TO 140                                  U   540
        PROD = 2.455                                                  U   550
        GO TO 160                                                     U   560
140     PROD = 0.0263 * GR ** 0.38                                    U   570
        GO TO 160                                                     U   580
150     CALL SMDEV (NAT,T(N1),T(N2),L,HTCOEF)                         U   590
        GO TO 170                                                     U   600
160     HTCOEF = 2.0 * (KA / L) * PROD                                U   610
170     RETURN                                                        U   620
        END                                                           U   630-
        SUBROUTINE NSCAN (NCND,IRCNT)                                 V    10
C       MULTI-SURFACE RADIATION EXCHANGE ROUTINE                      V    20
C       SCAN CTYPE FOR -2, FILL IF, JF, AM MATRIX                     V    30
        COMMON/BLK1/T(650),C(4000),NA(4000),NB(4000),Q(650)          V    40
        COMMON/BLK3/RF(50),CTYPE(4000),CURVE(20)                      V    50
        COMMON/BLK9/IF(2500),JF(2500),AM(50,50)                       V    60
        INTEGER CTYPE,CURVE                                           V    70
C       PLACE F*A FROM C(K) INTO AM MATRIX                            V    80
        J = 1                                                         V    90
        I = 1                                                         V   100
        II = 1                                                        V   110
        DO 10 K = 1,NCND                                              V   120
        IF (CTYPE(K) .NE.  - 2) GO TO 10                              V   130
        IF(II) = IABS(NA(K))                                          V   140
        JF(II) = IABS(NB(K))                                          V   150
        AM(I,J) = C(K)                                                V   160
        J = J + 1                                                     V   170
        II = II + 1                                                   V   180
C       IF J=IRCNT + 1 GO TO FIRST ELEMENT, NEXT ROW IN AM           V   190
        IF (J .EQ. IRCNT + 1) I = I + 1                               V   200
        IF (J .EQ. IRCNT + 1) J = 1                                   V   210
10      CONTINUE                                                      V   220
C       ARRANGE ELEMENTS OF EACH AM ROW IN ASCENDING ORDER OF NODE NO. V  230
        DO 40 I = 1,IRCNT                                             V   240
        J2 = IRCNT - 1                                                V   250
        DO 30 J = 1,J2                                                V   260
        K = (I - 1) * IRCNT + J                                       V   270
        L1 = J + 1                                                    V   280
        DO 20 L = L1,IRCNT                                            V   290
        M = (I - 1) * IRCNT + L                                       V   300
        IF (JF(M) .GT. JF(K)) GO TO 20                                V   310
        ASAVE = AM(I,J)                                               V   320
        AM(I,J) = AM(I,L)                                             V   330
        AM(I,L) = ASAVE                                               V   340
        JSAVE = JF(K)                                                 V   350
        JF(K) = JF(M)                                                 V   360
        JF(M) = JSAVE                                                 V   370
20      CONTINUE                                                      V   380
30      CONTINUE                                                      V   390
40      CONTINUE                                                      V   400
        RETURN                                                        V   410
        END                                                           V   420-
        SUBROUTINE AADJ (IRCNT)                                       W    10
C       MULTI-SURFACE RADIATION EXCHANGE ROUTINE                      W    20
C       CHANGE AM ELEMENTS FROM F*A TO TRUE AM VALUES                 W    30
C       AM IS MATRIX DEFINED FOLLOWING T.C.E.E., E3.15.               W    31
        COMMON/BLK9/IF(2500),JF(2500),AM(50,50)                       W    40
        J = 1                                                         W    50
        DO 20 L = 1,IRCNT                                             W    60
```

```
        N = IF(J)                                             W    70
        CALL AESRCH (N,IRCNT,AREA,EMIS)                       W    80
        FAC = (1.0 - EMIS) / AREA                             W    90
        DO 10 K = 1,IRCNT                                     W   100
        RKD = 0.0                                             W   110
        IF (L .EQ. K) RKD = 1.0                               W   120
        AM(L,K) = RKD - FAC * AM(L,K)                         W   130
10      CONTINUE                                              W   140
        J = L * IRCNT + 1                                     W   150
20      CONTINUE                                              W   160
        RETURN                                                W   170
        END                                                   W   180-
        SUBROUTINE AESRCH (N,IRCNT,AREA,EMIS)                 X    10
C       MULTI-SURFACE RADIATION EXCHANGE ROUTINE              X    20
C       SEARCH AND RETURN AREA, EMIS FOR NODE N               X    30
        COMMON/BLK8/IAE(50),EM(50),AR(50)                     X    40
        ITEST = 0                                             X    50
        DO 10 I = 1,IRCNT                                     X    60
        IF (IAE(I) .NE. N) GO TO 10                           X    70
        ITEST = 1                                             X    80
        AREA = AR(I)                                          X    90
        EMIS = EM(I)                                          X   100
        GO TO 20                                              X   110
10      CONTINUE                                              X   120
        IF (ITEST .EQ. 0) EMIS = 99.0                         X   130
20      RETURN                                                X   140
        END                                                   X   150-
        SUBROUTINE INVAM (IRCNT)                              Y    10
C       MULTI-SURFACE RADIATION EXCHANGE ROUTINE              Y    20
C       INVERT AM MATRIX                                      Y    30
        COMMON/BLK9/IF(2500),JF(2500),AM(50,50)               Y    40
        DIMENSION JC(50),VV(2)                                Y    50
        LLTEST = 0                                            Y    60
        VV(1) = 1.0                                           Y    70
        CALL MATINV (AM,50,50,IRCNT,IRCNT,LLTEST,JC,VV)       Y    80
        IF (LLTEST .NE. 1) GO TO 20                           Y    90
        WRITE (2,10)                                          Y   100
        STOP                                                  Y   110
10      FORMAT(22H MATRIX SOLUTION ERROR)                     Y   120
20      RETURN                                                Y   130
        END                                                   Y   140-
        SUBROUTINE SETBM (IRCNT)                              Z    10
C       MULTI-SURFACE RADIATION EXCHANGE ROUTINE              Z    20
C       SET UP BM MATRIX (EMIS*T**4*SIGMA)                    Z    30
C       BM IS MATRIX DEFINED FOLLOWING T.C.E.E., E3.15.       Z    31
        COMMON/BLK4/TZ,SIGMA,TFAC,TCON,KFAC,KVFAC,FLFAC,DFAC,HFAC   Z 40
        COMMON/BLK9/IF(2500),JF(2500),AM(50,50)               Z    50
        COMMON/BLK1/T(650),C(4000),NA(4000),NB(4000),Q(650)   Z    60
        COMMON/BLK10/BM(50),RJM(50)                           Z    70
        REAL KFAC,KVFAC                                       Z    80
        DO 10 I = 1,IRCNT                                     Z    90
C       FIND NODE NO. CORRESPONDING TO I IN FIRST NSURFS NUMBER OF JF   Z 100
        NN = JF(I)                                            Z   110
        CALL AESRCH (NN,IRCNT,AREA,EMIS)                      Z   120
        BM(I) = SIGMA * EMIS * ((T(NN) + TZ) ** 4) * 1.0E - 8 Z   130
10      CONTINUE                                              Z   140
        RETURN                                                Z   150
        END                                                   Z   160-
        SUBROUTINE RADIOS (IRCNT)                             AA   10
C       MULTI-SURFACE RADIATION EXCHANGE ROUTINE              AA   20
C       SOLVE FOR RADIOSITY MATRIX                            AA   30
        COMMON/BLK9/IF(2500),JF(2500),AM(50,50)               AA   40
        COMMON/BLK10/BM(50),RJM(50)                           AA   50
        REAL JSUM                                             AA   60
```

```
            DO 20 I = 1,IRCNT                                        AA   70
            JSUM = 0.0                                               AA   80
            DO 10 J = 1,IRCNT                                        AA   90
            JSUM = AM(I,J) * BM(J) + JSUM                            AA  100
      10    CONTINUE                                                 AA  110
            RJM(I) = JSUM                                            AA  120
      20    CONTINUE                                                 AA  130
            RETURN                                                   AA  140
            END                                                      AA  150-
            SUBROUTINE REX (N1,N2,IRCNT,FA,COND)                     AB   10
      C     MULTI-SURFACE RADIATION EXCHANGE ROUTINE                 AB   20
      C     CALCULATE CONDUCTANCE FOR MULTIPLE SURFACE RADIATION EXCHANGE  AB   30
      C     CONDUCTANCE IS T.C.E.E., E3.15.                          AB   31
            COMMON/BLK9/IF(2500),JF(2500),AM(50,50)                  AB   40
            COMMON/BLK1/T(650),C(4000),NA(4000),NB(4000),Q(650)      AB   50
            COMMON/BLK10/BM(50),RJM(50)                              AB   60
            T2 = T(N2)                                               AB   70
            T1 = T(N1)                                               AB   80
            TDIF = T1 - T2                                           AB   90
            IF (TDIF .EQ. 0.0) TDIF = TDIF + 0.0001 * T1             AB  100
      C     GET RADIOSITIES J1,J2                                    AB  110
            DO 10 I = 1,IRCNT                                        AB  120
            IF (N1 .EQ. JF(I)) RJ1 = RJM(I)                          AB  130
            IF (N2 .EQ. JF(I)) RJ2 = RJM(I)                          AB  140
      10    CONTINUE                                                 AB  150
            COND = FA * (RJ1 - RJ2) / TDIF                           AB  160
            RETURN                                                   AB  170
            END                                                      AB  180-
            SUBROUTINE MATINV (A,NC,NR,N,MC,LTEST,JC,V)              AC   10
      C     MATRIX INVERSION ROUTINE                                 AC   20
            DIMENSION A(NR,NC),JC(50),V(2)                           AC   30
            IW = V(1)                                                AC   40
            M = 1                                                    AC   50
            S = 1.                                                   AC   60
            L = N + (MC - N) * (IW / 4)                              AC   70
            KD = 2 - MOD(IW / 2,2)                                   AC   80
            IF (KD .EQ. 1) V(2) = 0.                                 AC   90
            KI = 2 - MOD(IW,2)                                       AC  100
            GO TO (10,30),KI                                         AC  110
      C     INITIALIZE JC FOR INVERSION                             AC  120
      10    DO 20 I = 1,N                                            AC  130
      20    JC(I) = I                                                AC  140
      C     SEARCH FOR PIVOT ROW                                    AC  150
      30    DO 160 I = 1,N                                           AC  160
            GO TO (50,40),KI                                         AC  170
      40    M = I                                                    AC  180
      50    IF (I .EQ. N) GO TO 100                                  AC  190
            X =  - 1.                                                AC  200
            DO 60 J = I,N                                            AC  210
            IF (X .GT. ABS(A(J,I))) GO TO 60                         AC  220
            X = ABS(A(J,I))                                          AC  230
            K = J                                                    AC  240
      60    CONTINUE                                                 AC  250
            IF (K .EQ. I) GO TO 100                                  AC  260
            S =  - S                                                 AC  270
            V(1) =  - V(1)                                           AC  280
            GO TO (70,80),KI                                         AC  290
      70    MU = JC(I)                                               AC  300
            JC(I) = JC(K)                                            AC  310
            JC(K) = MU                                               AC  320
      C     INTERCHANGE ROW I AND ROW K                             AC  330
      80    DO 90 J = M,L                                            AC  340
            X = A(I,J)                                               AC  350
            A(I,J) = A(K,J)                                          AC  360
```

```
90       A(K,J) = X                                                   AC 370
C        TEST FOR SINGULARITY                                         AC 380
100      IF (ABS(A(I,I)) .GT. 0.) GO TO 110                           AC 390
C        MATRIX IS SINGULAR                                           AC 400
         IF (KD .EQ. 1) V(1) = 0.                                     AC 410
         JC(1) = I - 1                                                AC 420
         LTEST = 1                                                    AC 430
         RETURN                                                       AC 440
110      GO TO (120,130),KD                                           AC 450
C        COMPUTE THE DETERMINANT                                      AC 460
120      IF (A(I,I) .LT. 0.) S = - S                                  AC 470
         V(2) = V(2) + ALOG(ABS(A(I,I)))                             AC 480
130      X = A(I,I)                                                   AC 490
         A(I,I) = 1.                                                  AC 500
C        REDUCTION OF THE I-TH ROW                                    AC 510
         DO 140 J = M,L                                               AC 520
         A(I,J) = A(I,J) / X                                          AC 530
140      CONTINUE                                                     AC 540
C        REDUCTION OF ALL REMAINING ROWS                              AC 550
         DO 160 K = 1,N                                               AC 560
         IF (K .EQ. I) GO TO 160                                      AC 570
         X = A(K,I)                                                   AC 580
         A(K,I) = 0.                                                  AC 590
         DO 150 J = M,L                                               AC 600
         A(K,J) = A(K,J) - X * A(I,J)                                AC 610
150      CONTINUE                                                     AC 620
160      CONTINUE                                                     AC 630
C        AX=B AND DET.(A) ARE NOW COMPUTED                            AC 640
         GO TO (170,220),KI                                           AC 650
C        PERMUTATION OF THE COLUMNS FOR MATRIX INVERSION              AC 660
170      DO 210 J = 1,N                                               AC 670
         IF (JC(J) .EQ. J) GO TO 210                                  AC 680
         JJ = J + 1                                                   AC 690
         DO 180 I = JJ,N                                              AC 700
         IF (JC(I) .EQ. J) GO TO 190                                  AC 710
180      CONTINUE                                                     AC 720
190      JC(I) = JC(J)                                                AC 730
         DO 200 K = 1,N                                               AC 740
         X = A(K,I)                                                   AC 750
         A(K,I) = A(K,J)                                              AC 760
200      A(K,J) = X                                                   AC 770
210      CONTINUE                                                     AC 780
220      JC(1) = N                                                    AC 790
         IF (KD .EQ. 1) V(1) = S                                      AC 800
         RETURN                                                       AC 810
         END                                                          AC 820-
         SUBROUTINE NDET (IRADFLG,IRCNT,NCND,LCT,RMXDT,MODE)          AD   10
C        THIS SUBROUTINE PRINTS NODE AND CONDUCTOR DETAILS            AD   20
         REAL L                                                       AD   30
         INTEGER CTYPE,BBTEST,AATEST,ATYPE,BTYPE,BB4                  AD   40
         COMMON /BLK1/ T,C,NA,NB,Q/BLK3/RF,CTYPE,CURVE/BLK5/AA1,BB1,BB2, AD   50
        1      BB3,BB4,ATYPE,BTYPE                                    AD   60
         COMMON/BLK7/TP(650),CAP(650),STAB(650),CAPCUR(650),NQ(20),  AD   70
        1NQPRS(20),TM(100),QT(100)                                   AD   80
         DIMENSION T(650),C(4000),NA(4000),NB(4000),Q(650),RF(50),AA1(50), AD   90
        1      BB1(50),BB2(50),BB3(50),BB4(50),CTYPE(4000),CURVE(20), AD  100
        2      ATYPE(50),BTYPE(50)                                   AD  110
         COMMON/BLK8/IAE(50),EM(50),AR(50)                           AD  120
         F(NAARG,NBARG) = (T(NAARG) - T(NBARG)) * CND                AD  130
         WRITE (2,170) LCT,RMXDT                                     AD  140
         NTEST = 0                                                   AD  150
         SUM2 = 0.0                                                  AD  160
         DO 160 I = 1,NCND                                           AD  170
         IF (NA(I) .EQ. NTEST) GO TO 20                              AD  180
```

```
         SUM = 0.0                                                        AD 190
         NTEST = NA(I)                                                    AD 200
         NNA = IABS(NA(I))                                                AD 210
         NNB = IABS(NB(I))                                                AD 220
         CALL CAPC (NNA,RDVCP)                                            AD 230
         WRITE (2,180) NA(I),T(NNA),Q(NNA),STAB(NNA),RDVCP               AD 240
         IF (NA(I) .LT. 0) WRITE (2,190)                                  AD 250
         IF (IRADFLG .EQ. 0) GO TO 10                                     AD 260
         CALL AESRCH (NNA,IRCNT,AREA,EMIS)                                AD 270
         IF (EMIS .GT. 2.0) GO TO 10                                      AD 280
         WRITE (2,210) AREA,EMIS                                          AD 290
  10     WRITE (2,200)                                                    AD 300
  20     NNA = IABS(NA(I))                                                AD 310
         NNB = IABS(NB(I))                                                AD 320
         IF (CTYPE(I)) 40,30,50                                           AD 330
  30     CND = C(I)                                                       AD 340
         GO TO 90                                                         AD 350
  C      RADIATION COND                                                   AD 360
  40     IF (CTYPE(I) .EQ.  - 2) CALL REX (NNA,NNB,IRCNT,C(I),CND)        AD 370
         IF (CTYPE(I) .EQ.  - 1) CALL RCOND (NNA,NNB,C(I),CND)            AD 380
         GO TO 90                                                         AD 390
  50     IF (CTYPE(I) .GT. 100) GO TO 60                                  AD 400
         LIM1 = CTYPE(I)                                                  AD 410
         CALL COND (CURVE(LIM1),C(I),T(NNA),T(NNB),CND)                  AD 420
         GO TO 90                                                         AD 430
  60     CALL CONV (I,CND,RE,RKINV,L)                                     AD 440
         IF (CTYPE(I) .LT. 201 .OR. CTYPE(I) .GT. 300) GO TO 90          AD 450
         BBTEST = CTYPE(I) - 200                                          AD 460
         NBT = BTYPE(BBTEST)                                              AD 470
         GO TO (70,80,70,80,70,80),NBT                                    AD 480
  C      BOUNDARY LAYER THICKNESS FOR LAMINAR FLOW. T.C.E.E., E2.24.      AD 490
  70     V = BB1(BBTEST)                                                  AD 500
         DEL = 5.0 * L / SQRT(V * L / RKINV)                             AD 510
         GO TO 90                                                         AD 520
  C      BOUNDARY LAYER THICKNESS FOR TURBULENT FLOW. T.C.E.E., E2.26.    AD 530
  80     V = BB1(BBTEST)                                                  AD 540
         DEL = 0.376 * L / ((V * L / RKINV) ** 0.2)                      AD 550
  90     FLUX = F(NNA,NNB)                                                AD 560
         SUM = SUM + FLUX                                                 AD 570
         IF (C(I) .EQ. 0.0) GO TO 150                                     AD 580
         H = CND / C(I)                                                   AD 590
         IF (CTYPE(I) .EQ. 0 .OR. CTYPE(I) .GT. 0) GO TO 100            AD 600
         WRITE (2,220) NB(I),CTYPE(I),C(I),CND,FLUX,H                    AD 610
         GO TO 150                                                        AD 620
  100    IF (CTYPE(I) .GT. 0) GO TO 110                                   AD 630
  C      CONSTANT CONDUCTOR                                               AD 640
         IF (C(I) .EQ. 0.0) GO TO 150                                     AD 650
         WRITE (2,230) NB(I),CTYPE(I),C(I),CND,FLUX                      AD 660
         GO TO 150                                                        AD 670
  110    IF (CTYPE(I) .GT. 100) GO TO 120                                 AD 680
  C      TEMP DEPENDENT COND                                              AD 690
         IF (C(I) .EQ. 0.0) GO TO 150                                     AD 700
         WRITE (2,240) NB(I),CTYPE(I),C(I),CND,FLUX                      AD 710
         GO TO 150                                                        AD 720
  120    IF (CTYPE(I) .GT. 200) GO TO 130                                 AD 730
  C      NATURAL CONVECTION COND                                          AD 740
         AATEST = CTYPE(I) - 100                                          AD 750
         NAT = ATYPE(AATEST)                                              AD 760
         WRITE (2,250) NB(I),CTYPE(I),NAT,C(I),CND,FLUX,H               AD 770
         GO TO 150                                                        AD 780
  130    IF (CTYPE(I) .GT. 300) GO TO 140                                 AD 790
  C      FORCED CONVECTION COND                                           AD 800
         IF (C(I) .EQ. 0.0) GO TO 150                                     AD 810
         WRITE (2,260) NB(I),CTYPE(I),NBT,C(I),CND,FLUX,RE,DEL,BB2(BBTEST), AD 820
```

```
          1H                                                           AD 830
          GO TO 150                                                    AD 840
C         FLUID CONDUCTOR                                              AD 850
140       IF (C(I) .EQ. 0.0 .AND. CTYPE(I) .LT. 401) GO TO 150         AD 860
          WRITE (2,270) NB(I),CTYPE(I),C(I),CND,FLUX                    AD 870
150       II = I + 1                                                   AD 880
          IF (NA(I) .EQ. NA(II)) GO TO 160                             AD 890
          WRITE (2,280) SUM                                            AD 900
          IF (NA(I) .LT. 0) GO TO 160                                  AD 910
          SUM2 = SUM2 + SUM - Q(NNA)                                   AD 920
160       CONTINUE                                                     AD 930
          WRITE (2,290) SUM2                                           AD 940
          RETURN                                                       AD 950
C                                                                      AD 960
170       FORMAT (1H0,40X,7HLOOPCT=,I5,10X,6HMAXDT=,E11.4/)            AD 970
180       FORMAT (1H0,4X,14HDETAIL OF NODE,I4,5X,12HTEMPERATURE=,E10.4,5X, AD 980
         1      6HPOWER=,E10.4,5X,20HSTABILITY CONSTANT =,E10.2,5X,4HCAP=, AD 990
         2 E10.4)                                                      AD1000
190       FORMAT(1H0,27X,35HTHIS IS A CONSTANT TEMPERATURE NODE)       AD1010
200       FORMAT(1H0,6X,4HNODE,2X,5HCTYPE,2X,5HCMODE,7X,1HC,           AD1020
         1      7X,11HCONDUCTANCE,8X,4HFLUX,8X,8HREYN NO.,3X,          AD1030
         2      11HBNDRY LAYER,5X,8HHYDR DIA,4X,13HHT TRANS COEF)      AD1040
210       FORMAT(1H0,27X,38HMULTI-SURFACE RADIATION EXCHANGE AREA=,    AD1050
         1      E10.4,6X,11HEMISSIVITY=,F5.3)                          AD1060
220       FORMAT(1H ,6X,I4,2X,I4,9X,E10.4,4X,E10.4,4X,E10.4,46X,E10.4) AD1070
230       FORMAT (1H ,6X,I4,2X,I4,9X,E10.4,4X,E10.4,4X,E10.4)          AD1080
240       FORMAT (1H ,6X,I4,2X,I4,9X,E10.4,4X,E10.4,4X,E10.4)          AD1090
250       FORMAT (1H ,6X,I4,2X,I4,3X,I3,3X,E10.4,4X,E10.4,4X,E10.4,46X, AD1100
         1      E10.4)                                                 AD1110
260       FORMAT (1H ,6X,I4,2X,I4,3X,I3,3X,E10.4,4X,E10.4,4X,E10.4,4X,E10:4, AD1120
         1      4X,E10.4,4X,E10.4,4X,E10.4)                            AD1130
270       FORMAT (1H ,6X,I4,2X,I4,9X,E10.4,4X,E10.4,4X,E10.4)          AD1140
280       FORMAT (1H ,42X,11HNET TOTAL =,E10.4)                        AD1150
290       FORMAT (1H+,76X,16HENERGY BALANCE =,E10.4/)                  AD1160
          END                                                         AD1170-
          SUBROUTINE BAL (NCND,IRCNT)                                  AE  10
C         THIS SUBROUTINE DETERMINES THE TOTAL SYSTEM ENERGY           AE  20
          REAL L                                                       AE  30
          INTEGER CTYPE,ATYPE,BTYPE,BB4                                AE  40
          COMMON/BLK1/T,C,NA,NB,Q/BLK3/RF,CTYPE,CURVE/BLK5/AA1,BB1,BB2, AE  50
         1 BB3,BB4,ATYPE,BTYPE                                         AE  60
          DIMENSION T(650),C(4000),NA(4000),NB(4000),Q(650),RF(50),AA1(50), AE  70
         1 BB1(50),BB2(50),BB3(50),BB4(50),CTYPE(4000),CURVE(20),      AE  80
         2 ATYPE(50),BTYPE(50)                                         AE  90
          F(NAARG,NBARG) = (T(NAARG) - T(NBARG)) * CND                 AE 100
          NTEST = 0                                                    AE 110
          SUM2 = 0.0                                                   AE 120
          DO 70 I = 1,NCND                                             AE 130
          IF (NA(I) .EQ. NTEST) GO TO 10                               AE 140
          SUM = 0.0                                                    AE 150
          DUMMY = 1.0                                                  AE 160
          RE = DUMMY                                                   AE 170
          RKINV = DUMMY                                                AE 180
          NTEST = NA(I)                                                AE 190
10        NNA = IABS(NA(I))                                            AE 200
          NNB = IABS(NB(I))                                            AE 210
          IF (CTYPE(I)) 30,20,40                                       AE 220
20        CND = C(I)                                                   AE 230
          GO TO 60                                                     AE 240
30        IF (CTYPE(I) .EQ.  - 1) CALL RCOND (NNA,NNB,C(I),CND)        AE 250
          IF (CTYPE(I) .EQ.  - 2) CALL REX (NNA,NNB,IRCNT,C(I),CND)    AE 260
          GO TO 60                                                     AE 270
40        IF (CTYPE(I) .GT. 100) GO TO 50                              AE 280
          LIM1 = CTYPE(I)                                              AE 290
```

```
         CALL COND (CURVE(LIM1),C(I),T(NNA),T(NNB),CND)         AE 300
         GO TO 60                                                AE 310
50       CALL CONV (I,CND,RE,RKINV,L)                            AE 320
60       FLUX = F(NNA,NNB)                                       AE 330
         SUM = SUM + FLUX                                        AE 340
         II = I + 1                                              AE 350
         IF (NA(I) .EQ. NA(II) .OR. NA(I) .LT. 0) GO TO 70       AE 360
         SUM2 = SUM2 + SUM - Q(NNA)                              AE 370
70       CONTINUE                                                AE 380
         WRITE (2,80) SUM2                                       AE 390
         RETURN                                                  AE 400
80       FORMAT(1H0,36X,16HENERGY BALANCE =,E10.4/)              AE 410
         END                                                     AE 420-
         SUBROUTINE FLOW (NCND,UU0)                              AF  10
C        THIS SUBROUTINE CALCULATES VELOCITY                     AF  20
C        FROM THE POTENTIAL NETWORK                              AF  30
C        T.C.E.E., SECTION 6.7.                                  AF  31
         INTEGER CTYPE,CURVE                                     AF  40
         DIMENSION T(650),C(4000),NA(4000),NB(4000),Q(650),CTYPE(4000)  AF  50
        1 ,RF(50),CURVE(20)                                      AF  60
         COMMON /BLK1/T,C,NA,NB,Q/BLK3/RF,CTYPE,CURVE            AF  70
         VX = 0.0                                                AF  80
         VY = 0.0                                                AF  90
         WRITE (2,80)                                            AF 100
         WRITE (2,70) UU0                                        AF 110
         DO 50 I = 1,NCND                                        AF 120
         NNA = IABS(NA(I))                                       AF 130
         NNB = IABS(NB(I))                                       AF 140
         VEL =  - (T(NNB) - T(NNA)) * SQRT(C(I))                 AF 150
         IF (VEL .LT. 0.0) GO TO 20                              AF 160
         IF (CTYPE(I) .EQ. 0) GO TO 10                           AF 170
         VY = VEL                                                AF 180
         GO TO 20                                                AF 190
10       VX = VEL                                                AF 200
20       II = I + 1                                              AF 210
         IF (NA(I) .EQ. NA(II)) GO TO 50                         AF 220
         IF (VX .EQ. 0.0) GO TO 30                               AF 230
         THETA = ATAN2(VY,VX)                                    AF 240
         THETA = THETA * 180.0 / 3.141592654                     AF 250
         GO TO 40                                                AF 260
30       THETA = 90.0                                            AF 270
40       V = SQRT(VX ** 2 + VY ** 2)                             AF 280
         NNNA = NA(I)                                            AF 290
         WRITE (2,60) NNNA,VX,VY,V,THETA                         AF 300
         VX = 0.0                                                AF 310
         VY = 0.0                                                AF 320
50       CONTINUE                                                AF 330
60       FORMAT(1H0,6X,4HNODE,I6,5X,4HVX =,E11.4,5X,4HVY =,E11.4, AF 340
        1   5X,3HV =,E11.4,5X,7HTHETA =,F7.2)                    AF 350
70       FORMAT(1H , 8X,46HPOTENTIAL FLOW CALCULATION - ANGLES IN DEGREES// AF 360
        1/,1H ,4HV0 =,   E11.4//)                                AF 370
80       FORMAT(1H1)                                             AF 380
    120  FORMAT(1H ,4X,4HNODE,I6,5X,3HVX=,E11.4,5X,3HVY=,E11.4/,  AF 390
        15X,3HV =,E11.4,5X,6HTHETA=,F7.2)                        AF 400
         RETURN                                                  AF 410
         END                                                     AF 420-
         SUBROUTINE PRESS (JJJ,CNDDD)                            AG  10
C        THIS SUBROUTINE CALCULATES PRESSURE CONDUCTANCE         AG  20
         DIMENSION T(650),C(4000),NA(4000),NB(4000),Q(650)      AG  30
         DIMENSION RF(50),CTYPE(4000),CURVE(20)                 AG  40
         COMMON/BLK1/T,C,NA,NB,Q/BLK3/RF,CTYPE,CURVE            AG  50
         INTEGER CTYPE,CURVE                                     AG  60
         N1 = IABS(NA(JJJ))                                      AG  70
         N2 = IABS(NB(JJJ))                                      AG  80
```

```
        ITEST = CTYPE(JJJ) - 400                                      AG   90
        GO TO (10,20),ITEST                                           AG  100
C       FLOW RESISTANCE BASED ON LINEAR DEPENDENCE OF PRESSURE LOSS   AG  101
C       VS. AIRFLOW.                                                  AG  102
10      CNDDD = 1.0 / C(JJJ)                                          AG  110
        GO TO 30                                                      AG  120
C       CFM REQUIRED FROM PREVIOUS ITERATION                          AG  130
C       FLOW RESISTANCE BASED ON QUADRATIC DEPENDENCE OF PRESSURE     AG  131
C       LOSS VS. AIRFLOW. T.C.E.E., E6.9.                             AG  132
20      DELP = T(N1) - T(N2)                                          AG  140
        IF (ABS(DELP) .LT. 0.00001) DELP = 0.00001                    AG  150
        CNDDD = 1.0 / SQRT(ABS(DELP) * C(JJJ))                        AG  160
30      CONTINUE                                                      AG  170
        RETURN                                                        AG  180
        END                                                           AG  190-
```

The following comments refer to the TNETFA Source Listing, Appendix VI:

The main program (lines A 560, A 2440, A 2470, A 3720, A 4095, A 4170, A 4250, A 4310) and subprograms STSTA (lines B 20, B 280, B 290, B 870, B 900, B 970), TRANF lines C 10, C 410, C 420, C 740, C 750, C 830), NDET (lines AD 10, AD 260) contain a seven character variable name IRADFLG. If this causes difficulty with your compiler, you may truncate the variable name to the six character name IRADFL.

The main program contains a reference to the subprogram NEGNODE. Both the reference (line A 2390) and the subprogram (line F 10) itself may be truncated to six characters, NEGNOD.

Page 379, line E 60: insert into the blank spaces preceding AA1, RF(50),
Page 390, line AD 40: insert into the six blank spaces following BB4,
,CURVE

The author strongly recommends that if a computer other than a Control Data Corp. machine is used, both TAMS and TNETFA be modified to use double precision. The following code additions would then be required:

TAMS:
1. Add before first executable statement in main program and all subroutines -

    IMPLICIT REAL*8 (A-H,O-Z)

2. Sequence # A 410, A 430, A 440 -
    Change REAL to REAL*8

TNETFA:

1. Add before first executable statement in main program and all subroutines -

    IMPLICIT REAL *8 (A-H, O-Z)

2. Add before first executable statement -

    INTEGER CTYPE, CURVE in SUBROUTINE RCOND
    REAL KFAC, KVFAC IN SUBROUTINES CONV, PRAIR, GRAS
    INTEGER CAPCUR in SUBROUTINE NDET

3. Change all type statements in all routines
    from REAL to REAL*8

4. Sequences # C240
    Change third argument from 0.0 to 0.0D + 00

5. Sequence # L 1230
    Change second argument from 1.0 to 1.0D + 00

Index